Constanze Stockhammer

Why Some Research Insights Find Their Way to Market and Some Do Not

AF092666

Constanze Stockhammer

Why Some Research Insights Find Their Way to Market and Some Do Not

The Relevance of Social Capital for Academic Entrepreneurship

Südwestdeutscher Verlag für Hochschulschriften

Impressum/Imprint (nur für Deutschland/ only for Germany)
Bibliografische Information der Deutschen Nationalbibliothek: Die Deutsche Nationalbibliothek
verzeichnet diese Publikation in der Deutschen Nationalbibliografie; detaillierte bibliografische
Daten sind im Internet über http://dnb.d-nb.de abrufbar.
Alle in diesem Buch genannten Marken und Produktnamen unterliegen warenzeichen-, marken-
oder patentrechtlichem Schutz bzw. sind Warenzeichen oder eingetragene Warenzeichen der
jeweiligen Inhaber. Die Wiedergabe von Marken, Produktnamen, Gebrauchsnamen,
Handelsnamen, Warenbezeichnungen u.s.w. in diesem Werk berechtigt auch ohne besondere
Kennzeichnung nicht zu der Annahme, dass solche Namen im Sinne der Warenzeichen- und
Markenschutzgesetzgebung als frei zu betrachten wären und daher von jedermann benutzt
werden dürften.

Verlag: Südwestdeutscher Verlag für Hochschulschriften Aktiengesellschaft & Co. KG
Dudweiler Landstr. 99, 66123 Saarbrücken, Deutschland
Telefon +49 681 37 20 271-1, Telefax +49 681 37 20 271-0, Email: info@svh-verlag.de
Zugl.: Wien, Wirtschaftsuniversität, Dissertation, 2009

Herstellung in Deutschland:
Schaltungsdienst Lange o.H.G., Zehrensdorfer Str. 11, D-12277 Berlin
Books on Demand GmbH, Gutenbergring 53, D-22848 Norderstedt
Reha GmbH, Dudweiler Landstr. 99, D- 66123 Saarbrücken
ISBN: 978-3-8381-0994-7

Imprint (only for USA, GB)
Bibliographic information published by the Deutsche Nationalbibliothek: The Deutsche
Nationalbibliothek lists this publication in the Deutsche Nationalbibliografie; detailed
bibliographic data are available in the Internet at http://dnb.d-nb.de.
Any brand names and product names mentioned in this book are subject to trademark, brand or
patent protection and are trademarks or registered trademarks of their respective holders. The
use of brand names, product names, common names, trade names, product descriptions etc.
even without
a particular marking in this works is in no way to be construed to mean that such names may be
regarded as unrestricted in respect of trademark and brand protection legislation and could thus
be used by anyone.

Publisher:
Südwestdeutscher Verlag für Hochschulschriften Aktiengesellschaft & Co. KG
Dudweiler Landstr. 99, 66123 Saarbrücken, Germany
Phone +49 681 37 20 271-1, Fax +49 681 37 20 271-0, Email: info@svh-verlag.de

Copyright © 2008 Südwestdeutscher Verlag für Hochschulschriften Aktiengesellschaft & Co. KG
and licensors
All rights reserved. Saarbrücken 2008

Produced in USA and UK by:
Lightning Source Inc., 1246 Heil Quaker Blvd., La Vergne, TN 37086, USA
Lightning Source UK Ltd., Chapter House, Pitfield, Kiln Farm, Milton Keynes, MK11 3LW, GB
BookSurge, 7290 B. Investment Drive, North Charleston, SC 29418, USA
ISBN: 978-3-8381-0994-7

Abstract

In today's knowledge-based society academic spin-offs are considered as one important form of technology transfer. However, most related entrepreneurial efforts fail. When looking at determining factors, researchers have to be regarded as members of a complex social network consisting not only of academic peers but also of participants from industry, policy, and supporting institutions, and friends and family members. This network is the academic entrepreneur's general framework of operation characterised by resource and communication flows of differing nature.

The thesis tries to answer the question under what conditions technological spin-offs emerge from the academic community with special emphasis on the entrepreneurs' social capital expressed in terms of their social networks. Thus, a detailed analysis of the characteristics of the social network academic entrepreneurs operate in at the different stages of the spin-off process is effected. By means of social network analysis effected by a survey among Austrian academic entrepreneurs, it is examined which network structures favour the recognition of entrepreneurial opportunities and their realisation in the form of a spin-off company. Thus generated insights are intended to contribute to nascent academic entrepreneurs' awareness how to optimally use their social contacts and networks in setting up their new ventures. Moreover, the work intends to provide valuable insights for political decision-makers as to the provision of necessary framework conditions supporting respective social dynamics required for the creation of innovative knowledge- und technology-intensive companies in Austria.

Kurzzusammenfassung

In unserer heutigen wissensbasierten Gesellschaft werden akademische Spin-offs als wichtige Form des Technologietransfers angesehen. Diesbezügliche unternehmerische Anstrengungen schlagen jedoch häufig fehl. Analysiert man die maßgeblichen Einflussfaktoren so ist es erforderlich, die unternehmerischen ForscherInnen als Mitglieder komplexer sozialer Netzwerke zu betrachten, die sich nicht nur aus wissenschaftlichen KollegInnen sondern auch aus VertreterInnen aus Wirtschaft, Politik und unterstützenden Institutionen sowie Freunden und Familienmitgliedern zusammensetzen. Diese Netzwerke sind der allgemeine Handlungsrahmen der potentiellen Spin-off-GründerInnen, charakterisiert durch Ressourcen- und Kommunikationsströme unterschiedlichster Natur.

Die gegenständliche Arbeit intendiert, einen Beitrag zur Beantwortung der Frage zu leisten, unter welchen Bedingungen technologische Spin-offs aus dem akademischen Umfeld heraus entstehen, wobei ein besonderer Schwerpunkt auf dem Konstrukt des sozialen Kapitals der UnternehmerInnen operationalisiert in Form ihrer sozialen Netzwerke liegt. Zu diesem Zweck wurde entlang der identifizierten Phasen des Spin-off Prozesses eine umfassende Analyse der Charakteristiken der sozialen Netzwerke, in denen akademische UnternehmerInnen agieren, durchgeführt. Eine auf der Methode sozialer Netzwerkanalyse basierende Erhebung bei österreichischen Spin-off Gründern und Gründerinnen lieferte Einsichten in die Zusammensetzung begünstigender Netzwerkstrukturen für erfolgreiche Chancenerkennung und Ausgründung. So generierte Erkenntnisse sollen dazu beitragen, das Bewusstsein junger akademischer UnternehmensgründerInnen für eine optimale Nutzung ihrer sozialen Kontakte und Netzwerke zur erfolgreichen Gründung eines Unternehmens zu schärfen. Darüber hinaus sollen die Ergebnisse der Arbeit politischen EntscheidungsträgerInnen als wertvolle Informationsgrundlage dienen, wie sie in Österreich entsprechende Rahmenbedingungen zur Förderung der für die Gründungsdynamik im wissens- und technologieintensiven Bereich notwendigen sozialen Interaktionen schaffen können.

Acknowledgement

At this occasion I would like to thank all people that have contributed to the realisation of this work with both their scientific and emotional support, and unbowed patience.

First of all, to my scientific advisor Prof. Rainer Hasenauer, who has taken the risk of supervising a work with a rather unconventional methodological approach. Secondly, to Prof. Gerhard Wührer, who accepted to co-advise a thesis from an other university, even in an other city. A willingness that is rarely seen in the Austrian academic community.

Special thanks to my former colleagues from HiTec Marketing, above all to Susanne Fuchs, who was always my encouraging model, and Walter Aigner, who inspired me with his unconventional thinking.

Thank you very much also to all acedmic founders that undertook the effort to complete a rather long and unusual questionnaire. Without their co-operation this work would have never been possible. I hope some of them will draw valueable inputs out of the results.

To my family and friends, who never stopped believing in me and were never tired to build me up with their unshakable trust. Special thanks have to be dedicated to Bernhard, who sacrificed more than one joint weekend and bore my temper as fluctuating with the ups and downs of the scientific progress.

Last, but just not least, boundless thanks to my mother. She is the one that rendred all this possible by making me the person that I am, and always gave me the feeling that whatever I do, in her eyes I am worth a laureateship that cannot be measured in scientific terms.

Table of Contents

Abstract i
Kurzzusammenfassung ii
Acknowledgement... iii
Table of Contents v
List of Tables vii
List of Figures x

1 Introduction 1
   1.1 BACKGROUND 1
   1.2 RESEARCH QUESTION 3
   1.3 FOCUS AND SIGNIFICANCE OF RESEARCH..................................... 3
   1.4 RESEARCH PARADIGM 5
      1.4.1 Critical Rationalism.. 5
      1.4.2 Scientific Explanation... 6
      1.4.3 Implications on the Research Design 7
   1.5 ORGANIZATION OF RESEARCH.. 8

2 Theoretical Framework 9
   2.1 ENTREPRENEURSHIP 9
      2.1.1 General Theoretical Perspective 9
      2.1.2 Entrepreneurship in the Field of High Technology 15
      2.1.3 Academic Spin-offs as Entrepreneurial Outcomes 18
   2.2 SOCIAL CAPITAL.. 23
      2.2.1 Definition 23
      2.2.2 Features of Social Capital 24
      2.2.3 Dimensions of Social Capital 26
      2.2.4 Network Aspects of Social Capital........................... 28
   2.3 RESOURCE-BASED THEORY 30
      2.3.1 Background 30
      2.3.2 Resources and Capabilities 30
      2.3.3 Resources Properties for Sustained Competitive Advantage 33
      2.3.4 Strategy and Value Creation 34
      2.3.5 New Venture Creation and Innovation..................... 36
   2.4 SUMMARY.. 38

3 Research Model and Hypotheses... 40
   3.1 THE SPIN-OFF PROCESS AND ITS PHASES 41
   3.2 FORMS OF SOCIAL CAPITAL.. 45
   3.3 INTEGRATION WITH RESOURCE-BASED PERSPECTIVE 46
   3.4 MODELLING AND GENERATION OF FIRST HYPOTHESES................. 48
      3.4.1 Opportunity Recognition... 49
      3.4.2 Formation 50
      3.4.3 Establishment 51

4 Methodology 55
   4.1 RESEARCH DESIGN.. 55
      4.1.1 General Considerations 55
      4.1.2 Methods of Empirical Social Research..................... 57
      4.1.3 Applied Research Design 58
   4.2 QUALITATIVE EXPLORATORY STUDY 61
      4.2.1 Introduction 61
      4.2.2 The Qualitative Interview 61
      4.2.3 Sampling.. 64

	4.2.4	Data Collection 65

    4.2.4   *Data Collection* 65
    4.2.5   *Analysis* 66
  4.3   QUANTITATIVE NETWORK ANALYSIS.. 72
    4.3.1   *Introduction* 72
    4.3.2   *Social Network Analysis* 72
    4.3.3   *Ego-centred Networks as Unit of Analysis*...................... 74
    4.3.4   *Network Measures of Social Capital* 76
    4.3.5   *Sampling*.. 82
    4.3.6   *Data Collection* 88
    4.3.7   *Variables*... 93
    4.3.8   *Descriptive Statistics* 101
    4.3.9   *Statistical Power* 121
    4.3.10 *Methodological Limitations*... 123

5   Research Results and Interpretation 126
  5.1   QUALITATIVE EXPLORATORY STUDY 126
    5.1.1   *Presentation of Results* 126
    5.1.2   *Reappraisal of Hypotheses*.. 129
  5.2   QUANTITATIVE ANALYSIS 136
    5.2.1   *Comparison of Means* 136
    5.2.2   *Exploring the Data*... 137
    5.2.3   *Testing for Interrelations*... 179
    5.2.4   *Testing of Hypotheses*... 240
    5.2.5   *Additional Findings* 247

6   Conclusions.. 249
  6.1   DISCUSSION OF EMPIRICAL RESULTS 249
    6.1.1   *General*.. 249
    6.1.2   *Information* 250
    6.1.3   *Resources* 252
    6.1.4   *Support* 253
    6.1.5   *Synthesis* 254
  6.2   LIMITATIONS 257
    6.2.1   *Sampling*... 257
    6.2.2   *Lack of Control Group* 258
    6.2.3   *Cross-sectional Analysis* 258
  6.3   CONTRIBUTIONS FOR THEORY AND PRACTICE 260
    6.3.1   *Theory* 260
    6.3.2   *Practice*... 261
  6.4   RESEARCH AGENDA 264

7   References 266

8   Appendices 283
  8.1   DATA SOURCES.. 283
    8.1.1   *Academia plus Business* 283
    8.1.2   *Life Science Austria* 283
    8.1.3   *Universities* 284
  8.2   DATA SHEETS.. 286
    8.2.1   *Ego* 286
    8.2.2   *Alter* 289
    8.2.3   *Alter-Alter* 294
  8.3   QUESTIONNAIRE... 298

List of Tables

Table 2-1: Spin-off Characteristics (Zahn et al. 2003 p. 167) 19
Table 4-1: Differences between Exploratory and Conclusive Research (Malhotra and Birks 2003 p. 63) 56
Table 4-2: Comparison of Quantitative and Qualitative Research (Bortz and Döring 1995 p. 274) 57
Table 4-3: Composition of the Qualitative Sample 65
Table 4-4: Code System for Qualitative Data Analysis 71
Table 4-5: Network Measures of Social Capital 76
Table 4-6: Overview of Questions 92
Table 4-7: Original Variables 95
Table 4-8: Aggregated Variables 98
Table 4-9: Compound Variables 100
Table 4-10: Descriptives – Personal Variables 101
Table 4-11: Frequencies of Binary Personal Variables 104
Table 4-12: Frequencies of Egos' Universities of Origin 104
Table 4-13: Frequencies of Egos' Position at University 106
Table 4-14: Frequencies of the Spin-offs' Industries 107
Table 4-15: Frequencies of Binary Environmental Variables 109
Table 4-16: Frequencies of Reaction University and Department 109
Table 4-17: Frequencies of Sources of Reputation 110
Table 4-18: Frequencies of Sources of Personnel 111
Table 4-19: Descriptive Statistics – Relational Roles 113
Table 4-20: Descriptive Statistics – Academic Contacts 114
Table 4-21: Descriptive Statistics – Informational Content 115
Table 4-22: Descriptive Statistics – Resources 116
Table 4-23: Descriptive Statistics – Support 117
Table 4-24: Descriptive Statistics – Tie Strength 118
Table 4-25: Descriptive Statistics – Multiplexity 118
Table 4-26: Descriptive Statistics – Network Size 119
Table 4-27: Descriptive Statistics – Sparsity 119
Table 4-28: Descriptive Statistics – Density 120
Table 4-29: Descriptive Statistics – Centrality 120
Table 4-30: Descriptive Statistics – Closure 120
Table 4-31: Non-parametric Tests 122
Table 5-1: Original Hypotheses 131
Table 5-2: Means of Structural Network Parameters 136
Table 5-3: Comparison of Means (Friedman Test) 137
Table 5-4: Spearman Correlations for Relational Content in Phase 1 143

Table 5-5: Spearman Correlations for Relational Content in Phase 2 149

Table 5-6: Spearman Correlations for Relational Content in Phase 3 155

Table 5-7: Spearman Correlations for Informational Content in Phase 1 158

Table 5-8: Spearman Correlations for Informational Content in Phase 2 160

Table 5-9: Spearman Correlations for Informational Content in Phase 3 163

Table 5-10: Spearman Correlations for Resource Content in Phase 1 166

Table 5-11: Spearman Correlations for Resource Content in Phase 2 168

Table 5-12: Spearman Correlations for Resource Content in Phase 3 170

Table 5-13: Spearman Correlations for Supportive Content in Phase 1 173

Table 5-14: Spearman Correlations for Supportive Content in Phase 2 175

Table 5-15: Spearman Correlations for Supportive Content in Phase 3 177

Table 5-16: Multiple Regression Analysis - Phase 1: Network Size with Functional Roles .. 182

Table 5-17: Multiple Regression Analysis - Phase 1: Network Size with Relational Content 183

Table 5-18: Multiple Regression Analysis Phase 1: Informational Content with Functional Roles 185

Table 5-19: Multiple Regression Analysis Phase 1: Informational Content with Information Types 187

Table 5-20: Multiple Regression Analysis Phase 1: Informational Content with Structural Parameters 188

Table 5-21: Multiple Regression Analysis Phase 1: Sources of Economic Information 189

Table 5-22: Regression Analysis Phase 1: Sources of Information about Potential Employees 190

Table 5-23: Regression Analysis Phase 1: Sources of Other Information 190

Table 5-24: Multiple Regression Analysis Phase 1: Resources with Functional Roles 191

Table 5-25: Multiple Regression Analysis Phase 1: Resources with Resource Types 193

Table 5-26: Multiple Regression Analysis Phase 1: Resources with Structural Parameters 193

Table 5-27: Multiple Regression Analysis Phase 1: Sources of Financial Resources 194

Table 5-28: Multiple Regression Analysis Phase 1: Sources of Infrastructure 195

Table 5-29: Multiple Regression Analysis Phase 1: Support with Functional Roles 196

Table 5-30: Multiple Regression Analysis Phase 1: Support with Support Types 197

Table 5-31: Multiple Regression Analysis Phase 1: Support with Structural Parameters 199

Table 5-32: Multiple Regression Analysis Phase 1: Sources of Emotional Support 200

Table 5-33: Multiple Regression Analysis Phase 1: Sources of Institutional Support 200

Table 5-34: Multiple Regression Analysis Phase 2: Network Size with Functional Roles 202

Table 5-35: Multiple Regression Analysis Phase 2: Network Size with Relational Content . 203

Table 5-36: Multiple Regression Analysis Phase 2: Informational Content with Functional Roles 204

Table 5-37: Multiple Regression Analysis Phase 2: Informational Content with Information Types 205

Table 5-38: Multiple Regression Analysis Phase 2: Informational Content with Structural Parameters 206

Table 5-39: Regression Analysis Phase 2: Sources of Economic Information 207

Table 5-40: Multiple Regression Analysis Phase 2: Resources with Resources Types 209
Table 5-41: Multiple Regression Analysis Phase 2: Resources with Structural Parameters 210
Table 5-42: Regression Analysis Phase 2: Sources of Financial Resources 210
Table 5-43: Regression Analysis Phase 2: Sources of Infrastructure 211
Table 5-44: Multiple Regression Analysis Phase 2: Support with Functional Roles 212
Table 5-45: Multiple Regression Analysis Phase 2: Support with Support Types 213
Table 5-46: Multiple Regression Analysis Phase 2: Support with Structural Parameters..... 214
Table 5-47: Regression Analysis Phase 2: Sources of Emotional Support 214
Table 5-48: Regression Analysis Phase 2: Sources of Institutional Support 215
Table 5-49: Multiple Regression Analysis Phase 3: Network Size with Functional Roles 217
Table 5-50: Multiple Regression Analysis Phase 3: Network Size with Relational Content . 218
Table 5-51: Multiple Regression Analysis Phase 3: Informational Content with Functional Roles.. 220
Table 5-52: Multiple Regression Analysis Phase 3: Informational Content with Information Types 222
Table 5-53: Multiple Regression Analysis Phase 3: Informational Content with Structural Parameters 223
Table 5-54: Regression Analysis Phase 3: Sources of Economic Information 224
Table 5-55: Regression Analysis Phase 3: Sources of Information about Potential Customers 225
Table 5-56: Regression Analysis Phase 3: Sources of Other Information 225
Table 5-57: Multiple Regression Analysis Phase 3: Resources with Functional Roles 227
Table 5-58: Multiple Regression Analysis Phase 3: Resources with Resource Types 228
Table 5-59: Multiple Regression Analysis Phase 3: Resources with Structural Parameters 229
Table 5-60: Regression Analysis Phase 3: Sources of Personnel 230
Table 5-61: Regression Analysis Phase 3: Sources of Infrastructure 231
Table 5-62: Regression Analysis Phase 3: Sources of Material Resources 232
Table 5-63: Multiple Regression Analysis Phase 3: Support with Functional Roles 233
Table 5-64: Multiple Regression Analysis Phase 3: Support with Support Types 234
Table 5-65: Multiple Regression Analysis Phase 3: Support with Structural Parameters..... 235
Table 5-66: Regression Analysis Phase 3: Sources of Emotional Support 235
Table 5-67: Dependent Variable: Network Size – Overview Regression Coefficients.......... 236
Table 5-68: Dependent Variable: Informational Content – Overview Regression Coefficients 237
Table 5-69: Dependent Variable: Resources – Overview Regression Coefficients............. 238
Table 5-70: Dependent Variable: Support – Overview Regression Coefficients.................. 238
Table 5-71: Sources of Different Content Types – Overview Regression Coefficients........ 239
Table 5-72: Final Status of Hypotheses – Overview 246
Table 8-1: Overview Science Parks of AplusB 283
Table 8-2: Universities in Austria... 285

List of Figures

Figure 1-1: Phases of the Research Process (Schnell et al. 1999 p. 8) 8
Figure 2-1: From a market need to a successful enterprise (Ardichvili et al. 2003 p. 112).... 12
Figure 2-2: Entrepreneurial Context (Bloodgood et al. 1995 p. 133) 14
Figure 2-3: The Chasm (Moore 1999 p. 17) 17
Figure 2-4: Two Basic Forms of Technology Transfer (Gibson and Stiles 2000 p. 198) 21
Figure 2-5: Technology Transfer and Diffusion Process (Source: Beer 2000 p. 44) 22
Figure 2-6: Model of Social Capital and Value Creation (Source: Tsai and Ghoshal 1998) ... 27
Figure 2-7: Structural Holes and Weak Ties (Burt 1992 p. 27) 28
Figure 2-8: Types of Resources (Thudium 2005 p. 275) 31
Figure 2-9: Relationship between Resource Properties and Sustained Competitive Advantage (Barney 1991 p. 112) 33
Figure 2-10: Resource-based Approach to Strategy Analysis (Grant 1991 p. 115)................ 35
Figure 2-11: Resource Pyramid of Value Creation (Brush et al. 2001 p. 71) 36
Figure 2-12: Dynamic Model of Resource Acquisition, Development, and Effects in New Ventures (Lichtenstein Bergmann and Brush 2001 p. 38) 37
Figure 3-1: "Dialectic" Spiral of Theory Development (Roth and Heidenreich 1993 p. 334)... 40
Figure 3-2: The Spin-off Funnel (Clarysse et al. 2005 p. 187) 42
Figure 3-3: The Global Process of Valorisation by Spin-off (Ndonzuau et al. 2002 p. 283) ... 42
Figure 3-4: Causal Model for the Explanation of the Entrepreneurial Success Path (Christensen et al. 2000 p. 10) 47
Figure 3-5: Model of Social Capital Impact on Entrepreneurial Success 48
Figure 4-1: Classification of Research Designs (Malhotra and Birks 2003 p. 62).................. 55
Figure 4-2: Building of Theories and Hypotheses (Atteslander 1993 p. 55) 59
Figure 4-3: Research Design 60
Figure 4-4: Forms of Qualitative Interviews (Kepper 1994 p. 39) 62
Figure 4-5: Process Model of Deductive Application of Categories (Mayring 2000)............. 68
Figure 4-6: Personal Network (Schenk 1995 p. 16) 75
Figure 4-7: Spin-off Population in Austria (Federal Ministry of Science and Research et al. 2007) 84
Figure 4-8: Gender Distribution of Sample 85
Figure 4-9: Year of Foundation of Sampled Companies 85
Figure 4-10: Distribution of Incubating Organisations 86
Figure 4-11: Regional Distribution of Spin-offs 87
Figure 4-12: Distribution of Industries 87
Figure 4-13: Histogram – Respondent's Age 102
Figure 4-14: Histogram – Year of Foundation 102
Figure 4-15: Histogram – Number of Co-Founders 103
Figure 4-16: Distribution of Universities of Origin 105
Figure 4-17: Distribution of Egos' Position at University 106

Figure 4-18: Distribution of Spin-offs' Industries 107
Figure 4-19: Distribution of Scientific Disciplines among Egos 108
Figure 4-20: Distribution of Binary Environmental Variables.. 109
Figure 4-21: Distribution Reaction of University and Department 110
Figure 4-22: Distribution of Sources of Reputation 111
Figure 4-23: Distribution of Sources of Personnel.. 112
Figure 5-1: Social Capital Measures along the Spin-off Process.. 133
Figure 6-1: Final Model of the Social and Environmental Factors of the Academic Spin-off Process 254
Figure 8-1: Start Page of Online Questionnaire 298
Figure 8-2: Name Generator 299
Figure 8-3: Question on Relational Types 299
Figure 8-4: Question on the Content of Relations 300
Figure 8-5: Relations of Alters 300
Figure 8-6: Questions on Co-founders 301

1 Introduction

1.1 Background

In today's knowledge-based society and knowledge economy universities and members of academic communities are facing a transition from merely educational and research-focussed activities towards an increased relevance of entrepreneurial orientation. Academic science has become increasingly entrepreneurial, not only with respect to business firms asking for research support or technology transfer but also in its internal dynamic (Etzkowitz et al. 2000 p. 43)[1]. Especially with respect to the enhanced penetration of technology in almost all sectors of economy and the dynamics it implicates, not only businesses feel the force to permanently adapt to a constantly changing environment, but also research has to cope with the necessity of leaving its "ivory tower" and turning its focus more towards "real needs" of markets and society. Yet, there is also increasing pressure from governments to make the benefits of public investments in university-based research activities not only available to the scientific community, but also to the wider economy. As a result, following the United States also in Europe increasing emphasis is put on the transfer of technological know-how in the form of academic spin-offs (Wright et al. 2004a).

In this context, high-technology spin-offs as particular knowledge-intensive ventures have been put in the focus of policy and scientific interest. High-tech start-ups are commonly regarded as main motor for an economy's innovative potential. Yet, their establishment usually requires a high level of scientific knowledge and considerable research background to build upon. In most cases, respective know-how can be primarily found at university faculties and comparable publicly-funded research institutions. In this context, the commercialisation of technological research findings in the form of academic ventures is increasingly considered to be preferential over the licensing alternative generating more benefits and directly fostering regional innovativeness (Christensen et al. 2001 p. 12; Ndonzuau et al. 2002).

Particularly in highly industrialised economies it pays off when technological innovations arrive at the market beeline by means of high technology spin-offs as main drivers of technological progress and employment. Besides the general positive effect on employment, high-tech start-ups significantly contribute to net job creation, since they operate on markets with huge growth potentials so that a successful market entry not automatically entails crowding out of competitors and thus displacement of labour (Beer 2000 p. 3-5).

A study of the German Centre for European Economic Research (ZEW) commissioned by the German Ministry for Education and Research (Egeln et al. 2002) dealing with the integration of science and economy and its effect on innovation diffusion and transfer, commercialisation and employment in Germany, confirms the great relevance of academics for structural change and growth. Academics not only provided for 60% of the start-ups in research- and knowledge-intensive sectors, their new ventures were also significantly more fast-growing and successful, thus assuming the lion's share in economic and employment growth

---

[1] Citations aim at facilitated retrieval of referenced text passages. Pager numbers are indicated in case of citations from books, studies, reports or long articles, i.e. normally longer than 20 pages. Citations from shorter articles or texts are deemed as sufficiently easy to retrieve and are thus referred to without numbers of pages. The same applies if general positions of authors developed and presented throughout the whole publication are reproduced and not individual selective passages.

(Egeln et al. 2006 p. 10). In this context, close scientific contact to the research institution represented a decisive success factor. In particular start-ups of academics with a commercialisation focus had very dense contacts and thus a very high transfer level of research results into economy. In Germany 3% of all former researchers in public research take part in new ventures per year, with more than half of them having collected several years of experience in business between academic career and start-up formation (Egeln et al. 2002 p. 12).

However, most efforts in commercialising technologies fail. There are great ideas and academic findings with high potential and relevance for economy, which never find their way to the market. The question arises why some academic entrepreneurs succeed in their efforts to transform their work into marketable products and why some research findings, though promising, are never realised?

When looking at potential influencing factors and determinants for success, special emphasis is put on the social context academic entrepreneurs operate in. This approach has been motivated by the increasing recognition of the significance of social networks in the creation and sustaining of new ventures as reflected in recent entrepreneurship literature (Aldrich and Zimmer 2001; Anderson and Jack 2002; Christensen et al. 2000; Elfring and Hulsink 2001; Greve and Salaff 2003; Hoang and Antoncic 2003). The main argument of respective research on this issue is that entrepreneurs' personal social networks provide them with vital resources, be them tangible or intangible, they would not have at their disposal internally, thus significantly fostering the entrepreneurship process.

Following this perspective, entrepreneurial researchers can no longer be seen as isolated "geniuses" operating on their own in secret back rooms suddenly emerging with a resourceful idea; rather they are members of a complex social network consisting not only of academic peers but also of participants from industry, policy, and supporting institutions as well as friends and family members. This network is the general framework of operation characterised by resource and communication flows of differing nature. It provides the researcher with access to a broad range of input resources and enables a targeted distribution of his / her findings and outcomes.

When it comes to social networks, an important explanatory element of this phenomenon can be found in the social capital concept (Anderson and Jack 2002). The reason for this lies in the central proposition of social capital theory that networks of relationships constitute a valuable resource for the conduct of affairs (Nahapiet and Ghoshal 1998 p. 243). Consequently, the concepts of social capital and social networks are highly intertwined with social network analysis providing an effective tool for quantitative measurement and statistical analysis.

1.2 Research Question

Against the background of the considerations elaborated in Chapter 1.1, the thesis primarily centres on the following main research question:

How does the social capital of successful academic entrepreneurs develop along the process of spinning off high-technology ventures from universities?

The research thus focuses on answering the question under what conditions technological spin-offs emerge from the academic community. Special emphasis is put on the role an entrepreneur's social capital, in the form of his or her surrounding social networks, plays in this context. Thus, a detailed analysis of the structure and composition of the social networks academic entrepreneurs operate in as well as their development in time is carried out. It will be examined, which structures of a researcher's personal network favour the discovery and recognition of entrepreneurial opportunities as well as their realisation in the form of a successful spin-off company and how these structures change in the course of spin-off formation.

In answering this research question a process perspective is adopted for analysing the spin-off establishment. The process of new venture creation is split up into various phases serving as basis and explanatory variable for the set up of related hypotheses.

1.3 Focus and Significance of Research

Both academic and public authorities are paying increased attention to the commercialisation of scientific and technical knowledge produced within publicly funded research institutions as a key element in sustaining regional economic development (Ndonzuau et al. 2002). The creation of university spin-offs is seen as one of the most promising approaches in this context, since it can provide significantly higher revenues to universities than the licensing alternative (Nicolaou and Birley 2003b p. 335). Moreover for governments, it has appeared to provide an effective means for public policy to have a direct and considerable impact on economic development (Bower 2003).

However, despite the changes in policies and new public resources, a recent OECD survey shows that, outside the United States, spinning off new ventures from research institutions has remained a process of technology transfer with minimal impact (Callan 2001).

Due to its R&D extensive nature, high technology is a good indicator of the growth process in a knowledge-based economy. The key roles played by continuous innovation and short time to market in high technology today are increasingly coming to characterize the rest of the economy. Thus in the years ahead, the dynamics of high tech industries will increasingly shape the process of development in all industries (Cortright and Mayer 2001 p. 2).

This research project intends to provide a more detailed insight into the black box, in which academic research is transformed into economic value by the creation of new ventures. While the importance of social networks to entrepreneurship has been increasingly recognised in existing research (Aldrich and Zimmer 2001; Birley 1985; Christensen et al. 2000; Dubini and Aldrich 1991; Elfring and Hulsink 2001; Greve and Salaff 2003; Hulsink and Elfring 2003; Larson and Starr 1993; Singh et al. 2000), respective studies on the very particular topic of academic spin-offs remain limited.

And even the increasingly emerging research on social networks or social capital in entrepreneurship in general is still faced with a couple of limitations as clearly summarised by Hoang and Antoncic (2003).

As the authors (Hoang and Antoncic 2003) stress, there are only few process-oriented studies accounting for the dynamic nature of networks and taking them as dependent variables to answer the question how entrepreneurial processes and outcomes influence network development over time. The main part of research rather focuses on networks as independent variables and how they affect the entrepreneurial process and its outcomes. So far, only partial empirical confirmation exists for a theory of network development. Many questions remain regarding how network content, governance, and structure emerge and develop over time.

Moreover according to Hoang and Antoncic (2003), current work on entrepreneurial success is limited by considerable conceptual vagueness regarding the rare resources vital to success, and how to measure the networks that provide those resources. Mapping networks of general information flows seems to be too far removed from resource flows linked to outcomes such as business establishment. Currently, a standard question to gather network data asks the entrepreneur to whom he or she would turn for advice or information (Burt 1984). This may not be fine-grained enough to reveal meaningful differences in network structure, and if differences are observed, there is little insight into the nuances of the entrepreneurial process that would explain them (Hoang and Antoncic 2003).

Although the value of networks as integral part of the explanation of entrepreneurial success is widely acknowledged, there remains a number of unresolved issues (Elfring and Hulsink 2001 p. 2). For example, the network perspective lacks to be specific about the context and the timing of the role of network relations (Bloodgood et al. 1995 pp. 125-26). And there is little specification of the various dimensions of a network and their impact on the early development of a venture.

Given these limitations, this thesis will focus on providing additional empirical contributions to the theory of network development. In particular, insights will be generated on the variation of network content, structure and composition as dependent variables at the different stages of the academic spin-off formation process.

As mentioned despite the general notion of the importance of social networks and social capital for the entrepreneurial process, little research has been performed so far transferring these concepts to the analysis of the academic spin-off process. Within the broad range of literature on university spin-offs recently emerging in Europe (Callan 2001; Chiesa and Piccaluga 2000; Clarysse et al. 2002; De Coster and Butler 2005; Degroof and Roberts 2004; Fontes 2005a; Ndonzuau et al. 2002; Steffensen et al. 2000), there are only very few studies (Johansson et al. 2005; Nicolaou and Birley 2003a; Nicolaou and Birley 2003b) taking a closer look at the influence of social networks on academic spinout creation. Yet, even those remain limited to investigating the relationship between an academic's personal network structure and the outcome of the spin-off efforts in terms of the researcher's involvement in the newly founded firm and his affiliation with the mother university. Again a rather static perspective is chosen, taking social networks as independent variable and presuming that the network structure remains unchanged in the course of the spin-off process.

Taking this research gap into account, the research project exclusively focuses on analysing the process and the conditions of academic start-up formation in terms of the founding academic's social network and its development over time.

Finally, much of the existing literature on academic spin-offs tends to be heavily US-biased, which means that it is based on an entrepreneurial culture which is recognized for its unique social, historical and institutional settings. As a result, most of the empirical findings cannot be transferred to European countries that, for example, do not have the same history of 'pro-entrepreneurship environment' as the United States (Ulhoi 2005).

While there is meanwhile an increasing number of studies on university spin-offs in different European countries, they concentrate on Anglo-Saxon and Scandinavian regions such as the Netherlands (Hulsink and Elfring 2003; van Geenhuizen 2003), Sweden (Dahlstrand 1997; Klofsten and Jones-Evans 2000), the United Kingdom (De Coster and Butler 2005; Druilhe and Garnsey 2004; Lockett et al. 2003; Vohora et al. 2004; Wright et al. 2004b) or Italy (Bellini and Zollo 1997; Chiesa and Piccaluga 2000), and are consequently not apt to be applied to the Austrian situation. Respective research (Egeln et al. 2006; Egeln et al. 2003) dealing with Austrian peculiarities still remains rare.

To overcome this deficiency this thesis explicitly focuses on the conditions and features academic start-ups have to face in Austria. In doing so, it aims at providing Austrian policy makers and the scientific establishment with useful insights on the Austrian situation as basis for targeted measures to promote spin-off creations.

1.4 Research Paradigm

To provide for an appropriate placement of the research project in the reference frame of philosophy of science and epistemology the basic philosophical and methodological concepts related to are briefly described in the following section. Starting with the epistemic understanding of critical rationalism, one of the most widely applied models of scientific explanation, the Hempel-Oppenheim-scheme, is introduced. Finally, resulting implications for the structure and design of the research project are highlighted.

1.4.1 Critical Rationalism

The dissertation project is bound to the critical rationalism of Karl Raimund Popper, who strongly adheres to the principle paradigm of empiricism (Schülein and Reitze 2002 p. 143). Criticizing logical positivism and its problems related to applying induction as method of theory generation, Popper departs from the idea that the truth of theories can be proved by finite observations. According to him, induction theoretically requires infinite trials, since each additional observation imposes the possibility of finally contradicting the theory. Yet, this is not practically realisable. Instead Popper postulates that theories are preliminarily accepted as long as they prove themselves in the observations made so far. In case an observation or an experiment contradicts a theory, it is rejected or falsified. Thus, he obverts the procedure of taking evidence applied by logical positivism. Instead of taking observations in order to prove the correctness of a theory, the aim of scientific work should consist in demonstrating the falsity of a theory by means of even these observations, i.e. despite of verifying a theory, theories should be falsified.

For that purpose, at the beginning a theory has to be developed and then subjected to systematic examination. If the predictions of the theory correspond to the results of the experiments or observations, it is deemed as preliminarily proved, which does not mean that it is true, but just that there is no evidence of its falsity so far. In case of contradictory results, the theory is rejected and a new theory has to be established or the original theory has to be redesigned in a way that it is not disproved by present experiments or observations.

In this process of theory development, testing and in most cases subsequent rejection theories are increasingly improved. They converge towards reality. However, at no point in time the researcher may state that they are definitely true, since they can always be disproved by a further observation. Yet, one of the core statements of critical rationalism is that we can only obtain uncertain knowledge.

According to Popper the demarcation criterion distinguishing between empirical-scientific and metaphysical propositions consists in falsification. Only those propositions predicate something about experiential reality that may fail at it, i.e. that can be subjected to a methodological review which holds the possibility of refuting them (Schülein and Reitze 2002 p. 146).

Following Popper the only logically justified method of inferring required in order to achieve a formal-logically grounded science is the method of deduction. Based on the principles of logic the deductive conclusion starts from the general sentence, since an a priori statement about the general has to cover each special case. This general sentence constitutes the first premise. The second sentence is an assigning sentence. It represents the second premise of the logical conclusion. From the general sentence and the assigning sentence a formally founded conclusion can be drawn. Thus, deduction is the derivation of a sentence (the conclusion) from one or more other sentences (premises).

For example:

If it rains, it is wet.	(1st premise)
It rains.	(2nd premise)
It is wet.	(conclusion)

However, it has to be noted that the logical foundation of a coherent conclusion not necessarily implies the truth. If one premise is not right, the conclusion can not be right either. The validity of a deduction is based on the logical relationship between premises and conclusion, but not on the actual logical values of the premises and the conclusion.

1.4.2 Scientific Explanation

Popper's fundamental considerations are perpetuated in Carl Gustav Hempel's and Paul Oppenheim's theory of scientific explanation, also know as the Hempel-Oppenheim-scheme or H-O-scheme. In their original article "*Studies in the Logic of Explanation*" (1948) Hempel and Oppenheim analyzed what has come to be known as "deductive-nomological" (lat. *deductio*, deduction, and hell. *nomos*, law) explanation. It explains a phenomenon by showing that the phenomenon can be deducted from a general law and a series of special conditions (antecedents). According to them a scientific explanation has the following structure:

$A_1, A_2, ..., A_k$	antecedent conditions	⎫ explanans
$L_1, L_2, ..., L_k$	general laws (hypotheses)	⎭
E	explanandum	

Antecedent and general laws together form the explanans. The explanandum, i.e. a sentence describing a singular phenomenon, logically results from the explanans. These conditions result in a potential explanation / explication, if in addition the following adequacy conditions are fulfilled:

- The explanandum has to result from the explanans in a deductive manner.
- The explanans has to contain a general law.

A *true explanation* arises from adding following condition:
- The law has to be true.

In other words, there exists a *general law* stipulating that if events a_1 of the type A_1, a_2 of the type A_2; ... and a_k of the type A_k occur, that an event e of the type E will occur. As the explanation consists in demonstrating that a series of events a_1, a_2, ... a_k is connected to another event e in that the connector is a law covering the respective events, the model is also referred to as covering-law model.

In social and business sciences the H-O-scheme is also applied to stochastic and incomplete explanations. In case of stochastic explanations instead of all-sentences probability sentences are used.

1.4.3 Implications on the Research Design

Following the fundamental idea of Popper's critical rationalism, the dissertation project emanates from the specification of general hypotheses and antecedent conditions on the academic spin-off process and its social structures. Respective hypotheses and related basic assumptions are derived from existing scientific literature on the subject and from the insights obtained in the course of a preparatory explorative study preceding quantitative examination.

In a next step these hypotheses are subjected to an extensive examination regarding their potential falsification through a large number of empirical observations made within the scope of a quantitative statistical analysis. Given no contradictory observations of statistical relevance are made, the hypotheses are deemed as preliminarily accepted and at contributing to scientific explanation in the sense of the deductive-nomological model of Hempel and Oppenheim.

1.5 Organization of Research

The structure of the dissertation project and accordingly the layout of this publication as its primary output basically follow the main phases of empirical research processes as depicted in Figure 1-1.

Referring to the first step, the subject of research formulated as research problem and the underlying motivation for its selection have already been presented in Sections 1.1, 1.2 and 1.3 of this chapter. Departing from an identification of appropriate theories for the treatment of the research problem, existing work in the field is introduced and analysed in more detail in Chapter 2. This theoretical framework serves as basis for the establishment of a preliminary model and related hypotheses (Chapter 3) subject to empirical validation. To this end the core components of the model have to be operationalised accordingly as basis for subsequent measurements in the field. The overall procedure of these measurements is defined and presented in Chapter 4. As the research design consists of a mixed-method approach combing qualitative and quantitative research phases, the subsequent steps of selecting suitable research units, and gathering, entering and analysing data are presented under the respective heading of the research phase concerned, i.e. the qualitative phase in Chapter 4.2 and the quantitative phase in Chapter 4.3. Chapter 5 contains an extensive presentation and interpretation of the results of the data analysis, as basis for deducting conclusions and an outlook on future research in the light of prevailing limitations in Chapter 6.

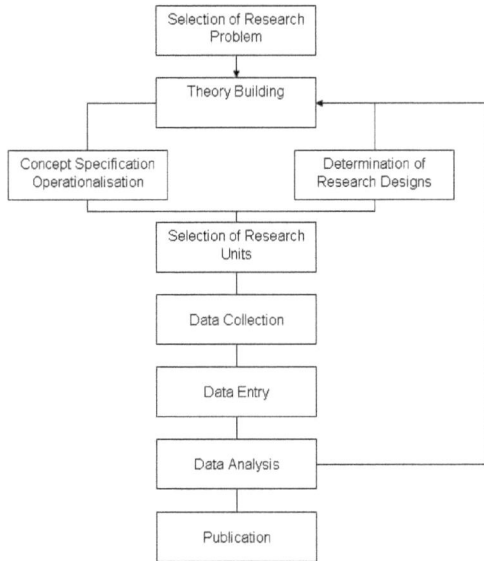

Figure 1-1: Phases of the Research Process (Schnell et al. 1999 p. 8)

2 Theoretical Framework

The following section provides for a theoretical placement of the research on-hand in the light of relevant literature on the topic. In a first step the two concepts dominating the theoretical considerations and associated hypothetical implications of the dissertation project are presented in more detail, i.e. the entrepreneurial process and the social capital concept. The perspective drew on to join these two conceptual positions is the one of the resource-based theory. It provides for the theoretical umbrella under which the two concepts can be reasonably integrated.

2.1 Entrepreneurship

Starting with a theoretical view of the entrepreneurship concept in general, attention is turned towards the specific conditions imposed on entrepreneurial activities in the field of high technology and the typical characteristics of academic spin-offs, as one occurrence of entrepreneurial endeavours. This focus is deliberately set to provide the ground for a profound scientific approach to the subject of this thesis dealing with the establishment of high tech start-ups from the university.

2.1.1 General Theoretical Perspective

Richard Cantillon (1680-1734) was the first economist to investigate the role of entrepreneurship in the economy. He considered entrepreneurs as actors that buy goods and services at a certain price to resell them at a possibly higher price.

In the course of time the term became customary to denote risk-oriented persons, who stimulate economic processes by introducing new and improved processes. In this notion the concept is commonly attributed to the French economist Jean Baptiste Say (1776–1832). According to Say an entrepreneurs is someone, who co-ordinates the employment of different production factors. In his words *"The entrepreneur shifts economic resources out of an area of lower and into an area of higher productivity and greater yield"* (Say 1803 p. 66).

Despite these intellectual developments in the 18th century, it was not until 1911 that Schumpeter (1883-1950) gave us the modern version of the entrepreneur as creative visionary willing and able to convert a new idea or invention into a successful innovation (Grebel et al. 2003). Schumpeter suggested that entrepreneurs are market participants that engage in the process of 'creative destruction' by applying new combinations of economic value creating activities such as (Schumpeter 1934 p. 66):

(1) introducing new products or qualities of products

(2) opening of new markets

(3) tapping new sources of supply of raw materials

(4) carrying out the new organisation of any industry.

These new combinations of value creating activities transform the market, essentially destroying the status quo and creating a whole new wave of innovation (Park 2005). In doing so, the entrepreneur creates disequilibrium in the economy (Schumpeter 1934 p. 66).

This Say-Schumpeter-approach can be seen as main basis of today's definitions. The contemporary economist Peter Drucker (1909-2005), for example, extends this concept with a focus on seizing existing opportunities. He defines an entrepreneur as a person that does not necessarily create something new, but takes existing knowledge as opportunity, which he realises in founding an organisation. According to Drucker (1985 p. 28) *"the entrepreneur always searches for change, responds to it, and exploits it as an opportunity"*. As the following sections will show this opportunity-oriented perspective has become the focal point of many current definitions.

An important contemporary economist born in 1930s, who seized the element of entrepreneurial opportunities, is Kirzner (1973; 1979). Kirzner opposes neoclassic economics as to the existence of an equilibrium which is based on the assumption of complete information (Kirzner 1973 p. 137). In his opinion, by pursuing opportunities the entrepreneur contributes to a movement towards economic equilibrium. Yet, an equilibrium situation will be never reached. A vital point of Kirzner's theory is the imperfect distribution of information. According to him, the economic process is characterized by discovery and learning. The entrepreneur benefits of the imperfect distribution of information as a result of the superior information and knowledge he possesses (Philipsen 1998 p. 5).

The focus of entrepreneurship research changed in the late 1980s and early 1990s with authors (Bygrave and Hofer 1991; Gartner 1985) proposing a more holistic approach to the study of entrepreneurship (Park 2005).

2.1.1.1 Definitions

When it comes to defining the phenomenon of entrepreneurs and entrepreneurship, unfortunately, still no single definition of the entrepreneur exists. Yet, a good version in particular suitable to this research project is provided by Kurtako and Hodgetts.

According to them (Kurtako and Hodgetts 1994 p. 6),

> "*an entrepreneur is an innovator or developer, who recognises and seizes opportunities; converts those opportunities into workable/marketable ideas; adds value through time, effort, money, or skills; assumes the risk of the competitive marketplace to implement these ideas; and realises the rewards from these efforts*".

Analogous Bygrave (1997; 1991) defines an entrepreneur as someone who perceives an opportunity and creates an organisation to pursue it. As a result, the entrepreneurial process involves all the functions, activities, and actions associated with perceiving opportunities and creating organisations to pursue them.

Similarly, Morris et al. (1993 p. 595) see entrepreneurship as the process of creating value by bringing together a unique package of resources to exploit an opportunity, which is detailed by Harwood's definition of entrepreneurship as (Harwood 1982 p. 92)

> "*the process of assembling resources to create and build an independent enterprise, encompassing creativity, risk taking and innovation*".

Shane and Venkatamaran (2000 pp. 218-19) also put the entrepreneurial opportunity in the centre of their considerations. According to them and following Kirzner (1997; 1979) entrepreneurship is concerned with two related processes:

(1) the discovery of profitable opportunities, and

(2) their exploitation.

New venture creation is explained in terms of *opportunities and resources* that are combined in various ways. Thus, the heart of entrepreneurship lies in the identification of new opportunities, the exploitation of the opportunities, the identification of needed resources, and the acquisition of the resources. In the following this postulated relevance of resources will provide for the crossover to the resource-based perspective in Chapter 2.3.

2.1.1.2 Entrepreneurial Opportunity

As these definitions reveal the *opportunity* is a vital element in the entrepreneurial process. It constitutes the starting point and the core of each entrepreneurial venture. In this context, an important contribution to entrepreneurship research has been Kirzner's (1979) entrepreneurial discovery theory focusing on opportunities and their realisation. According to him, ideas become an opportunity when their commercial value is recognized. The underlying theory is basically grounded in three interrelated key concepts:

(1) the entrepreneurial role,

(2) the role of discovery, and

(3) rivalrous competition.

Following the first concept, the entrepreneur grasps opportunities for entrepreneurial profit created by temporary absence of full adjustment between input and output markets caused by prior entrepreneurial errors. The *alert entrepreneur* discovers these earlier errors and acts upon them. In this regard, entrepreneurial alertness refers to an attitude of receptiveness to available (but so far) overlooked opportunities. Yet, the capacity of market participants to discover earlier errors is central.

Coming to the role of *discovery* there is a tendency for profit opportunities to be discovered and grasped by routine-resisting entrepreneurial market participants. This process implies elements of surprise and discovery: one discovers one's previous (unknown) ignorance involving the surprise that one had overlooked something readily available (Kirzner 1997 pp. 71-72). This feature of discovery characterises the entrepreneurial process of the market.

The last concept of *rivalrous competition* assumes that the above-described process is made possible by the freedom of entrepreneurs to enter the market in which they see opportunities for profit. In being alert to such opportunities and in grasping them, they are competing with other entrepreneurs. However, rivalrous competition also features a potential for discovery by revealing information, which no one was aware of its having been lacking. In addition the competitive process is an entrepreneurial one in that it depends crucially on the incentives provided by the possibility of entrepreneurial profit.

Following Kirzner's perspective of the value creating aspect of opportunities, a couple of definitions can be identified in existing literature on this topic. Dorf and Byers (2005 p. 4) for example view an opportunity as a favourable juncture of circumstances with a good chance fur success or progress. Hulbert et al. (1997) for example state that a business opportunity is the chance to meet an unsatisfied need that is potentially profitable. DeBono (1978), in turn, defines opportunity as a course of action that is possible and worth pursuing. According to him, recognising opportunities requires non-linear or lateral creative thinking.

Based upon the concept of value creation, Ardichvili et al. (2003) created a conceptual framework depicting the development opportunities undergo on their way from initial idea to a successful enterprise (see Figure 2-1).

According to them (Ardichvili et al. 2003) and following Kirzner (1997), in its most elemental form an opportunity comes either as an "imprecisely-defined market need" or due to "un- or under-employed resources or capabilities". In the first case, prospective customers may not be able to articulate their needs or (von Hippel 1994), but may still be able to recognize the value of something new presented to them. Thus, opportunities seen from the perspective of prospective customers represent *value sought* (Ardichvili et al. 2003).

Underutilized or unemployed resources, as well as new capabilities or technologies may offer possibilities to create and deliver new value for potential customers, even though the precise form the new value will take may be undefined. Respective opportunities arising from underutilized or unemployed resources, from technology or other types of knowledge or abilities, can be denoted as *value creation capability* (Ardichvili et al. 2003).

As the market need becomes more precisely defined in terms of benefits and value sought, and resources become more precisely defined in terms of potential uses, the opportunity transforms into a business concept. This concept contains the core notions of how the market need might be served or the resources deployed.

Over time this business concept grows into a business model matching market needs and resources. If the concept originated as market need, i.e. value sought, now the type and amount of resources, required to address that need, are identified. If the concept arose from underemployed resources, i.e. value creation capability, that capability's benefits and value to particular users and uses are more explicitly detailed at this point.

As the business model advances, cash flows, schedules of activities and resource requirements are added. These additions enable the business concept to transform into a fully detailed business plan as foundation for the subsequent new venture creation. However, some businesses may also be started with incomplete or unarticulated business plans.

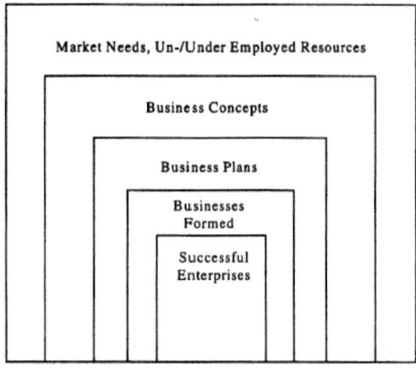

Figure 2-1: From a market need to a successful enterprise
(Ardichvili et al. 2003 p. 112)

Opportunities are evaluated at each stage of their development, although the evaluation may be informal or even unarticulated. Individuals may informally pursue investigations of presumed market needs or resources (including inventions) until concluding either that these warrant no further consideration, or that more formal pursuit of the possibility is appropriate.

As a result, entrepreneurship can be seen as dynamic process that evolves over time and comprises various phases. Existing literature on this topic provides a number of phase models of the entrepreneurial process ranging from three to four and more stages (Bhave 1994; Bygrave 1997; Galbraith 1982; Kazanjian and Drazin 1990; Morris 1998; Roberts 1991; Van de Ven et al. 1984). In principle, these approaches can be reduced to have the following components in common (Hansen and Bird 2001): entrepreneurs typically first gain access to technology; next, they market-test their newly developed products; and finally, they enter their products in designated markets and start business operations.

While having its limitations, a major strength of the literature on stage models is that it adds to the understanding of the rather complex phenomenon of entrepreneurship, and is thus applied in the course of this research. Following Hite and Hesterly (2001), the key assumption is that each stage represents more than mere changes over time, but rather functions as proxy for many strategic issues. The presumption of the underlying life cycle approaches is that each stage stands for a unique strategic context that influences the nature and extent of a firm's external resource needs, and related resource acquisition challenges that have to be overcome at each stage to survive.

2.1.1.3 Network Approaches to Entrepreneurship

Drawing on the Austrian economics argument that entrepreneurship exists because of information asymmetry between different actors (Hayek 1945), which is also followed by Kirzner (1997; 1979) taking it as important prerequisite of entrepreneurial opportunities, a crucial element is the information entrepreneurs **both** already possess, such as technological knowledge or experiences from past business activities, and obtain in the course of time as a result of purposeful search or by chance, such as information about the market conditions.

In this regard, the phenomenon of entrepreneurial behaviour cannot be meaningfully separated from the social and economic context in which that behaviour occurs (Bloodgood et al. 1995 p. 123). Contacts to and exchange with other people play a decisive role. Information is not only relevant for discovering and identifying potential entrepreneurial opportunities, but also enables cost-efficient access to the resources needed to realise these opportunities in concrete products and services to cover the gap between problems and solutions identified on the market. As Mahoney and Michael (2004 p. 7) note, with respect to entrepreneurship, searching for information is the entrepreneurial activity par excellence. Successful entrepreneurs are seen to have information benefits over other market participants. Benefits which might have been obtained as a result of the favourable position they occupy in their reference network.

According to Brüderl and Preisendörfer (1998), there are two different network approaches to entrepreneurship. The first one refers to the personal network of entrepreneurs, i.e. to individual relations of business founders as focal persons; the second one concentrates on the organizational network of businesses, i.e. on collective relations new firms are embedded in (Dubini and Aldrich 1991; Uzzi 1996).

The relevance of personal networks was first analysed in greater detail by Aldrich and Zimmer (2001), who viewed the entrepreneurial process as embedded in a changing network of continuous social relations that facilitate and constrain linkages between entrepreneurs, resources and opportunities. This network approach has become a popular theoretical perspective in the debate about entrepreneurship and small business formation (Brüderl and Preisendörfer 1998), and is also applied in the scope of this research.

The personal network approach is based on the common premise that entrepreneurship is a social role, embedded in a social, political, and cultural context.

Thus, entrepreneurs should not be viewed as isolated and autonomous decision makers, but as actors involved into a special micro-context (Brüderl and Preisendörfer 1998). The role-set of an entrepreneur may, for example, comprise of family members, venture capitalists, bankers, investors, partners, distributors, and customers (Bloodgood et al. 1995).

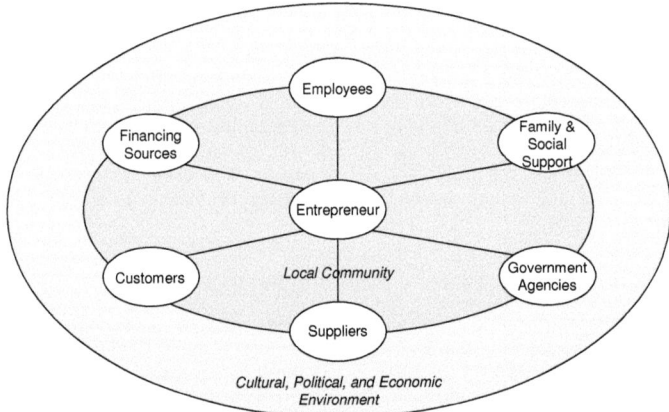

Figure 2-2: Entrepreneurial Context (Bloodgood et al. 1995 p. 133)

If entrepreneurs are conceived as organizers and coordinators of resources (Bygrave 1997; Morris 1998; Shane and Venkataraman 2000), social networking is directly connected with the very idea of an entrepreneur. The organization and coordination of resources require social activity and social interactions. Establishing a new business means that existing social relationships have to be activated and new ones created.

However, as Birley (1985) notes, entrepreneurial ventures not only require access to information, resources and opportunities, but also access to network members who are motivated to support the entrepreneur. In this context, networks are often used as much for assurance and validation as for economic motives (Birley 1985). While economic necessities often provide the impetus for inter-organisational alliances, non-economic ties often bind and maintain relationships essential to the survival of the entrepreneurial venture (Larson 1992 p. 98). Network members, in particular those with relevant past experience, often serve as sources of advice to entrepreneurs thus contributing to avoid typical pitfalls (Bloodgood et al. 1995 p. 126).

Both the specific resource needs of entrepreneurs and their established set of social contacts should be considered if one wants to predict how future business relations will develop. As access to resources is critical for entrepreneurs and their new firms, successful entrepreneurs are eager to direct their networking activities towards likely channels of resources and motivated network members (Bloodgood et al. 1995 p. 129). However, each person is subject to time and energy constraints that limit their ability to build and maintain extremely large networks. As networks expand, more time must be devoted to external network maintaining activities at the sacrifice of internal network functions necessary for success. Thus, entrepreneurs have to aim at network efficiency (Bloodgood et al. 1995 p. 129), meaning that they have to strive for a maximum return of invested networking time, or, to put it the other way round, minimal networking efforts to achieve a certain output, such as information or resources obtained.

In this context, it is assumed that network support improves the entrepreneurial process by at least three basic mechanisms (Brüderl and Preisendörfer 1998):

(1) Social relations and social contacts are important for gaining access to information; compared with information received from formal sources, information received from network ties is often more useful, reliable, exclusive, and less redundant.

(2) Network contacts provide access to customers and suppliers; a diverse network contributes to spreading information on the new firm.

(3) Social networks open the possibility to broaden the financial basis of a new firm; given the financial restrictions many new firms are confronted with, informal credits received from network contacts are particularly helpful in the start-up phase.

These imputed functions of entrepreneurs' social networks are an essential part of the research model to be tested within the scope of this work.

2.1.2 Entrepreneurship in the Field of High Technology

Starting up a business in the field of high technology imposes specific requirements to the nascent entrepreneur. To understand the conditions entrepreneurial researchers have to face when founding a start-up in the high technology domain, the characteristics of the industry are briefly depicted.

When trying to find a common definition of high technology, one encounters unexpected difficulties. While the term is frequently used, there exists no generalisable definition in technical or management research (Gardner et al. 2000).

Starting with the notion of technology itself, in the context of high technology it is frequently taken in the narrow sense of natural science and engineering or technical knowledge (Granstrand 1998).

The broadest definition of technology has probably been launched by Frances Stewart (1977 p. 1), where it includes

> "all skills, knowledge and procedures required for making, using and doing useful things. Technology therefore, includes the software of production – managerial and marketing skills, and extended to services- administration, health, education and finance."

Similarly Rogers (1995 p. 12) tends to link the term to innovation and to view technology, including social technology, as

> "a design for instrumental action that reduces the uncertainty of cause–effect relationships involved in achieving a desired outcome".

Burgelman et al. (1987 p. 4) put more emphasis on the process perspective by defining technology as

> "the practical knowledge, know-how, skills, and artefacts that can be used to develop a new product or service and/or a new production/delivery system. Technology can be embodied in people, materials, cognitive and physical processes, plant, equipment, and tools."

In this regard, technologies represent potentials for solving problems. Technology and the incorporated knowledge can be classified as capital (Hasenauer et al. 1994).

Accordingly Hasenauer et al. (1994 p. 14) distinguish between:

- Product technology
- Process technology
- Management technology

The concepts underlying most definitions of high technology use one or a combination of three factors (Riche et al. 1983):

(1) the utilization of scientific and technical workers,

(2) expenditures for research and development, and

(3) the nature of the product.

Regarding the first factor, Hodson and Parker (1988) for example consider high-technology industries as those with large numbers of engineers, scientists, and highly skilled technicians.

The OECD (2000) focuses more on the second factor by characterising high-technology industries by their intensity of research and development expenditures, and explicitly quoting aerospace, pharmaceuticals, computers and office machinery, communications equipment, and scientific instruments as high-technology sectors.

Tyson (1993 p. 124), in turn, combines the first two factors by defining a high technology industry as one in which knowledge is a prime source of competitive advantage for producers, who in turn make large investments in knowledge creation. Reflecting this definition, high technology industries are usually identified as those with above-average spending on research and development, above-average employment of scientists and engineers, or both.

Similarly, the National Science Foundation (Roessner et al. 2001) attributes high technology industries a great dependence on science and technology innovation that lead to new or improved products and services with substantial economic impact, fuelled both by large research and development spending, and a higher than industry average sales growth.

Following Shanklin and Ryans (1984) businesses must meet three criteria to be labelled 'high technology':

1) the business requires a strong scientific-technical basis,

2) new technology can quickly make existing technology obsolete and

3) as new technologies come on stream their applications create or revolutionize markets (supply as well as demand).

This corresponds to Viardot's (2004 p. 6-7) characterisation of high tech products according to the three features most frequently cited by marketing managers interviewed by the author: incorporation of sophisticated technology, short product life cycle, and the integration of innovation.

According to McKenna (1985) high-tech industries are characterized by complex products, large numbers of entrepreneurial competitors, customer confusion, and rapid change.

Summing up these varying notions of high technology, it may be stipulated that high tech industries are characterised by the following common features : a high degree of market and technological (Moriarty and Kosnik 1989; Shanklin and Ryans 1984).

Market uncertainty emanates from the ambiguity as to the type and extent of customer needs that can be satisfied by the technology. It results, firstly, from customers' uncertainty regarding which needs the technology will satisfy and to what extend. As a result users may delay adopting an innovation requiring increased product information and awareness building. Secondly, customer needs may be subject to rapid and unpredictable changes in high technology environment. Thirdly, customers' anxiety is supported by the uncertainty whether the market will eventually establish technical standards with which the products must be compatible if the buyer hopes to use them with other products, people, or organizations. Fourthly, also against the background of the previous three factors, predicting how fast a high-tech innovation will spread is difficult for both costumers and suppliers. And finally, the eventual size of the market is hard to assess resulting from the difficulties in forecasting whether and to what extent the mainstream market will adopt a technology (Moriarty and Kosnik 1989).

Moore explicitly addresses the last problem in his book "Crossing the Chasm" (Moore 1999). When presented to the market for the first time, such products primarily attract innovators and early adopters, or so called visionaries (see Figure 2-3). These visionaries are willing to adopt the new technology despite its associated uncertainties and high price due to the particular psychological and substantive benefits they obtain. However, the majority of the market consists of pragmatists, who require different incentives to accept such an offer. In contrast to visionaries they need a stronger and more immediate stimulus to be convinced to adopt an innovation. For this purpose they refer to the experiences of previous adopters (Fuchs 2005 p. 55). The chasm now refers to the gap that exists between these two worlds. In particular, visionaries are not good references for pragmatists. They provide tales of heroics and not stories of smooth, predictable adoption. Pragmatists want references from other pragmatists. Hence, the transition between the two markets is difficult, with many high-tech firms never crossing the chasm and succeeding in getting through to the more support requiring majority.

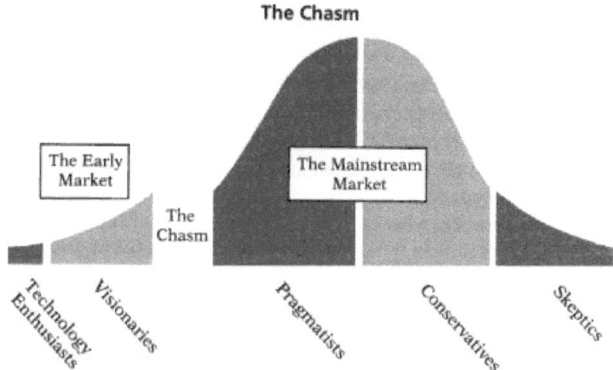

Figure 2-3: The Chasm (Moore 1999 p. 17)

The second characteristic feature of high technology markets is the *technological uncertainty*. While market uncertainty stems from the lack of knowledge about customers' needs and wishes, technological uncertainty refers to the uncertainty whether the technology can deliver on its promise to meet the needs, once they have been articulated (Moriarty and Kosnik 1989). There are five sources of technological uncertainty. The first originates from a lack of information about whether a technology and thus a product will function as promised. Second, due to the complexity involved in the development of high technology products, the timetable for product availability is not very predicable. Third, there is uncertainty about whether the supplier of a high-tech product will be able to provide prompt, effective service. Fourth, the technology may have unanticipated side effects. Finally, in high technology markets technological uncertainty arises because of questions about technological obsolescence, whether and when the market will turn to another technology to replace the current generation of products (Moriarty and Kosnik 1989).

Following Lin et al. (2006 p. 2), high-tech entrepreneurship can thus be understood as the pursuit of opportunities beyond resources currently controlled and a willingness to face the challenge of enormous technology and market uncertainty. According to the authors, high technology entrepreneurs pursue business opportunities by introducing technological innovations to the world. A perspective that is also followed in the course of this research project.

2.1.3 Academic Spin-offs as Entrepreneurial Outcomes

Academic spin-offs as specific form of new entrepreneurial ventures are companies established on the basis of research outcomes produced at universities. Also referred to as "start-ups" or "spin-outs", university spin-offs represent an important mechanism of technology transfer from public research to market, to be developed into useful products and services.

2.1.3.1 Spin-offs in General

The term spin-off usually denotes a new company that arises from a parent organisation (Carayannis et al. 1998).

More specifically and following the definition proposed by Steffensen et al. (2000), a spin-off is a new company formed (1) by individuals who were former employees of a parent organisation, and (2) a core technology that is transferred from the parent organisation to the new company. Spin-offs thus represent the transfer of a technological innovation to a new entrepreneurial company that is formed around that technological innovation (Rogers et al. 2001). To be a spin-off, this must also include the transfer of some rights, e.g., assets or knowledge, from the existing legal body to the new firm or body (Lindholm Dahlstrand 1997). Depending on the maturity of the know-how to be transferred emerging intellectual property right (IPR) issues have to be taken into account and represent increasingly relevant factors in the process of spinning-off from the mother company. Yet, in the worst case the incubating organisation, i.e. the university, may hamper the creation of the young venture by retaining control of the required technology, if it is sufficiently developed to be secured by property rights that are then kept by the university.

In their definition of spin-outs Carayannis et al. (1998) also consider the opportunity that the technology may be transferred to the new company while the inventor/founder remains with the parent organisation. The parent organisation may be a private company, a research institution, or a public organisation.

To be classified as a spin-off, a new venture has to comprise both

(1) the transfer of persons (founder of the spin-off) and

(2) the transfer of ideas, technologies or products developed by the spin-off founder in the course of his former employment

form the parent organisation to a newly founded firm that has to be economically and legally independent (Beer 2000 p. 3).

		Technology Transfer	
		No	Yes
Person Transfer	No	Traditional new venture formation	Formation by technology adoption
	Yes	Formation by experienced persons	**Spin-off formation**

Table 2-1: Spin-off Characteristics (Zahn et al. 2003 p. 167)

Consequently, the academic entrepreneur is the key in this transfer. He is both carrier of the transferred ideas, products or technologies, and object of transfer, thus establishing the link between parent organisation and spin-off (Zahn et al. 2003 p. 166).

2.1.3.2 Defining Academic Spin-offs

A common two-dimensional definition of academic spin-offs (Clarysse et al. 2002; Steffensen et al. 2000) is the one of Smilor et al. (1990), who define an academic spin-off as an enterprise of which

(1) the entrepreneur is an academic, a research worker or a student, who left the university to start a company or who started a company while he or she was still attached to the university and of which

(2) the business activity is founded on a technological development or innovative concept developed at the university.

Roberts and Malone (1996) identified four principal entities involved in the spin-off process:

(1) the *technology originator(s)*, who bring(s) the technological innovation through the innovation development process to the point at which the transfer of this technology to the commercial sector can begin;

(2) the *entrepreneur* (or the entrepreneurial team), who takes the technology created by the originator and attempts to create a new business venture centred on the technology;

(3) the *parent organisation*, in which R&D activities to create the technological innovation take place, and which may assist or restrict the spin-off process by controlling intellectual property rights; and

(4) the *venture investor*, who provides the financial resources to establish the spin-off and may also provide for business management expertise.

In the academic spin-off, both the roles of originator and the entrepreneur may coincide. Most studies on academic spin-off companies have identified models of the academic entrepreneur as an individual who is the technology originator but also assumes the role of the entrepreneur.

In this respect, Nicolaou and Birley (2003a) provide the following definition, which is both encompassing and parsimonious. Spinouts involve:

1. The transfer of a core technology from an academic institution into a new company.

2. The founding member(s) may include the inventor academic(s) who may or may not be currently affiliated with the academic institution.

Compared to non-academic spin-offs and start-ups, university spin-offs are usually faced with particular difficulties and particularities resulting from the environment and conditions they emerge from. According to Vohora et al. (2004) compared to other high-tech start-up ventures they have to deal with two completely different problems.

First, typical problems occur as they develop from an initial idea in a non-commercial environment to becoming an established competitive profit-generating firm. Universities in particular often lack resources and academic entrepreneurs may lack commercial skills to set-up new ventures in an attempt to commercialise technological assets.

Second, during its whole establishment process and even beyond the academic spin-off has to deal with conflicting objectives of key stakeholders such as the university, the academic entrepreneur, the venture's management team and financial providers.

2.1.3.3 Academic Spin-offs as Technology Transfer Mechanisms

As mentioned, spin-offs are increasingly considered as important means of technology transfer transforming scientific output into commercially applicable products and services (Chiesa and Piccaluga 2000; OECD 2000). While the process of transferring university technology and knowledge into application can take place through many channels, such as publications, cooperation and licensing to existing companies, the creation of new business entities is gaining importance due its particularly favourable effects on the economic growth and technological competitiveness of the innovation system (Rasmussen and Borch 2004). University spin-offs not only promote the fast dissemination and application of scientific know-how, but also significantly contribute to job creation, quality of life, and the innovation potential in general.

Technology transfer is the application of information – a technological innovation – into use (Rogers et al. 2001). It usually involves a source of technology that possesses specialized technical skills, and the transmission to receptors who do not posses these specialised skills and who cannot or do not want to create the technology themselves (Williams and Gibson 1990).

Due to its complex, systemic, context-related, tacit and person-embodied nature (Cohen and Levinthal 1990; Nonaka and Takeuchi 1995), and the differences in scope and purpose between academic and business research (Dasgupta and David 1994), knowledge transfer from public research to the productive sector is a difficult undertaking. It requires a transformation process that involves devising applications for new scientific concepts and/or turning technologies and prototypes into viable products and services. Academic spin-offs appear particularly suitable for this task (Fontes 2005b).

According to Gibson and Stiles (2000) there are two possibilities to transfer technology from research to commercial application:

(1) spinning-out technologies into start-up companies (dashed line), and

(2) transferring innovative technologies to established firms (solid line).

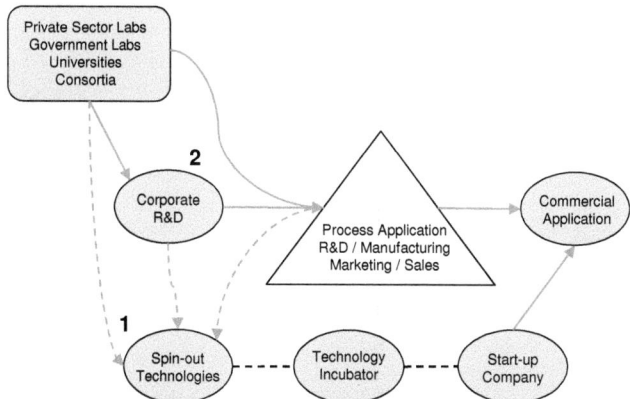

Figure 2-4: Two Basic Forms of Technology Transfer (Gibson and Stiles 2000 p. 198)

In the first case, technologies or knowledge apt to form the basis of a spin-off may originate in the private sector, government labs, universities, and consortia. These spin-off companies may or may not be nurtured by an incubator. With regard to transfer of technologies to established firms researchers conduct state-of-the-art, pre-competitive research, and transfer these results by such varied means as research publications, licensing, patents, students graduated, and personnel transfer (Gibson and Stiles 2000).

According to Beer (2000 p. 39-40) the technology transfer process consists in the transmission of technological knowledge from a technology provider to a technology taker representing two economically and legally independent organisations. The transfer can also occur indirectly by means of technology intermediaries.

As mentioned university spin-offs represent one possible form of technology transfer. Yet, their possible role in the technology transfer process is ambiguously. As Beer (2000 p. 41) highlights, they may be technology recipients, since they take up new technologies developed at universities and draw on them as basis of their business activities. Moreover, they can be seen as technology givers that hand on new technologies to companies, i.e. their customers. Finally, spin-offs may be interpreted as technology intermediaries between academia and economy, as they are often considered as important transfer media for disseminating new technological know-how (Sandberger 1986).

As already practised by Beer (2000 p. 42) in the course of this thesis Carayannis et al.'s (1998) understanding of technology transfer as taking place from the originating organisation to the spin-off is pursued. This relates to Rogers (1995) notion of technology as basis for innovation. As spin-offs represent innovative organisations they require new technologies to create innovation. Thus, technologies have to be transferred from the university to the spin-off.

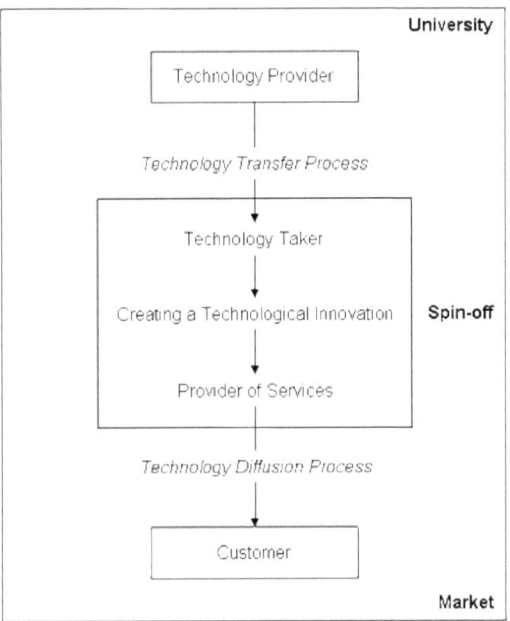

Figure 2-5: Technology Transfer and Diffusion Process (Source: Beer 2000 p. 44)

As Beer summarises in Figure 2-5, the production and provision of technology takes place in the course of the basic research at the university. At the same time the transfer process is initialised, since in most cases researcher and spin-off founder are the same person or in close contact to each other, e.g. as teacher and student. Together with applied research the crossover of persons and know-how to the spin-off usually starts. In this phase the technology is further transferred and adopted. As soon as the innovation process has passed on to the spin-of, processing of the technology starts in the spin-off. This may already be the case during applied research, or at the time of development. In this context, the technology is adopted and processed by the spin-off, i.e. it represents the basis of the services the spin-off offers on the market. These may be product or process innovations, material goods or services or information goods (software) as immaterial goods. The important point is that the technology manifests itself in the goods and services of the new venture. This is where the actual technology transfer process ends. Through the products of the spin-off the technology is passed on to all, who employ the services of the spin-off. However, it no longer represents a technology transfer process in the actual sense between two participants, but a technology diffusion process, in the course of which the technology diffuses through the goods and services of the spin-off to the market. As such the notion holds true that spin-offs can be considered as transfer media for disseminating new technological know-how: they are technology recipients in the technology transfer process, create their own service offer, and initialise the diffusion process in the market (Beer 2000 p. 42-43).

2.2 Social Capital

The central proposition of social capital theory is that networks of relationships constitute a valuable resource for the conduct of social affairs. The value of social capital is its ability to make possible achievements that would not be possible without it, or only be achieved at extra cost (Nahapiet and Ghoshal 1998). Social networks consisting of extended family, community-based or organisational relationships are theorised to complement the effects of education, experience, and financial capital (Bourdieu 1985; Coleman 1990; Loury 1987). Researchers of social capital, thus, are primarily concerned with the significance of relationships as a resource for social action (Burt 1992; Coleman 1990). As Davidsson and Honig (2003) summarise, social capital theory accounts to actors' ability to draw benefits from their social structures, networks, and memberships.

As Burt (2000) notes, social capital is a metaphor about competitive advantage originating in a better connection of parties involved. People, who do better, are somehow better connected (Burt 2000). Social relations and interactions improve actors' access to and control of information, knowledge and resources, and enable the establishment of trust, norms, and collective attitudes facilitating transactions in the widest sense. Social capital constitutes of both the network of relations and the assets and resources this network provides (Burt 1992 pp. 11-12).

A problem with the term social capital is its widely differing definitions. A comprehensive definition of social capital stresses its informal nature, subsumes social networking and accommodates reciprocity. An ample definition should also state that the concept is not limited to individuals and can be applied in many relationships, be they economic, social/personal, political, etc. (Wikipedia 2007b). Social capital is multidimensional, and occurs at both the individual and the organisational levels (Nahapiet and Ghoshal 1998).

2.2.1 Definition

The concept of social capital was originally used in community studies to describe relational resources embedded in personal ties that are useful for the development of individuals in community social organisation (Jacobs 1965; Loury 1977).

Bourdieu and Coleman further advanced and introduced and developed the social capital framework in modern sociology (Suvanto 2000). The first cohesive elaboration of the term was by Pierre Bourdieu. In *The Forms of Capital* (1985) he contrasts social capital to two other forms of capital: economic capital and cultural capital. Building on this original work, Bourdieu and Wacquant (1992 p. 119) define that

> "*social capital is the sum of resources, actual or virtual, that accrue to an individual or group by virtue of possessing a durable network of more or less institutionalised relationships of mutual acquaintance and recognition*".

This definition is instrumental, emphasising the benefits an individual obtains when participating in social networks and the deliberate construction of social relationships in order to achieve these benefits (Portes 1998).

James Coleman adopted Loury's (1977) definition in developing the general wide-scale social capital theory that scholars mostly use today. According to him (Coleman 1990 p. 302),

> "*social capital is defined by its function. It is not a single entity, but a variety of different entities having two characteristics in common: They all consist of some aspect of social structure, and they facilitate certain actions of individuals who are within the structure. Like other forms of capital, social*

> capital is productive, making possible the achievement of certain ends that would not be attainable in its absence. ... Social capital is embodied in the relations among persons."

In the late 1990s, the concept became very popular, with the World Bank devoting a research program to it and its theoretical review in Robert Putnam's book *Bowling Alone*.

According to the World Bank (1999),

> "social capital refers to the institutions, relationships, and norms that shape the quality and quantity of social interactions. In this view, social capital is not only the sum of the institutions, which underpins the society, it is the glue that holds them together".

Putnam, in turn, grounds his work in Coleman's metaphor by maintaining the focus on action facilitated by social structure (Burt 2001 p. 2):

> "social capital .... refers to features of social organisation, such as trust, norms and networks, that can improve the efficiency of society by facilitating coordinated actions" (Putnam 1993 p. 167).

Fukuyama has further investigated the link between trust, social capital and national economic success. He defined social capital as:

> "the existence of a certain set of informal values or norms shared among members of a group that permit cooperation among them" (Fukuyama 1995 p. 16).

Nan Lin's (1999 p. 30) concept of social capital has a more individualistic approach:

> "Investment in social relations with expected returns".

According to him (Lin 1999 p. 31), four elements – information, influence, social credentials and reinforcement – explain why social capital works in instrumental and expressive actions that are not accounted for by other forms of capital such as economic capital or human capital.

Nahapiet and Ghoshal (1998 p. 243) have further elaborated the concept and defined it as,

> "the sum of the actual and potential resources embedded within, available through, and derived from the network of relationships possessed by an individual or social unit. Social capital thus comprises both the network and the assets that may be mobilized through that network."

Due to its particular appropriateness for the objectives of this research resulting from its emphasis on resources and structures in the form of social networks, **the last definition is the one primarily adopted when talking about social capital in the course of this work.**

2.2.2 Features of Social Capital

Generally speaking, social capital is a form of capital in the sense of resources available to actors to achieve their goals (Bourdieu 1985) like physical and human capital. However, it also has some specific features. All forms of capital are productive, i.e. they are indispensable for certain outcomes, and not completely fungible but specific to particular actions (Suvanto 2000). Yet, social capital unlike other forms of capital is embedded in the relationship between and among actors thus being jointly owned and providing no exclusive ownership rights (Burt 1992). Moreover, although it has value in use, it cannot be traded easily (Nahapiet and Ghoshal 1998). Rather it is a form of capital that can change as relationships and rewards change, and it disappears when the relationship disappears (Suvanto 2000). As opposed to other forms of capital, social capital increases rather than decreases with its use (Leana and Van Buren III 1999).

Social capital can be subdivided into individual and collective capital (Christensen et al. 2000). Individual social capital is defined as a set of social relations (social ties) surrounding the actor (here, the entrepreneur), and which can be mobilised more or less consciously when needed. Analysis focusing on individual social capital primarily addresses the individual benefits resulting from the inclusion of the individual within his social environment. Collective social capital, in turn, refers to relationships at another level, that is, between groups. Research dealing with collective social capital thus refers to collective benefits that usually arise from participatory and associative dynamics among or within communities (Franke 2005 p. 11).

An entrepreneur's social capital consists of all social relationships and social structures used to achieve his or her goals. Social capital is therefore the result of a dynamic interaction. It becomes "capital" if it is used by actors in concrete situations (Coleman 1990; Pizzorno 1999). As Jansen notes (2000 pp. 37), social capital is used up, like in every capital cycle, by being transformed into other forms of capital, goods or services that in turn may be applied to increase social capital. In essence, five benefits or values provided by social capital can be distinguished (Burt 1997; Burt 1992; Coleman 1990; Putnam 1993):

1. family and group solidarity based on social closure
2. trust in the validity of general social norms
3. information
4. profit opportunities through structural autonomy
5. social influence

In case of the first two types the collective character dominates, while in case of the last three the individual benefit is in the forefront. In this context, different social structures provide different services. What may be useful for the one value, e.g. collective solidarity, may be detrimental for another purpose, e.g. extensive information. Moreover, the position of individual actors differs in the same social structure: the same structure may yield high profits for one actor that have to be paid by the other (Jansen 2000 p. 38).

Social capital manifests itself in different forms, primarily in trust, norms, and networks. As Landry et al. specify (2000), trust is developed over time through repeated series of interactions. Norms of appropriate behaviour, too evolve over time as a result of a series of interactions and exchange of resources. Finally, networks develop as actors develop reliable and effective communication channels across organisational boundaries (Landry et al. 2000). Social capital can be a useful resource both by enhancing internal organisational trust through the bonding of actors, as well as by bridging external networks in order to provide resources (Adler and Kwon 2002; Putnam 2000). As Jane Fountain and others have begun to report, there is a useful link between social capital and technological innovation (Fountain 1998; Tsai and Ghoshal 1998). Compared with social capital, economic capital in the traditional sense refers to material values, such as money, means of production, shares and property institutionalised by property rights.

2.2.3 Dimensions of Social Capital

Granovetter (1992) originally made the distinction between structural and relational embeddedness. While the structural dimension of social capital is about social interactions, the sum of relationships within a social structure, the relational dimension refers to direct relationships of a person to others and the assets rooted in these relationships, such as trust and trustworthiness (Tsai and Ghoshal 1998).

Drawing on Granovetter's (1992) distinction between structural and relational embeddedness, Nahapiet and Ghoshal (1998) argue for a third dimension of social capital: the cognitive dimension, which is about shared values or paradigms that allow a common understanding of appropriate ways of acting. According to them, these three dimensions of social capital – structural, relational, and cognitive – are highly inter-related.

Structural Dimension

According to Nahapiet and Goshal (1998 p. 251) the structural dimension of social capital refers to the overall pattern of connections between actors, which could be described as network ties, network configuration, and appropriable organisation. It reflects whom you reach and how (Burt 1992). According to Tsai and Ghoshal (1998), an actor's location in a social network captures the structural dimension of social capital. In essence, the more contacts you have with your environment, the more structural capital you have. An interesting concept in this context is the network morphology, which reflects the density, hierarchy, and connectivity of the linkages in a network (Arenius 2002 p. 53). Structural capital determines the potential or the possibilities of actors to access information and resources (Liao and Welsch 2001).

Relational Dimension

The relational dimension concerns the quality of personal relations people have developed over time, i.e. it focuses on the character of the connection between individuals. It has been interpreted as relational content, tie strength and relational trust. The relational dimension of social capital becomes evident in forms of behaviour, such as trust, willingness and the capacity to cooperate and coordinate. From a relational view, social capital exists in a relationship between two parties, who develop personal bonds, attachment, and trust (Arenius 2002 p. 14). The higher the degree of interaction, the more communication channels are available for use, the more easily actors are able to build up trust and trustfulness, the more easily information, resources and other forms of exchange can be effected within the network. Trust between parties promotes cooperative activity through which further trust can be generated. Thus, trust is the precursor to resource acquisition, knowledge combination and exchange (Liao and Welsch 2001).

Cognitive Dimension

Nahapiet and Goshal defined this dimension as those resources providing '*shared representations, interpretations and systems of meaning among parties*' (Nahapiet and Ghoshal 1998 p. 243). According to Arenius (2002 p. 15) it can be considered as second dimension of the quality of the relationship. It refers to the degree to which people share common codes, narratives or norms. The sharing may be among the dyadic partners or among the members of a network (Arenius 2002 p. 15). As Coleman (1990) highlighted a norm exists when the socially defined right to control an action is not held by the actor but by others. Thus, cognitive social capital provides a set of norms of acceptable behaviour (Anderson and Jack 2002).

Moreover, cognitive social capital supports the common understanding of collective goals and of ways to interact with one another, by helping parties to see the potential value of their exchange and combination of resources.

Referring to the formation and establishment of spin-offs, new venture formation is not only a process characterised by intentionality, resources, boundary and exchange, but also by the necessity to obtain organisational legitimacy in order to ensure acceptance of the market, and access and appropriation of external resources. Moreover, cognitive capital provides social support, a safety net, which allows the entrepreneur to break social norms in the process of risk-taking.

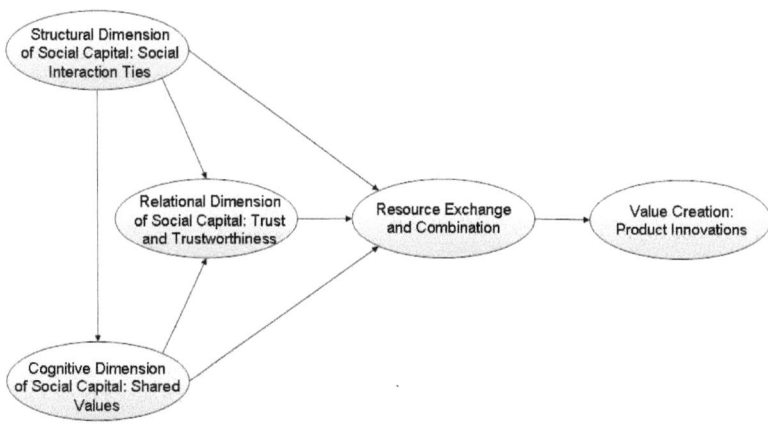

Figure 2-6: Model of Social Capital and Value Creation (Source: Tsai and Ghoshal 1998)

As Tsai and Ghoshal (1998) note all three dimensions of social capital interact with each other. Their analysis of the relation between the three dimensions and patterns of resource exchange and product innovation revealed that shared vision and centrality are sources of trust and trustworthiness. An actor occupying a central location in a network is likely to be perceived as trustworthy by other actors in the network. Trust and trustworthiness in turn increase resource exchange and combination, and resource exchange and combination increase product innovations. Their findings imply that social capital contributes to product innovation and value creation, and that dimensions of social capital reinforce the creation of other dimensions (Arenius 2002 p. 62).

2.2.4 Network Aspects of Social Capital

As the previous chapter reveals, the fundamental proposition of social capital theory is that network ties provide access to resources and information (Liao and Welsch 2003). As social capital is invisible, ubiquitous, and hard to pin down, it is usually studied in terms of its manifest structures and effects (Arenius 2002). This means that people, who research social capital most of the time study relationship networks (Nahapiet and Ghoshal 1998).

As a result social capital is often operationalised through the identification of networks and the respective network structure (Davidsson and Honig 2003). In this context, there are two major but somehow contradictory concepts referring to the impact of network structure on the generation of social capital: Coleman's (1990) closure argument and Burt's (1992) concept of structural holes.

2.2.4.1 Structural Holes

One way to operationalised social capital is the concept of structural holes coined by Burt (2000; 1992). It provides a good instrument to measure the efficacy of a person's social capital (Degenne and Forsé 1999 p. 117). According to Burt social capital inherent in network structure consists of information and control advantages due to the bridging of structural holes.

Based on Granovetter's (1973; 1982) strength of weak ties, Burt (1992 p. 28) defines structural holes as people or groups that are otherwise not connected thus representing sources of non-redundant information. Individuals whose relationships span such holes are in a unique position to obtain new information and broker the information flow between people as well as to control the activities that bring together people from opposite sides of the hole (Burt 2000). For Burt social capital is a function of the structural holes between connections. The term he uses for the measurement of social capital is "network constraint", which captures both the size and density effects of social networks with the size of a network decreasing constraints (i.e. adding structural holes) and density on the contrary increasing constraint (i.e. reducing structural holes).

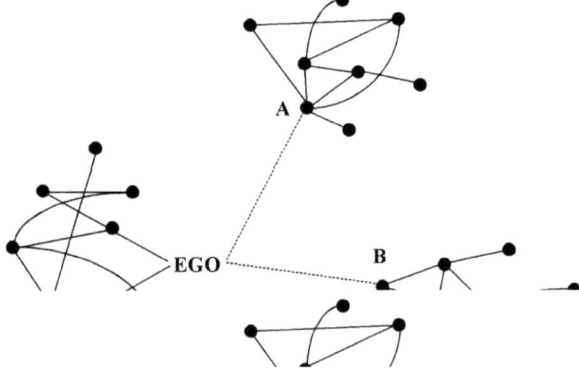

Figure 2-7: Structural Holes and Weak Ties (Burt 1992 p. 27)

Figure 2-7 illustrates Burt's notion of structural holes. A, B, and C are components of a network, so called clusters. Points or nodes represent network actors and the lines or edges are the relations between them with continuous lines relating to strong and dashed lines relating to weak ties. Ego, in the middle, serves as bridge over a structural hole, since without ego the clusters A, B, and C would not have been connected.

2.2.4.2 Closure

Whereas for Burt the source of social capital lies in sparse networks rich in structural holes, Coleman (1990) on the opposite sees the value of social networks in their cohesiveness. According to him dense networks with a high level of connection between actors, promote the development of social norms and sanctions that facilitate trust and co-operative exchanges. Mutual trust diminishes behavioural uncertainty and reduces opportunistic tendencies allowing for a freer and greater exchange of fine-grained, high quality information (Liao and Welsch 2003). Also actors' readiness to provide proprietary know-how is enhanced, if they have a closer relationship to alters. The amount of social capital accruing to an actor can therefore be defined as a function of the closure of the network surrounding him (Gargiulo and Benassi 2000). Granovetter approaches the same argument from a sociological perspective, by stressing the positive effect of common third parties making trust more likely by increasing the threat of sanctions (Granovetter 1985).

2.2.4.3 Combining the Two Approaches

Despite the seemingly contradictory character of Burt's and Coleman's theories, both concepts can be jointly applied when studying the impact of network structure in the spin-off process. On the one hand, an academic entrepreneur needs access to diverse and non-redundant information, in the sense of information not available from more than one source, as provided by bridging of structural holes. In particular relations to actors from other communities or institutions furnish the necessary novel inputs and knowledge and bring about multidisciplinarity required for the recognition and creation of entrepreneurial opportunities in an academic context. To be able to combine and aggregate different scientific and commercial know-how, the entrepreneur has to draw on a broad and multifaceted knowledge base allowing for an optimal leveraging of opportunities.

On the other hand, spinning off a new venture based on a recognised opportunity is characterised by a high degree of uncertainty and risk. To reduce these drawbacks resources of information whose reliability can be taken for granted are required. Moreover, in particular with respect to critical or proprietary information or knowledge, actors disposing of such assets will only be willing to disclose them to parties whom they can trust either because of frequent contact (strong ties) or because of trusted third parties warranting for the alter's trustworthiness. Furthermore, when it comes to searching for strategic partners or investors, trust as generated in dense networks increases an academic entrepreneur's attractiveness to potential allies.

Past research has revealed that networks either primarily consisting of structural holes or primarily being characterised by dense and connected relations, both have their drawbacks. Uzzi (1996) demonstrates the effect of varying network constraint in his study on the New York apparel industry. Opposing arm's-length ties (weak ties) to embedded ties (strong ties), he found that increased embeddedness, i.e. the presence of strong ties, promotes economic performance by enabling resource sharing, co-operation, and co-ordination among firms until a certain threshold is reached after which the positive effect of embeddedness reverses itself. Since an actor's capacity to maintain network ties is limited, too many strong ties compensate their advantages by paralysing firms in a way that they are sealed off in the network from new information or opportunities outside of the network.

From the above it may be concluded that the two different network specifications of social cohesiveness and structural holes indicate the necessity of both features in an entrepreneur's social network in order to successfully perceive and realise relevant entrepreneurial opportunities. In this respect and following the research design of Nicolaou and Birley (2003a) a contingency approach is chosen reconciling the two sources of social capital. The content of the network as a contingency factor examines how the value of social capital varies with different types of relationship (Burt 2000).

2.3 Resource-based Theory

The resource-based view of the firm is one of the theories in strategic management and entrepreneurship literature that has received considerable attention in recent years (Conner 1991). The reason for this development is that compared to the traditional focus on the external environment of the firm (industry structure and relative competitive position) an analysis from the viewpoint of the firm provides a more stable basis for long-term strategic planning in an increasingly dynamic and uncertain environment.

A resource-based approach to strategic management concentrates on costly to imitate attributes of the firm as sources of economic rents and, thus, as central drivers of performance and competitive advantage (Conner 1991).

2.3.1 Background

The perspective traces back to Edith Penrose (1959), who pioneered by founding the idea of viewing the firm as bundle of resources. The term "resource-based" as such was introduced by Wernerfelt (1984) in his characterization of firms as collections of resources rather than sets of product-market positions. Sustained competitive advantage is generated by the uniqueness of this bundle of resources at the core of the firm (Conner and Prahalad 1996). In other words, it is the heterogeneity, not the homogeneity, of the productive services available from its resources that give each firm its distinctive character (Kostopoulos et al. 2002). The theory addresses the central issue of how superior performance can be attained relative to other firms in the same market and contends that superior performance results from acquiring and exploiting unique resources of the firm (Saffu and Manu 2004).

2.3.2 Resources and Capabilities

Generally speaking, *resources* are tangible and intangible factors that are tied semi-permanently to the firm and that may trigger a strength or a weakness (Wernerfelt 1984). While tangible resources may be usually traded on factor markets and are consequently easily identifiable, such as financial and physical assets, intangible resources rarely find factor markets and are thus poorly tradeable, comprehensible, and exactly assessable (Hall 1993), e.g. employees' knowledge, experiences and skill, firm's reputation, brands, patents. They are the input to the production process.

According to Barney (1991) resources can be grouped according to three categories:

(1) *physical capital resources* such as physical technology, plant, equipment or access to raw materials,

(2) *human capital resources* including training, experience, judgement, relationships, and insight of employees and managers, and

(3) *organisational capital resources* consisting of a firm's formal reporting structure, formal and informal planning, controlling and coordinating systems, as well as informal relations among groups inside and outside of the firms.

Other authors expand this view by considering also financial and technological resources and reputation (Grant 1991; Hofer and Schendel 1978; Williamson 1975), and relational, legal, and informational resources (Hunt 2000).

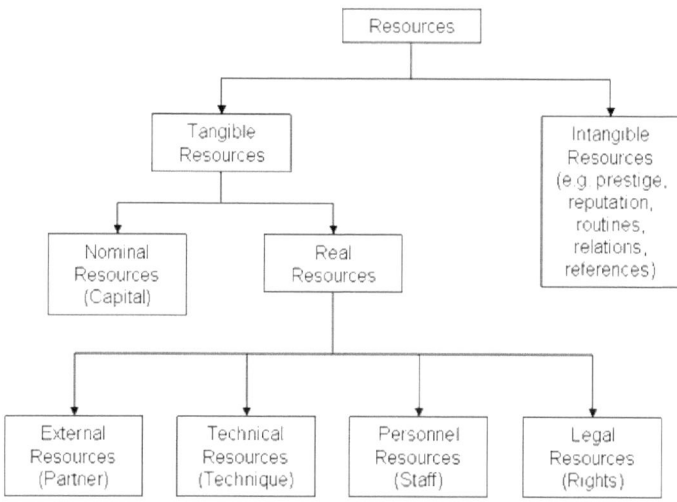

Figure 2-8: Types of Resources (Thudium 2005 p. 275)

Resources, however, are only one part of superior firm performance. Even though they are inputs into the production process, on their own, few resources are productive. Productive activity requires the cooperation and coordination of teams of resources (Grant 1991). As a result of the interaction and co-ordination of resource groups organisational capabilities emerge. They enable a resource group to perform a certain task or activity.

Capabilities thus refer to a firm's capacity to deploy and coordinate different resources to affect a desired end (Amit and Schoemaker 1993; Kostopoulos et al. 2002). They represent information-based, firm-specific processes developed over time as complex interaction of resources (Amit and Schoemaker 1993). Amit and Schoemaker (1993) also compare them with 'intermediate goods' generated by the firm to ensure enhanced productivity of its resources, strategic flexibility, and protection for its final product and services. Makadok (2001) advances this definition by specifying that a resource is an observable asset that can be valued and traded, such as a brand, a patent or a plant. A capability, in turn, is not observable, cannot be valued, and changes hands only as part of its entire unit. Further a capability, for example a marketing capability, can be valuable on its own or enhance the value of a resource, like a brand (Makadok 2001). A similar distinction is made by Foss and Eriksen (1995), who stipulate that resources can be both tangible (physical capital) and intangible (human capital), while capabilities are always intangible and emerge from the interaction between multiple agents (i.e. they are independent of individual agents). As Kostopoulos et al. (2002) summarise, the two characteristic features that distinguish capabilities from resources are that they are (1) firm-specific, since they are embedded in the organisation and its processes, and (2) intermediate, since their primary purpose is to enhance to effectiveness and productivity of a firm's resources.

As Grant (1991) notes, creating capabilities is not just a matter of combining a team of resources, but capabilities involve complex patterns of coordination among people and between people and other resources. To optimise such coordination learning by repetition is required. This refers to Nelson and Winter's (1982) concept of 'organisational routines'. Organisational routines in their understanding are regular and predictable patterns of activity, which consist of a sequence of coordinated actions by individuals. Consequently, a capability can be also regarded, in essence, as a routine, or a number of interacting routines, and the firm itself as large network of routines (Grant 1991).

Building on Nelson & Winter's (1982) view of the organization as a set of interdependent operational and administrative routines which slowly evolve on the basis of performance feedbacks, Teece, Pisano and Shuen (1997) define the concept of 'dynamic capabilities' as *"the firm's ability to integrate, build and reconfigure internal and external competencies to address rapidly changing environments"*.

Prahalad and Hamel (1990) use the term 'core competencies' to describe central, strategic capabilities, which refer to the collective learning in the organisation, particularly how to coordinate various production skills and combine different streams of technology. *Competencies* in turn differ from capabilities by being located on a higher level of aggregation. They do not refer to a singular or incidental capability to perform a task, but to the in-depth control of the interaction of various resources. To be core competencies, they have to be able to strategically position the organisation within its competitive environment (Thudium 2005 pp. 277-78). If, according to Prahalad and Hamel (1990), core competence refers to harmonising skills and streams of technology, it also has to do with the organisation of work and the delivery of value. Thus, core competence is communication, involvement, and a deep commitment to working across organisational boundaries. Similar to social capital, core competence does not deplete with use. Yet, competencies have to be cultivated and protected, as knowledge fades if it is not used (Prahalad and Hamel 1990).

2.3.3 Resources Properties for Sustained Competitive Advantage

As mentioned a basic assumption of the resource-based perspective is that the resource bundles and capabilities underlying production are heterogeneous across firms. This *heterogeneity* or asymmetry of resources is responsible for the differing success of companies within an industry.

Moreover, for the resulting competitive advantage to be sustainable, resources also have to be imperfectly mobile (Barney 1991). Resources are *imperfectly mobile* if they cannot be traded, for example because property rights are not well defined, or they are idiosyncratic to the extent that they have no use outside the firm (Peteraf 1993). Other resources are tradeable, but loose value while traded due to their firm-specific nature, bounded rationality or high switching or transactions cost involved.

In addition, to be a resource of sustained competitive advantage, resources and capabilities must be (Barney 1991; Hoopes et al. 2003):

- *Valuable*. Resources are valuable if they enable a firm to apply strategies that improve its efficiency and effectiveness, i.e. they improve its market position relative to competitors.

- *Rare*. A firm only enjoys a sustained competitive advantage when resources are available in short supply relative to demand.

- *Imperfectly imitable*. Valuable and rare resources can only be sources of sustained competitive advantage if firms that do not possess them cannot obtain them. Resources can be imperfectly imitable for one or a combination of three reasons:
  1. the ability of a firm to obtain them is dependent on *unique historical conditions*, which means that other firms without that particular path in history cannot get them;
  2. the link between the resources and the competitive advantage of the firm that possesses them is *causally ambiguous*, i.e. the relation between the resources of a firm and its competitive advantage is not or very poorly understood;
  3. the resource is *socially complex* and thus beyond a firm's ability to systematically manage and influence. Examples include interpersonal relations of managers, company culture or reputation.

- *Substitutability*. There must be no strategically equivalent resources or bundle of resources that can be exploited separately to implement the same strategies by different competitors.

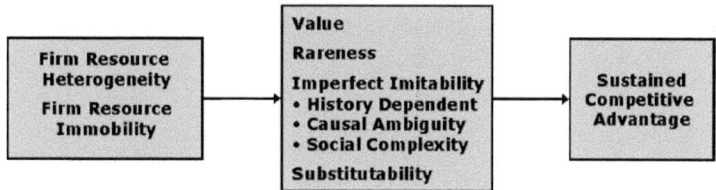

Figure 2-9: Relationship between Resource Properties and Sustained Competitive Advantage (Barney 1991 p. 112)

2.3.4 Strategy and Value Creation

As already mentioned resource-based theory has delivered important contributions to strategic management by turning the traditional focus on a firm's external environment (Porter 1980; Rumelt 1984) more towards its internal skills and resources as basis for the generation of sustainable economic rents. In the following, the impact of this resource-based perspective on strategy formulation is highlighted in more detail.

On a business level, strategy has been defined as *"the match an organisation makes between its internal resources and skills ... and the opportunities and risks created by its external environment"* (Hofer and Schendel 1978 p. 12). According to Mahoney and Pandian (1992) strategy can be also viewed as a continuing search for rent, where rent is defined as return in excess of a resource owner's opportunity costs.

As presented in the previous chapters, the resource-based approach argues that the firm's unique capabilities, such as technical know-how and managerial ability, are important sources of heterogeneity that form the basis of sustained competitive advantage (Mahoney and Pandian 1992). In particular, distinctive competence and superior organizational routines in one or more of the firm's value-chain functions may enable an organisation to generate rents from a resource advantage (Hitt and Ireland 1985). As Penrose puts it (1959 p. 54), a firm may achieve rents not because it has better resources, but rather because its distinctive competence involves making better use of its resources.

As Grant (1991) summarises, exploring these relationships between resources, competition, and profitability requires analysing competitive imitation (Lippman and Rumelt 1982; Reed and Defillippi 1990; Rumelt 1984), the appropriability of returns (Amit and Schoemaker 1993; Collis and Montgomery 1995), the role of imperfect information in creating profitability differences (Barney 1991), and the means by which the process of resource accumulation can sustain competitive advantage (Dierickx and Cool 1989). All these activities can be subsumed under the heading of "the resource-based view of the firm". Yet, Grant (1991) also locates a lack of a framework for a resource-based approach to strategy formulation that integrates a number of the key themes arising from this theoretical stream. To this end, he proposes the following organizing framework based on a five-stage procedure for strategy formulation (see Figure 2-10):

1) analysing the firm's resource-base,
2) appraising the firm's capabilities,
3) analysing the profit-earning potential of firm's resources and capabilities,
4) selecting a strategy, and
5) extending and upgrading the firm's pool of resources and capabilities.

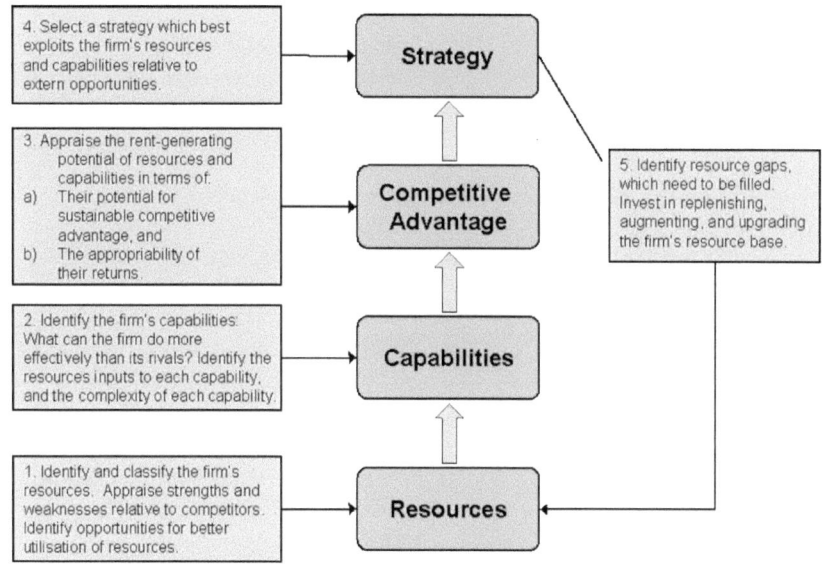

Figure 2-10: Resource-based Approach to Strategy Analysis (Grant 1991 p. 115)

An important addition to the considerations presented so far, is that Grant does not view the resource base of a firm as predetermined, with the primary task of strategy makers to deploy these resources in a way that maximises rents over time. He rather turns the focus on managing and developing the firm's resource portfolio. This involves replacing investments to maintain the firm's stock of resources and to augment resources to extend positions of competitive advantage and broaden the organisation's strategic opportunity set. To both fully exploit a firm's collection of resources, and to build up competitive advantage for the future, the external acquisition of complementary resources may be necessary (Grant 1991).

Departing from Grant's notion of developing a firm's resource portfolio, the resource-based approach may be also framed in a more dynamic context referring to a firm's effectiveness in deploying and working on its resources. This implies adopting a Schumpeterian perspective. Schumpeterian competition involves implementing 'new combinations' including new methods of production as well as organisational innovation (Makadok 2001). Mahoney and Pandian (1992) transfer this Schumpeterian competition into the resources-based framework by considering a firm's *"new combinations of resources"* (Penrose 1959 p. 85) as a means to achieve competitive advantage. This so-called 'dynamic-capability view' highlights the importance of an alternative rent-creation mechanism, i.e. capability building (Amit and Schoemaker 1993; Teece et al. 1997).

Another important contribution of resource-based theory to strategic management comes from Dierickx and Cool, who have coined the term of a 'strategic asset' (Dierickx and Cool 1989). According to them (Dierickx and Cool 1989) there are strategic factors that cannot be traded, but only be accumulated, such as reputation or R&D capabilities. Other examples are fast product development cycles, brand management, buyer-seller relationships or a firm's installed user base. As the authors put it (Dierickx and Cool 1989 p. 1506): *"The strategic asset is the cumulative result of adhering to a consistent set of policies over a period of time".*

Based upon this notion, Amit and Schoemaker (1993) further conclude that when making investment decisions concerning strategic assets, managers are basically confronted following three challenging tasks:

(1) anticipating possible futures,

(2) assessing competitive interactions within each projected future, and

(3) surmounting organizational inertia and internal discussion in order to realign the firm's bundle of strategic assets to provide for a sustained unique market value.

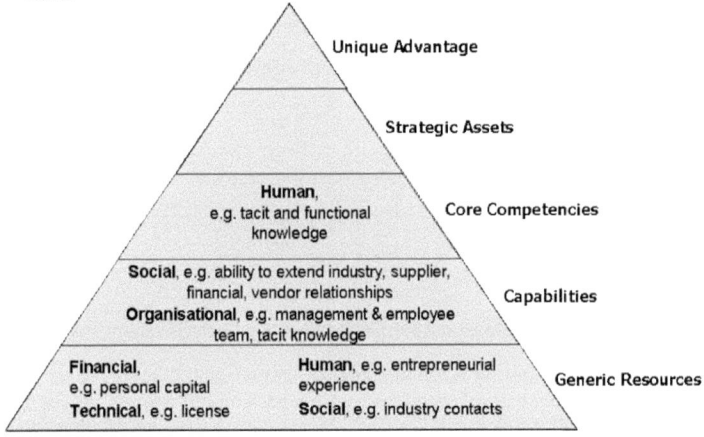

Figure 2-11: Resource Pyramid of Value Creation (Brush et al. 2001 p. 71)

Brush et al. (2001 p. 71) summarise the way of transforming generic resources into a sustained unique market value in their resource pyramid of value creation (see Figure 2-11). Following, the Schumpeterian notion of innovation, the development of basic resources into unique resources involves combining and configuring them into increasingly complex combinations. Thus, the initial resource combinations ascend the pyramid of resource development by being interrelated in a way that generates capabilities and competencies up to the accumulation of strategic assets as basis for sustained unique advantage. As Brush et al. note (2001 p. 76) novelty and flexibility are essential (Penrose 1959), in particular when business models adapt to new technology, as in the case of creating new ventures, which will be presented in more detail in the next chapter.

2.3.5 New Venture Creation and Innovation

Previous work (Conner 1991; Rumelt 1984) on resource-based theory considers entrepreneurship as integral part of the resource-based framework. Since then, an increasing number of entrepreneurial researchers have applied the resource-based theory of the firm to get a better understanding of the role resources play in new venture creation and establishment (Chandler and Hanks 1994; Mosakowski 1993). In line with this research, entrepreneurial actions are analysed as the creation and combining of resources that generate new heterogeneous resources. According to this perspective, the entrepreneurial act involves suggesting alternative uses of resources that have not been previously discovered leading to unique competitive advantage (Alvarez and Busenitz 2001). The entrepreneur may choose matching resource combinations to a specific industry environment, or building up combinations that are new (Miller and Shamsie 1996).

Following Brush et al. (2001), six types of resources can be identified as key to the spin-out process: human, social, financial, physical, technology, and organizational. Moreover, decisions about how to combine resources benefit from an assessment of the entrepreneur (Brush et al. 2001 p. 76). This is one of the reasons why the person of the founder is attributed such a central role in the operation of the new venture. To the extent that the business owner makes all the important decisions, his skills become a critical asset, on which the success of the firm depends (Saffu and Manu 2004). As Brush et al. (2001) further expound, when the firm comes into existence, resources unique to the founder have to be transformed into organisational resources, and to pass the combinatory process to form a unique advantage. Creating such an advantage involves formalising and transforming resources into specialised capabilities that can be integrated into higher order clusters of competencies, enabling the new venture to outmatch others in the industry (Brush et al. 2001).

According to Nelson and Winter (1982), firms typically possess pockets of excellence, i.e. certain activities they are good at, such as technological expertise or integrated procedures. The entrepreneur can advance or extend these starting endowments generally into the business. However, when looking at the creation of new ventures from a resource-based perspective, it becomes obvious that in the early stages of new venture development it is the identification and acquisition of resources, rather than deployment or allocation activities, that affect the long-term success of a young organisation (Bergmann Lichtenstein and Brush 2001). Especially start-ups depend for their resources upon their environment (Heirman et al. 2003). Yet, in addition entrepreneurs sometimes also have to spin off resources that are negative or not productive (Stevenson and Gumpert 1985), which is not always easy to realise. However, continually assessing whether the venture has the right resources for the opportunity, can help prevent resource weaknesses or inadequacies (West and De Castro 2001).

Although new ventures start with the entrepreneur's initial resource endowments, they cannot survive over time without acquiring and developing additional organisational resources (Chandler and Hanks 1998). As a new venture passes its phases of development, it becomes necessary to develop and transform resources (Penrose 1959) to preserve the right fit of resources to changes in the product-market strategy (Chandler and Hanks 1994) and in the environment. Thus, shifts in the types and combinations of resources are characteristic of the growth and development of new ventures (Mosakowski 1993; Penrose 1959). Related performance outcomes, in turn, influence which resources are identified, acquired, or developed next, as a means to extend growth or improve fit in dynamic environments (Lichtenstein Bergmann and Brush 2001).

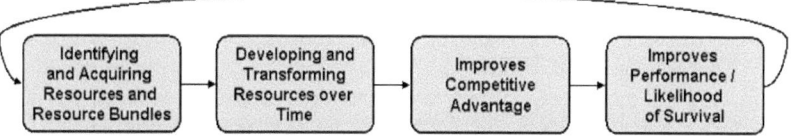

Figure 2-12: Dynamic Model of Resource Acquisition, Development, and Effects in New Ventures (Lichtenstein Bergmann and Brush 2001 p. 38)

2.4 Summary

Relating the previous theoretical discussion to the focus of this research, its main objective can be summarised as analysing the entrepreneurial act of spinning off new ventures from universities from a resource-based perspective with social capital as resource and access to resources.

The resource-based perspective turns the focus towards the skills and assets considered at large to constitute a firm. When these resources are combined in unique and inimitable manner they are assumed to form the basis of sustained competitive advantage. As depicted, resources need not necessarily be tangible, physical assets, but may be, and already are in an increasing number of cases, intangible assets, such as employees' knowledge, experiences and skill or a company's reputation on the market. Grant (1991) comes up with the increasing relevance of non-physical resources in the today's knowledge-based economy by accounting the knowledge-based view as an extension of the resource-based view to the extent that knowledge is viewed as the most important resource of the firm. The processes of knowledge acquisition, creation, dissemination and integration are in the focus of this perspective (Mäkinen 2001). A firm's knowledge base is therefore influenced and partly derived from its embeddedness in its local social and industrial context, and in the habitus of its managers and workers (Meso and Smith 2000).

However, what really makes the difference in the quest for profit in a competing environment is the company's ability to transform its resources into superior capabilities or organisational routines. As highlighted earlier, this requires learning procedures following repeated interactions also as important prerequisite for the creation of dynamic capabilities, integrating, building and reconfiguring internal and external competencies to address rapidly changing environments.

This is where the role of networks comes into play. It is Bouty (2000), who explicitly addresses the relevance of the exchange of informal resources between researchers across organisational boundaries as foundation for innovation due to the learning processes involved. As she notes in more detail: *"Researchers develop collaboration strategies and networks of allies through information and service exchanges. Thus, the social dimension is essential to the individual exchange decision."*

As Mäkinen (2001) points out, networks are of particular importance to small companies. Firms enter into networks or alliances for a wide range of different reasons. Access to external resources and new product ideas, risk sharing and inter-organisational learning are frequently cited reasons (Bellini 1999; Foss 1999; Sydow and Windeler 1998). Summing up, companies use networking as a way of closing the resource gap between available and needed resources (Mäkinen 2001).

Yet, while the network theory and the resource-based view of the firm have become important paradigms of management research, entrepreneurship researchers continue to show limited interest in studying resource strategies and the role of networks in building dynamic capabilities in entrepreneurial start-ups. Only little can be found in the literature about new venture resource strategy to explain how entrepreneurs continuously enhance, manage and develop resources to match the shifting product-market strategy by using networks on a sustainable basis (Ramachandran and Ray 2006).

This is where the underlying research project digs deeper. When looking at the networking capabilities of a single entity, i.e. the company or in this context the entrepreneur, it is the individual level of social capital that comes into play. Focusing on individual social capital the following definition of social capital is adopted in the course of the subsequent analysis:

> *Social capital is the amount and quality of the external relationships possessed by an individual.*

The quality of relationships encompasses both the relational and cognitive dimension of social capital as introduced by Nahapiet and Ghoshal (1998). In order to dispose of social capital, an individual must have external relationships. In addition to the mere existence of relations, also the quality of these relationships is of importance. Following Davidsson and Honig (2003), in this research social capital is applied in terms of social exchange (Emerson 1972), to examine the effects of exchange ties on performance.

In this research project the perspective of resource-based theory is chosen to incorporate the concepts of entrepreneurship, in particular when applied to the special case of academic spin-offs, and the social capital approach operationalised in terms of social networks in one common theoretical model to be empirically tested within the scope of the dissertation. The set-up of this model and its major components are highlighted in the next chapter.

3 Research Model and Hypotheses

Generally, theory is defined as a system of logically consistent propositions about social phenomena (Atteslander 1993 p. 43). In empirical research theories have to be limited to propositions that can be empirically validated. In the main, theories are decisions about the relevance and conditions of ascertainable occurrences of social reality.

On the basis of the theoretical concepts introduced in the previous chapter, in the next section a model is established to be subjected to empirical investigations within the scope of this research. In this regard the dissertation can be seen as further contribution to the production of scientific growth of cognition as depicted in the "dialectic" spiral of theory development of Figure 3-1.

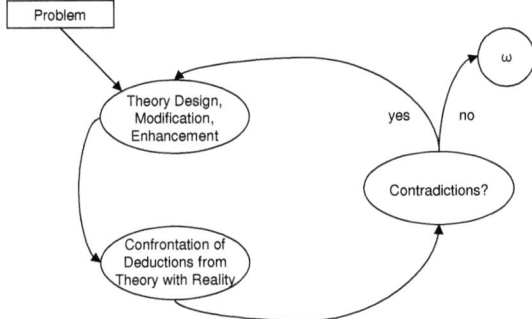

Figure 3-1: "Dialectic" Spiral of Theory Development (Roth and Heidenreich 1993 p. 334)

According to this spiral process, departing from a research problem a preliminary model of explanation is drafted and confronted with reality. In case of contradictions between the predictions of the theory and the real data, the theory is modified and expanded. Again it is confronted with reality, etc. In the course of this cycle, the theory is further advanced, and becomes more complete, differentiated and applicable (Roth and Heidenreich 1993 p. 333).

Following this process, a first model incorporating assumptions on the effect of social capital in the academic spin-off formation process is drafted and translated into empirically revisable statements in the form of deducted hypotheses. In this regard the main principle of empirical research methodology – as represented by the Critical Rationalism – is made allowance for: All propositions of an empirical science have to permit their examination in their confrontation with reality (Kromrey 1994 p. 33).

To fuse the concept of social capital with entrepreneurship theory in one model, resource-based theory is applied. Resource-based theory in the context of entrepreneurship stipulates that nascent entrepreneurs require a set of different resources for the establishment of their new venture. When combined in a unique way these resources provide the entrepreneur with the necessary capabilities or core competences to offer added value to the market. Resources can be tangible or intangible and in line with Burt (1992 p. 8) may be grouped in financial, human and social capital. The research on-hand focuses on the latter form of capital in both of its functions: a form of resource and access to other resources (Boari and Presutti 2004). In particular the effects of social capital on the opportunity recognition and new venture formation process are analysed. In this context a contingency approach is pursued assuming that different forms of social capital are important with respect to different relational contents and at different stages of the entrepreneurial process.

3.1 The Spin-off Process and Its Phases

Greve and Salaff (2003) apply Wilken's (1979) breakdown of the entrepreneurial establishment process comprising of:

(1) motivation phase

(2) planning phase

(3) establishment phase

In the first phase, the *motivation phase*, entrepreneurs discuss the initial idea and develop their business concept. In the *planning phase* provisions are made to set up a firm, while in the last phase, the *establishment phase*, the firm is actually established and run. Elfring and Hulsink (2001; 2003) also assume a trichotomy of the entrepreneurial process. They distinguish three sub-processes in which different network properties play a role:

1. The ability to discover an *opportunity*
   Networks and particularly weak ties provide access to a broad range of information that helps the entrepreneur to spot opportunities. They also raise the alertness of entrepreneurs.

2. The ability to secure *resources*
   Networks and in particular strong ties play an important role in gathering the necessary resource to exploit the identified opportunity. Network actors to whom the entrepreneur has strong tie relationships are in general more eager to provide the required resources than occasional weak tie relations. Social transactions support the entrepreneur in accessing the resources below the market price.

3. The ability to gain *legitimacy*
   When starting something considered innovative obtaining legitimacy is crucial. Faced with Stinchcombe's (1965) "liability of newness" a new venture has to organise institutional support and legitimacy.

Regarding the specific form of academic spin-offs also in this case a stage-based perspective prevails within relevant research literature. Vohora et al. (2004) for example identify five different phases university spin-outs have to pass during their development from initial research finding to a sustainable company. According to them, academic spin-offs emerge not so much through discrete stages of growth but rather through non-linear "phases" of development separated by critical junctures that have to be overcome in order to progress towards the next phase of development (Vohora et al. 2004).

Clarysse and Moray (2004) suggest that the founding of a spin-off can be seen as a process characterised by three different stages. The first, the so called *"invention"* phase, is a period of high research intensity during which technical uncertainty prevails. In the second, the *"transition"* phase, technical uncertainty decreases and the business idea is validated. The last phase is the *"innovation"* phase, where the validation of growth expectation is carried out. The whole spin-off process is represented as a funnel (see Figure 3-2), with a declining number of ideas from the invention phase to be validated as having an economic value for a new venture. During the transition phase, still fewer business ideas are deemed to exhibit growth expectations and consequently enter the innovation or business development stage (Branscomb and Auerswald, 2001).

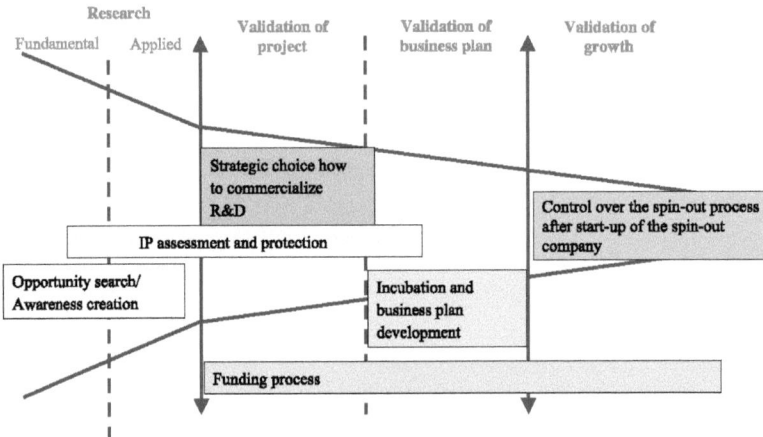

Figure 3-2: The Spin-off Funnel (Clarysse et al. 2005 p. 187)

To enable entrepreneurs to move through the different stages of the spin-off process, there may be a need for support from the parent research institution in the sense of an incubating role. Degroof (2002) has defined the following activities of a proactive spin-off management process that have been incorporated by Clarysee et al. (2005) in the conception their spin-off funnel. First, opportunity search consists of trying to identify technologies with a commercial potential. Second, intellectual property assessment involves investigating if patents have already been filed for the specific technology and, if not, perhaps seeking for according legal protection. This step also includes examining whether licensing or commercializing through a spin-off venture are the preferred options. Third, appropriate spin-off projects are selected based on their intrinsic potential and on the comparison with alternative projects. Fourth, business plans are developed. Fifth, research institutions have to direct their spin-offs towards potential sources of funding. Sixth, as soon as funding is secured, the venture can be formally incorporated, and spin-off coaching started.

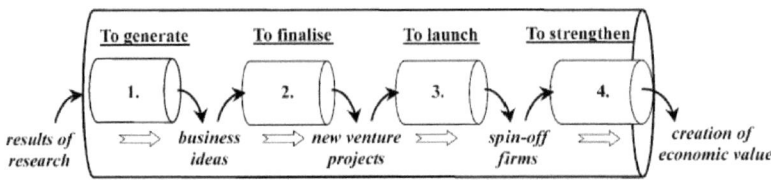

Figure 3-3: The Global Process of Valorisation by Spin-off (Ndonzuau et al. 2002 p. 283)

Ndonzuau et al. (2002) also propose a stage model approach to the academic spin-off process (see Figure 3-3). Yet, they identify four stages as relevant in explaining the transformation of academic research results to economic value:

Stage 1: to generate *business ideas* from research

Stage 2: to finalize *new venture projects* out of ideas

Stage 3: to launch *spin-off firms* from projects

Stage 4: to strengthen the *creation of economic value* by spin-off firms

According to them (Ndonzuau et al. 2002) the four stages are not wholly independent of each other. Rather is the process neither straightforward nor spontaneous, but interlard with obstacles, difficulties, impediments, and other sources of resistance.

Stage 1: generating business ideas

At the first stage ideas with a potential for commercialisation are generated and assessed. Typical problems the nascent spin-off has to face at this stage relate to the academic culture and weak internal identification and assessment capabilities. Most universities still adhere to the traditional "scientific" paradigm according to which the sole purpose of academic research consists in generating further knowledge in the form of publications and to pass on this knowledge by education. This attitude often significantly impairs recognising the commercial value of one's research results and even if recognised hampers their commercial realisation due to adverse effects of publishing on economic attractiveness and patenting. Moreover, academic entrepreneurs often lack the required know-how to assess the commercial potential of an idea. They may have the scientific-technological knowledge to perform an evaluation of its technological feasibility, yet this alone is not sufficient. From the perspective of business exploitation the commercial viability must also be assessed to verify the extent to which there might be a viable market (Ndonzuau et al. 2002).

Stage 2: finalising new venture projects

At the second stage the ideas generated in the first stage are considered and the most promising one translated into genuine entrepreneurial projects. Major issues to be covered at this stage comprise the protection and the development of the idea to transform it from a vague concept into a coherent and structured venture creation project. Following Ndonzuau et al. (2002) development has to be effected on a technological and on a commercial level. The purpose of the technological development consists in verifying the possibilities of industrial exploitation by means of a first prototype. In addition, the prototype may also be applied to demonstrate to potential customers and partners what the technology can achieve. In the course of commercial development a business plan is elaborated laying down the main business strategy and its main elements such as investments and turnover. In this regard, the business plan also serves as a selling document for bankers and investors by providing a structured picture of how the results are intended to be exploited. Certainly the most crucial point is financing this development process. While on the one hand public funding is essentially dedicated to fundamental research, on the other hand very few private financial backers (i.e. venture capitalists) invest so early in the process because of the unpredictability and instability of the high-technology market and the supposedly low entrepreneurial capabilities of researchers (Ndonzuau et al. 2002).

Stage 3: launching spin-off firms

The third stage deals with setting-up a new venture to exploit an opportunity managed by a professional team and supported by available resources. According to Ndonzuau et al. (2002) two problems have to be faced with at this stage: (1) the availability of resources, and (2) the relationship between the spin-off and its mother university. Besides material and financial resources, intangible, human resources also play a vital role. Given academic entrepreneurs' lack of commercial know-how, people disposing of the required management expertise have to be identified and attracted. One issue in this context is access to outside coaches for mentoring the management team (Radosevich 1995).

Stage 4: creating economic value

The purpose of the last stage is to consolidate and strengthen the economic value created by the new firm. This implicates providing for the necessary conditions and infrastructure to bar young spin-offs from leaving the region (Ndonzuau et al. 2002). Moreover, due to insufficient funding originally product-oriented spin-offs may decide to change their offer to mere service provision such as consulting, an activity initially set up to temporarily bridge financing gaps and reduce risks.

Taking the model of Ndonzuau et al. (2002) as basis and following the spin-off funnel of Clarysee et al. (2005), the dissertation project focuses on the first three phases of the academic spin-off process. Yet, for the purpose of this research the following three stages have been identified:

(1) **opportunity recognition**
(2) **formation**
(3) **establishment on the market**

The first stage refers to the process of identifying and evaluating research ideas with a potential for an entrepreneurial *opportunity* out of the academic context. In the course of their research activities academic entrepreneurs produce and accumulate a large pool of valuable intellectual property, which then generates the potential opportunity for commercialisation (Vohora et al. 2004). Yet, potential opportunities have to be recognised by the researcher and upon recognition assessed according to their technological and commercial feasibility and attractiveness to filter out the most promising ones. During this stage entrepreneurs have a heightened demand for extensive and diversified information to perform these tasks.

At the *formation* stage most promising opportunities are further elaborated and advanced to provide for more detailed technological and economic specifications. A prototype is produced demonstrating the technical feasibility of the intended development. Potential fields of application are identified and screened for their attractiveness. Most importantly a business plan is set up constituting a fundamental basis for negotiations with strategic partners and venture capitalists. A crucial point in this context is securing necessary funds to finance these development activities and provide for a solid foundation for the next stage. This comprises gathering resources and information about sources of resources.

In the *establishment* phase the new venture is actually spun off from its institutional university background and has to assert itself on the market. For that purpose in principle two types of resources are required: human capital and physical capital (money and material). In the first case personnel for the spin-off and external coaches and consultants have to be located and attracted that provide for the required management expertise often lacking in the person of the scientific entrepreneur. Referring to the second case start-up firms of course also need both material and financial resources. A key issue in this context is to gain the necessary legitimacy for long-term success. Since financial providers view start-ups, and in particular university spin-offs, as very risky due to their lack of track records, they ask for respective securities by explicit certification from well-established individuals and organisations (Hoang and Antoncic 2003; Wright et al. 2004b).

3.2 Forms of Social Capital

According to a sociological perspective of entrepreneurship social capital is considered as a main factor of success for the establishment and growth of young firms (Aldrich and Zimmer 2001). Firm formation is presented as a social phenomenon where individuals within a social network decide to found firms (Grebel et al. 2003). In this context, social capital serves as both as resource and as access to resources such as human and physical capital.

The development of the entrepreneur's social capital through the stages of the venture formation process is analysed with respect to its diverse occurrences in the form of different network properties. In this context networks have several useful properties for entrepreneurs (Greve and Salaff 2003). One of them is their *size*. With expanding personal networks entrepreneurs also broaden their access to useful information and resources from other persons. During the entrepreneurial process entrepreneurs are assumed to expand their networks as a result of their entrepreneurial activities (Nicolaou and Birley 2003a).

Another property is the *structure* of the network, whether it is sparse or dense. According to Coleman (1990) dense networks promote trust and co-operation. On the other hand are sparse networks more likely to feature structural holes, a source of non-redundant novel information (Burt 2000; Burt 1992 p. 28). Thus, depending on the content to be provided through the network relations as well as on the stage of the entrepreneurial process either one or the other structural property is more beneficial.

The personal network of an entrepreneur can be decomposed in different subsets regarding to the *content* of the interaction. According to Larson and Starr (1993) the emerging organisation basically requires three types of resources: information, physical and financial resources, and symbolic support (approval and legitimacy). Following this approach information networks, resource networks and support networks are analysed. While the first consist of the people the entrepreneurs address to discuss aspects of establishing and running a business (Greve and Salaff 2003), thus representing a valuable source of information, resource networks are persons that provide the entrepreneur with required assets such as property, capital and credit. And finally the people in the support network back and encourage the entrepreneurs with emotional and institutional support.

Relating to *tie strength* it is possible to differentiate between bridging and bonding capital (Putnam 2000). Bonding refers to relations among relatively homogenous groups such as family members and close friends and is similar to the notion of strong ties. Bridging social capital regards relations with distant friends, associates and colleagues. These ties tend to be weaker and more diverse but support progress.

Following Schenk (1983) relations can be also distinguished according to the nominal *affiliation* of an entrepreneur's network partners to a certain group. To account for this notion a general distinction between an entrepreneurial researcher's private, academic and business network is made. Whereas the private network consists of family members, friends and acquaintances, academic contacts can be classified into relationships to members of the same discipline as the entrepreneurial researcher and to members of other disciplines. Finally, the business network comprises of industry contacts, capital providers, and members of public authorities or of coaching and consulting institutions. The distinction in different actor groups is also applied to account for network heterogeneity or homogeneity (Jansen 2000).

3.3 Integration with Resource-based Perspective

To consolidate the entrepreneurial process of academic spin-off formation and the concept of social capital from the perspective of resource-based theory the causal model of Christensen et al. (2000 p. 10) is applied as provided in their article "The Entrepreneurial Process In a Dynamic Network Perspective".

In this model the authors integrate human, financial and social capital as forms of resources (Burt 1992 p. 8) to explain the success of an entrepreneurial venture. In doing so, they distinguish two levels of analysis: (1) the entrepreneur/individual level focusing on the entrepreneur as individual actor, and (2) the company level, which refers to the newly established firm.

Coming to the contribution of social capital, individual and collective social capital is differentiated. While the former is composed of the social relations surrounding the entrepreneur, the latter is the result of all social interactions taking place in a given society. Human capital is defined as resulting from an entrepreneur's experience and educational background both in general as well as specifically related to the entrepreneurial activity. Financial capital consists of personal funds and general funds provided by development agencies, government-backed loans, and venture capitalists.

According to Christensen et al. (2000 p. 12) all three types of capital relate to each other. While for example social capital in the form of contacts to resource providers helps to obtain external funding, later on, the firm's success may allow acquiring new human capital needed thus creating additional social capital. In general, social, human, and financial capital come into play both before the launch and during the initial phases of a new venture, since they are needed to establish and run a firm.

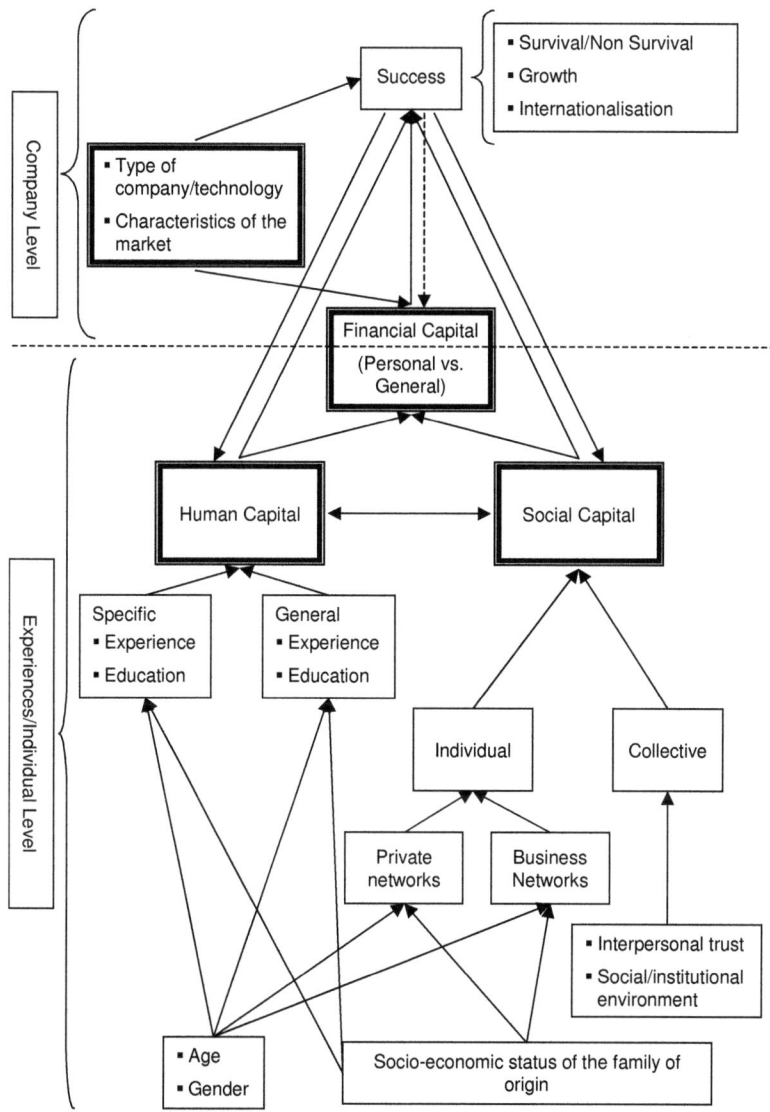

Figure 3-4: Causal Model for the Explanation of the Entrepreneurial Success Path
(Christensen et al. 2000 p. 10)

3.4 Modelling and Generation of First Hypotheses

When analysing the influence of social capital on opportunity recognition and formation and the establishment of academic spin-offs, focus is put only on a section of the above-described model expanding it with additional assumptions.

According to the focus of the dissertation on the social capital of academic entrepreneurs, research is restricted to the level of the entrepreneur. It is the individual person that is of main interest not the future company as an entity. In this context, object of investigation is the individual social capital of the academic entrepreneur not its collective realisation.

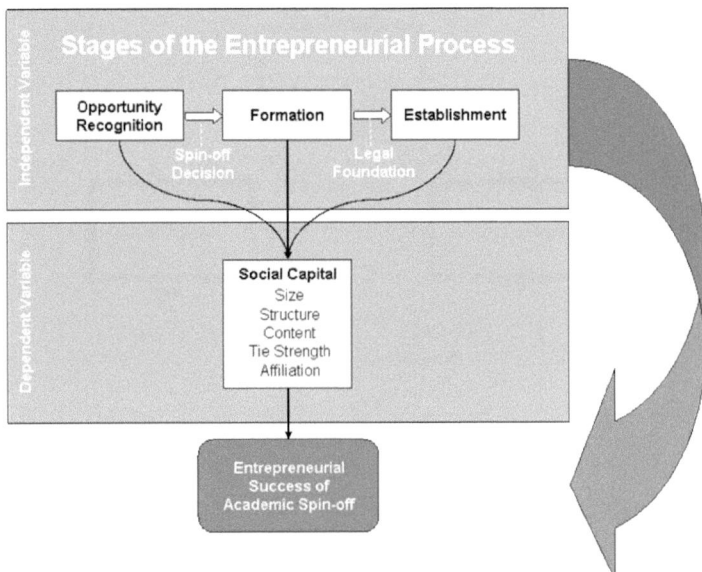

Figure 3-5: Model of Social Capital Impact on Entrepreneurial Success

Against the background of the considerations of Chapter 3.2 this individual social capital in its realisation as ego-centred networks is further analysed according to network size and structure, content and strength of relationships and categorical affiliation of network partners (see Figure 3-5). Financial capital flows into the model in the form of content provided through social networks. To account for the influences of human capital, the entrepreneur's experience and scientific background are levied for control purposes. Due to the relative homogeneity of the academic sample regarding the educational background of its members, no significant contributions are expected from this variable. Thus education is not included in the model. However, gender and age are considered.

For the generation of first hypotheses the preponderance of different realisations of social capital at different stages of the entrepreneurial process is postulated (Greve and Salaff 2003). Following the resource-based perspective it is assumed that successful academic entrepreneurs entertain specific network structures to mobilise the resources they require at the respective stages. Yet, depending on the phase of the research commercialisation process resource requirements and thus forms of social capital vary for the succeeding entrepreneurial researcher.

3.4.1 Opportunity Recognition

The recognition of possible entrepreneurial opportunities is the first prerequisite for the successful formation of an academic spin-off. For that purpose the academic entrepreneur has to put his / her technological research findings in the economic framework of the market comparing the fit of his supply with existing or emerging demand. This comparing analysis requires the entrepreneur to be able to draw on a broad range of information. However, individuals are limited in their ability to process and store information, which results in bounded rationality (Hills et al. 1997; Simon 1976; Singh et al.; Singh et al. 2000). The network of the entrepreneur can help to expand these boundaries by broadening his / her knowledge and information base (Hills et al. 1997).

Hills et al. (1997) found that entrepreneurs' networks are important for the opportunity recognition process. According to them entrepreneurs with extensive networks are more likely to identify opportunities than solo entrepreneurs. Especially weak ties as sources to non-redundant information (Granovetter 1973; Granovetter 1982) benefit the recognition of entrepreneurial opportunities, since they lead to a more varied set of information and resources. Academic entrepreneurs usually dispose of a large stock of very specialised knowledge in their respective discipline. This represents an important foundation to build upon, i.e. the basic idea. Yet, to make this idea a real opportunity additional knowledge coming from other scientific areas or even from beyond academia are required and can be accessed via the researcher's personal network.

Networks not only provide access to valuable information promoting the identification of opportunities, but also reduce uncertainty related to the their exploitation by delivering risk-reducing signals about a venture opportunity (Hulsink and Elfring 2003). Yet, they are valuable sources supporting the entrepreneur in locating and evaluating the opportunity.

Singh et al. (2000), too, found in their research that the size of an entrepreneur's network was positively related to the number of new venture opportunities recognised. However, although a large network offers more information in case the network is dense information will be rather redundant. Thus, Singh et al. (2000) investigated whether the structure of the network, i.e. the prevalence of structural holes, contributed to the opportunity recognition process by providing access to more expansive and diverse knowledge compared to dense networks.

Besides the availability of diverse information, the entrepreneur's alertness, defined as the ability to notice, without search, opportunities previously overlooked (Kirzner 1997), is another important prerequisite for identifying opportunities. In this context, weak ties raise the alertness of entrepreneurs and thus set into motion a chain of events started by spotting an opportunity (Elfring and Hulsink 2001). This is particularly vital with regard to recognising opportunities out of the university context. Due to the academic culture researchers are not trained and often even discouraged to view their research output in the light of an economic and more application-oriented perspective. Academic entrepreneurs may be at a disadvantage in their sensitivity to maker and user problems, unmet needs and interests, and novel combination of resources (Wright et al. 2004b). Often an external trigger from outside of the academic community is required to set the ball rolling.

Against the background of the previous considerations following preliminary *hypotheses* are derived for the research project:

The first stage of the entrepreneurial process is dominated by the provision of and access to information as major promoter for recognising entrepreneurial opportunities out of the scientific technological context. Yet, the crucial content provided by an academic's social network is information. Consequently *information networks* are those forms of social capital that are of vital importance at the first entrepreneurial phase.

H1.A: The opportunity recognition phase of the entrepreneurial process is characterised by information networks, i.e. the personal social network of the academic entrepreneur consists predominantly of information-related interactions.

To be provided with non-redundant information constituting the basis for novel and innovative ideas and opportunities, entrepreneurs' social networks have to be sparse and rich in *structural holes*.

H1.B: The opportunity recognition phase of the entrepreneurial process is characterised by sparse social networks of academic entrepreneurs that are rich in structural holes.

Accordingly, information variety is enriched if academic entrepreneurs' networks are featured by bridging social capital comprising infrequent relations to not so familiar alters (*weak ties*).

H1.C: The opportunity recognition phase of the entrepreneurial process is characterised by social networks dominated by weak ties.

To be able to recognise opportunities out of his / her rich stock of technical knowledge, the entrepreneurial researcher needs novel and complementary information to realise potential economic values. Such information can be best obtained via contacts to actors in other fields of research or even outside of the academic community. In the latter case existing research has found that in the first stage of the entrepreneurial process increasing emphasis is put on tapping pre-existing relationships such as ties to family and friends (*private network*) (Hoang and Antoncic 2003).

H1.D: During the opportunity recognition phase of the spin-off process the information networks of academic entrepreneurs are dominated by contacts to academic actors in different research areas and to private network members.

3.4.2 Formation

After successful opportunity recognition and evaluation, the academic entrepreneur enters the next stage, the stage of planning and preparing the ground for successfully spinning off a new venture. To transform an opportunity into a genuine entrepreneurial project, technological development, i.e. the production of a prototype, and commercial development, i.e. the elaboration of a business plan, are required (Ndonzuau et al. 2002) to prove the technological and economic viability of the entrepreneurial project. Moreover, both outputs can be used to impress and convince capital providers and potential business partners. To accomplish these tasks the entrepreneurial researcher needs both information and financial resources.

The first provides the researcher with the necessary (commercial) knowledge to set up the business plan, such as information on market conditions and developments, legal provisions, strategic partners, required investments, and possible turnover and operating costs. This is a crucial point, since developing a technology into a marketable product requires capabilities derived from prior industry and entrepreneurial experience, which the entrepreneurial scientist may not possess to a sufficient degree (Wright et al. 2004b). Moreover, information about where to obtain required financial backing, and relevant assets, such as machines and laboratory space, to set up a prototype can be accesses by virtue of personal networks.

Financial funds are required to finance the technological and commercial developments; in particular with regard to the production of the prototype a costly issue. In this context social networks play an important role in securing access to resources. According to Elfring and Hulsink (Elfring and Hulsink 2001; Hulsink and Elfring 2003) networks and in particular strong ties represent a valuable tool in getting the required funding to transform the spotted opportunities. Especially alters who are connected to the entrepreneur via strong ties are more motivated to help him / her than network members with whom he / she has only loose relationships.

Departing from these considerations following preliminary assumptions are derived for the dissertation project:

The second phase of the entrepreneurial process relating to the formation of the entrepreneurial opportunity basically in the form of a prototype and a business plan is characterised by the search for missing information and appropriate resources. Yet, this stage is dominated by information and resource flows and thus by both *information and resource networks*.

> H2.A: *The formation phase of the entrepreneurial process is characterised by both information and resource networks, i.e. the personal network of the academic entrepreneur contains more information- and resource-related interactions.*

Yet, as stipulated above to obtain novel and non-redundant information networks need not to be dense and composed of overlapping relations. Rather loosely-structured entrepreneurial networks, rich in *structural holes* are to be preferred.

> H2.B: *The information network of the formation phase is characterised by sparse social networks of academic entrepreneurs that are rich in structural holes.*

On the other hand, the provision of resources, in particular of financial resources, to a nascent entrepreneur calls for a high level of trust both in the person of the entrepreneur and in his idea. This trust is best established in *dense networks* due to the effect of social closure fostering the compliance with norms by enforced community pressure (Coleman 1990; Granovetter 1985). Both trust and norms are developed over time through repeated series of interactions being the characteristics of strong ties (Boari and Presutti 2004).

> H2.C: *The resource network of the formation phase is characterised by dense social networks of academic entrepreneurs.*

Regarding tie strength it is stipulated that again a difference can be observed between information and resource networks. While information networks rely more on *weak ties* for the generation of a wider range of different and novel inputs, resource networks require the prevalence of *strong ties* to foster mutual trust as basis for the provision of funds and assets. This leads to the following two hypotheses:

> H2.D: *The information network of the formation phase is characterised by weak ties.*

> H2.E: *The resource network of the formation phase is characterised by strong ties.*

3.4.3 Establishment

After having produced a first prototype and prepared the business plan, in the third phase the academic entrepreneur makes the final decision to start his business and actually spins off. For an entrepreneurial project to become a concrete economic reality it is essential to secure the required resources. This refers to both physical, i.e. financial and material resources, and human capital in the form of skilled employees (Ndonzuau et al. 2002).

The management of high-technology firms differs greatly from that of university laboratories, thus the newly established firm has to be managed by a professional team also skilled in the necessary management expertise (Ndonzuau et al. 2002).

Whereas the academic founder of the firm usually disposes of the required technological know-how, economic knowledge often has to be acquired form external sources. To this end appropriate personnel has to be identified and involved in the new venture.

Besides the above mentioned human capital, also a broad range of material resources is required such as assets, machines, office space, etc. that have to be financed in one or another way. Again securing access to resources can be best achieved by cultivating personal networks with strong relationships. (Elfring and Hulsink 2001; Hulsink and Elfring 2003) providing for the required trust. In addition, potential entrepreneurs assess their ability to secure the required resources at relatively low cost on the basis of their strong ties. Resources have to be obtained at minimum cost, since paying the market price is often to expensive (Elfring and Hulsink 2001). In this context social interactions play a vital role. Resources can be acquired far below market prices, if entrepreneurs employ social assets such as friendship, trust, and obligation (Starr and MacMillan 1990).

Founding a new firm is an incising step in the career of a researcher requiring lots of support and encouragement from his surroundings. Frustrating events that force down overly optimistic initial expectations are not seldom, especially during the market entry phase. In this situation, emotional support received from the spouse and other family members might be very helpful to sustain emotional stability (Brüderl and Preisendörfer 1998). As a result, entrepreneurs who receive much support from the family might be more successful (Sanders and Nee 1996). Hence, it is postulated that at this phase the network of the successful entrepreneur is coined by relations providing emotional support.

Once spun-off the new venture has to assert itself on the market. In this context, legitimacy in the form of institutional support of a powerful external actor, is often critical to the success of a new venture to overcome the so called "liability of newness" introduced by Stinchcombe (1965). Building network exchange structures with outsiders that are identified as critical resource suppliers that can stabilise the new firm as player in the targeted market is essential (Dubini and Aldrich 1991).

Generally, it could be observed (Hoang and Antoncic 2003) that relationships gain increased multiplexity with advancing entrepreneurial stage. Initially pure instrumental relations develop also social or affective traits, and contrariwise non-instrumental relations are also applied to business purposes with progressing interaction. Given that multiplexity may serve as indicator for relation intensity (Jansen 2003 p. 108), this implies a general tendency towards increasing tie strength within the entrepreneur's personal network.

In addition the third stage of the entrepreneurial process is characterised by information networks with higher quality information exchange between actors involved (Hoang and Antoncic 2003). To cope with the dynamic market conditions the newly created spin-off has to develop and acquire genuine routines and procedures (Vohora et al. 2004). Besides reliable information on qualified human resources, accumulation of internal and external capabilities related to product development, manufacturing, marketing, sales and distribution are vital to ensure that the spin-off can become established in the marketplace (Wright et al. 2004b). Related information is more tacit of nature and thus requires intensified interaction and closeness of partners to be transferred.

Summing up the above made considerations, the following first hypotheses can be established.

In the third phase of the entrepreneurial process the social network of the academic entrepreneur is characterised by exchanges of all three resources key to the success of the emerging organization: *information, resources, and support*.

H3.A: The establishment phase of the entrepreneurial process is characterised by information, resource and support networks, i.e. the personal network of the academic entrepreneur comprises of interactions transferring information as well as material and symbolic support.

Whereas in the earlier phases novel, non-redundant information was required to complement the researcher's knowledge and stimulate new thoughts, at the final stage of the entrepreneurial process emphasis is more on the reliability of the information provided and the trustworthiness of the informants. This requires the existence of a trusting environment for information exchange as generated in dense networks. In highly connected networks actors know each other providing the academic entrepreneur with referees on the information provider and his / her input. Moreover, dense networks enforce the compliance with norms and rules due to increased social pressure. Non-compliance and dishonesty can be uncovered and pursued more easily, due to the prevailing transparency and the better knowledge network actors have of each other. Yet, contrary to the information networks of the first two phases, in the establishment phase the entrepreneurs' information networks are *dense* and more interconnected.

H3.B: The academic entrepreneur's information network of the establishment phase is characterised by a dense and interconnected structure.

In addition, acquisition of non-codified knowledge of professional procedures and routines is more likely to be efficiently effected via *strong ties*. Such tacit knowledge is rather transferred by regular and intensive interactions determined by shared objectives.

H3.C: The academic entrepreneur's information network of the establishment phase is characterised by strong ties.

While securing resources is already an important issue in the formation phase, in the establishment phase it gains even more importance. Spinning-off and trying to get hold on the market usually requires huge investments in assets of all kind. However, letting financial funds to someone requires a certain amount of trust and security of reliable payback. Yet, due to a lack of track records and past experiences, young spin-offs find it rather difficult to prove the necessary level of trustworthiness. Being embedded in *dense networks* characterised by prevalence of mutual trust and social norms facilitates such efforts.

H3.E: The academic entrepreneur's resource network of the establishment phase is characterised by a dense and interconnected structure.

A critical issue for the new venture's success is to secure required assets at low cost, which can be best achieved by means of *strong ties*. In addition, network partners to whom the academic entrepreneur is connected by strong relationships are more likely to provide the required human and physical capital.

H3.F: The academic entrepreneur's resource network of the establishment phase is characterised by strong ties.

Besides material support, the entrepreneurial researcher also features an increased need for *symbolic support* in the form of emotional and institutional assistance. Spinning-off and leaving the comparably save university environment is an incising and risky decision that requires a lot of emotional encouragement from the part of the nascent entrepreneur's immediate environment. Such support is typically provided by *strong ties* to close friends and members of the family.

In addition, the young firm has to cope with the aforesaid liability of newness, which is particularly a problem for young high-technology ventures. Since the quality of a new venture often cannot be directly observed, potential evaluators have to judge the firm based on observable attributes expected to correspond to the unknown underlying quality (Stuart et al. 1999). One of these indicators is the exchange partners of the new venture. To overcome the legitimacy barrier, network actors, like competitors, distributors or customers, have to be mobilised to generate positive signalling effects (Hulsink and Elfring 2003). Strong relationships with prominent organizations convey the fact that the venture has earned a positive evaluation from experienced and influential actors (Stuart et al. 1999). These prior research findings allow the following conclusion:

H3.G: The academic entrepreneur's support network of the establishment phase is characterised by strong ties.

H3.H: Emotional support is predominantly obtained from members of the academic entrepreneur's private network.

H3.I: Institutional support is predominantly obtained from members of the academic entrepreneur's business network.

4 Methodology

The following chapter presents the methodology applied to empirically validate the research model and related hypotheses introduced in Chapter 3.

4.1 Research Design

The following chapter introduces the design of the research process subject of this thesis. After a brief introduction into the possible occurrences of research designs in general, an overview of common methods of social research is provided. The last section finally presents the specific research design applied to structure this study.

4.1.1 General Considerations

According to Malhotra and Birks (2003 p. 58) a research design is a framework or blueprint for conducting a research project detailing the procedures required to obtain the information needed to structure or solve the research problem. The main purpose of empirical research is the generation and examination of reasonable hypotheses about reality (Bortz and Döring 1995 p. 29). Following this emphasis Malhotra and Birks (2003 p. 62) broadly classify research designs into exploratory or conclusive designs. Figure 4-1 provides an overview of the two groups.

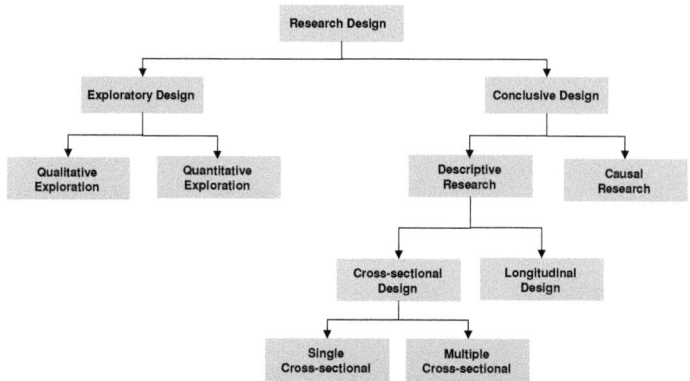

Figure 4-1: Classification of Research Designs (Malhotra and Birks 2003 p. 62)

Malhotra and Birks (2003 p. 63) stipulate that the main objective of *exploratory research* consists in providing insights into and an understanding of a phenomenon. This type of research is often applied in cases where problems need to be defined more precisely, relevant courses of action have to be identified, or additional insights have to be generated prior to confirming findings by means of a conclusive design. Required information is only loosely defined in the form of research questions rather than hypotheses and the research process applied is characterised as being flexible, loosely structured, and occasionally evolutionary in nature (Malhotra and Birks 2003 p. 63). Apart from data mining techniques, exploratory research is usually dominated by qualitative methods. Methods are applied flexibly and versatilely, since no formal research protocols and procedures are employed (Malhotra and Birks 2003 p. 64).

The purpose of *conclusive research*, on the other hand, is to describe specific phenomena, to test specific hypotheses and examine specific relationships. To this end, the required information has to be clearly specified. Thus, conclusive research is generally more formal and structured than exploratory research (Malhotra and Birks 2003 p. 65). Table 4-1 compares the main differences between the two types of research.

	Exploratory	Conclusive
Objectives	Providing insights and understanding of the nature of phenomena	Testing specific hypotheses and examining relationships
Characteristics	Required information is loosely defined	Information needed is clearly defined
	Research process is flexible, unstructured, and may evolve	Research process is formal and structured
	Samples are small	Sample is large and aims to be representative
	Data analysis can be qualitative or quantitative	Data analysis is quantitative
Findings / Results	Can be used in their own right	Can be used in their own right
	May feed into conclusive research	May feed into exploratory research
	May illuminate specific conclusive findings	May set a context to exploratory research
Methods	Expert surveys	Surveys
	Pilot surveys	Secondary data
	Secondary data	Databases
	Qualitative interviews	Panels
	Unstructured observations	Structured observations
	Quantitative exploratory multivariate methods	Experiments

Table 4-1: Differences between Exploratory and Conclusive Research (Malhotra and Birks 2003 p. 63)

As Figure 4-1 depicts, conclusive research can be divided into a *descriptive* and a *causal* branch. While the former aims at describing certain phenomena with respect to selected characteristics and their changes, the latter's main objective is to achieve evidence relating to cause-and-effect (causal) relationships. Both descriptive and causal research are characterised by the prior formulation of specific research questions or hypotheses. They require a planned and structured design and are typically based on large representative samples (Malhotra and Birks 2003 p. 65).

Similar to Malhotra's and Birk's distinction of research designs, Bortz and Döring (1995 p. 332) classify empirical research according to its purpose into:
- *Exploratory research* aiming at developing theories and hypotheses
- *Explanative research* aiming at examining theories and hypotheses
- *Descriptive research* aiming at describing populations

According to them (Bortz and Döring 1995 p. 332) exploratory studies are carried out ahead of explanative investigations and serve the development of scientifically testable hypotheses. In doing so, exploratory research is by no means devoid of any theory. Already the selection of variables to be incorporated in the explored data set, their operationalisation and measurement or the selection of research units is guided by partly implicit, partly explicit assumptions and theories. In contrast to explanative research, this prior theoretical understanding is not elaborated and focussed enough to allow for the formulation of operational and finally also statistical hypotheses that can be subjected to significance testing (Bortz and Döring 1995 p. 332).

4.1.2 Methods of Empirical Social Research

According to Atteslander (1993 p. 24) social data in the narrower sense are systematically collected aspects of social reality. Yet, social facts in empirical social research are only those extracts of social reality that are to be documented in the light of the theoretical assumptions. They can be expressed in verbal terms or by means of measured values. This corresponds to the distinction of qualitative and quantitative data, and the related underlying research paradigms of qualitative and quantitative research.

In this regard, quantitative research is often associated with methods based upon quantifications of observed reality. Whereas qualitative research refers to research methods that mainly do without measurements and focus upon interpretations of verbal material. However, qualitative and quantitative research not only differ in the form of data material processed, but also with respect to research methods, subject, and understanding of science (Bortz and Döring 1995 p. 271). These differences are often illuminated in the form of antithetic pairs (see Table 4-2).

Quantitative	Qualitative
Nomothetic	Idiographic
Deductive	Inductive
Particularistic	Holistic
Explanative	Exploratory
Ahistorical	Historical
Explain	Understand
"Hard" methods	"Soft" methods
Measure	Describe
Sample	Individual Case
Behaviour	Experience
Scientific	Humanistic

Table 4-2: Comparison of Quantitative and Qualitative Research (Bortz and Döring 1995 p. 274)

Bortz and Döring (1995 p. 274), however, recommend to understand these polarities not in the sense of dichotomies, but rather as bipolar dimensions, and to apply them cautiously.

Particular methodologies can be associated with particular social theories. Thus, research methods do not only refer to technical considerations, but also involve understanding theories of knowledge and philosophical standpoints (Spratt 2003). Qualitative and quantitative research roots back to two different research paradigms. The quantitative approach can be related to positivistic thinking, which stands for objective observation and precise measurement (Bazeley 2004 p. 143). Its main goal is to detect patterns in the experience and behaviour of humans. The existence of such legalities or general laws (nomos) is thereby postulated (nomothetic approach). The more knowledge about these legalities becomes available, the more observable events can be explained and future events predicted (Kromrey 1994 p. 24). Thus, the main purpose of the quantitative approach is to establish causal laws that enable the prediction and explanation of social phenomena (Malhotra and Birks 2003 p. 136). A mechanistic idea of man is pursued assuming that humans are primarily controlled by external influences (Bortz and Döring 1995 p. 274).

Opposed to this perspective is the interpretative paradigm, which associated with qualitative research. Its representatives object to the assumption of a predefined structure with fundamental and constant legalities ("social laws"). They posit that men create and thus constantly change the social structures, in which they live with each other, as a result of their actions. The at a time given current status is deemed to be the result of complex sequences of interactions. It is constantly put anew at the disposal, constantly interpreted and advanced anew (Kromrey 1994 pp. 24-26). Interpretivists hold the view of self-determined, reasonably acting humans, whose behaviour may not be explained by means of external, objectively observable factors, but only be understood by communicatively reproducing their subjective world outlook and internal motives. In this sense the objective of qualitative research is to describe individualising single historical events or cultural products (ideographic approach) (Bortz and Döring 1995 p. 274). The sense and events, i.e. the meaningfulness of the course of action as experienced by participants and observants, shall be reconstructed. The recorded history becomes traceable and understandable (Heinze 1995 p. 12).

Qualitative research methods are designed to help researchers understand people and the social and cultural contexts within which they live. Kaplan and Maxwell (1994) argue that the goal of understanding a phenomenon from the point of view of the participants and its particular social and institutional context is largely lost when textual data are quantified. While quantitative research aims at causal determination, prediction, and generalisation of findings, qualitative research seeks illumination, understanding, and extrapolation to similar situations. Thus, qualitative research results in a different type of knowledge than does quantitative inquiry (Hoepfl 1997).

Yet, it can be noted that although qualitative studies predominantly focus on understanding in the sense of a reconstruction of the actors' perspective, they do not generally abstain from explanations. Explanations come into play, whensoever theoretical concepts, which do not necessarily correspond to actors' daily understanding, are applied to analysis. Likewise the reconstruction of actors' perspectives is not completely factored out in quantitative research (Bortz and Döring 1995 pp. 274-77).

4.1.3 Applied Research Design

According to Atteslander (1993 pp. 55-56) independent of the motivation of research at the beginning of a research project the question of adequate theories has to be answered. He distinguishes between three types of situations as depicted in Figure 4-2. In case A theories are existent and also hypotheses have been formulated that can be subjected to empirical investigation. However, even if theories are available, it may happen that no or only to some extent hypotheses have been formulated with respect to the subject matter of research. Thus, terms have to be defined, checked as to their reference to reality, and first pre-tests with formulated hypotheses to be carried out. Only then the main research can be started. This situation refers to case B. In the last case (C) vague notions of theoretical interrelations exist, yet, they are not formulated. Single hypotheses may be subjected to exploratory tests as preliminary propositions. Linking these hypotheses to theoretical concepts is in most cases carried out simultaneously.

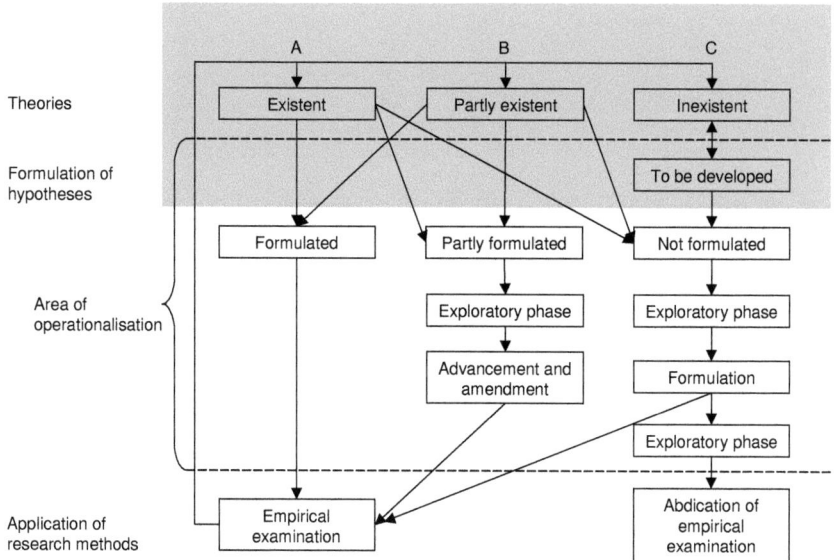

Figure 4-2: Building of Theories and Hypotheses (Atteslander 1993 p. 55)

Thus, when specifying an appropriate research design, it is necessary to check to what extent theories and related hypotheses are already available to be empirically examined within the scope of research. In case of this research, theories to the phenomenon of academic spin-offs and social networks in entrepreneurship are already available and presented in Chapter 2. Yet, hypotheses related to answer the imposed research question of how they can be jointly applied when looking at the process of academic spin-off formation are only partly present (see Chapter 3) and have to be further developed. For that reason the research design is split into two phases starting with an exploratory part for refining and validating existing assumptions and identifying additional hypotheses.

As described in Chapter 4.1.2, due to its characteristic features qualitative research is particularly appropriate to meet the purpose of exploratory investigations. As Hoepfl (1997) notes, qualitative methods are particularly appropriate in situations where one needs to first identify the variables that might later be tested quantitatively. A holistic understanding of phenomena and procedures that are so far only vaguely known shall be obtained. Especially when it comes to exploring social phenomena, the researcher is dependent on the perception and input of the relevant actors involved. Emphasis is not to explain certain events, since they are not yet known, but to identify and describe social patterns as they can be found in reality. In this context, qualitative data feature the strength that they focus on naturally occurring, ordinary events in natural settings (Miles and Huberman 1994). As a result a qualitative approach is applied in the course of the first phase of the project to gain a better understanding of the researched phenomenon from the perspective of the subjects involved.

To empirically validate derived propositions and examine detected relationships on a larger scale, the exploratory phase is followed by a conclusive phase as presented in Chapter 4.1.1. To this end, the required information has to be clearly specified. Involved concepts are operationalised by means of quantitative analysis, in particular social capital measures are captured via quantitative social network analysis, in the second phase.

As the main objective of the conclusive part of research consists in describing entrepreneurs' social structures at different stages of the academic spin-off formation process and the changes of these structural features in time, a descriptive approach was adopted. Due to pragmatic considerations and time constraints a cross-sectional design was chosen.

As a result the general research design of the dissertation project can be mapped as depicted in Figure 4-3.

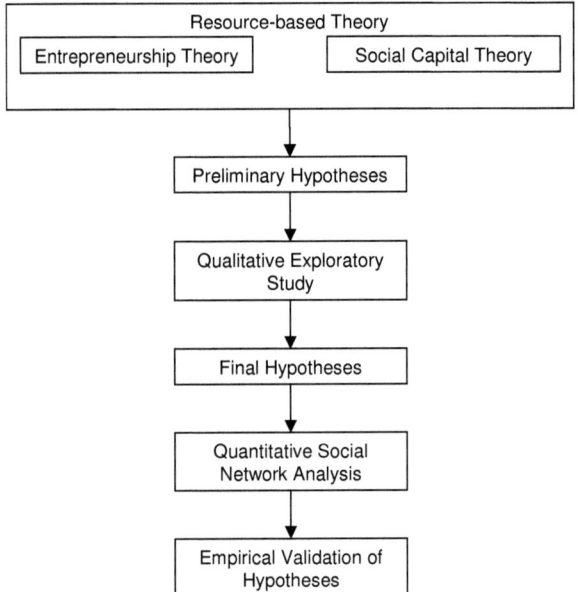

Figure 4-3: Research Design

Summing up the dissertation basically consists of the following two parts:

1. An **exploratory qualitative study**
2. A **quantitative social network analysis**

In doing so, the research design seizes the advantages of the complementary nature of qualitative and quantitative research. These advantages are attributed to the fact that qualitative and quantitative methods are not only characterised by different requirements as to the gathering and processing of data, but also by different general perspectives and attitudes providing for a differentiated access to reality. Mixed methods often combine nomothetic and idiographic approaches in an attempt to serve the dual purpose of generalisation and in-depth understanding (Bazeley 2004 p. 144); a path which is also adopted in this research project.

4.2 Qualitative Exploratory Study

The main aim of the qualitative approach is to gain explorative insights in the composition and structure of the researchers' personal networks at the different stages of opportunity recognition, spin-off formation and establishment. A general understanding of the academic venture creation process shall be provided to support the subsequent network approach. First hypotheses derived from existing literature in the field shall be preliminarily evaluated and modified and, if indicated, additional hypotheses generated to be subjected to statistical validation in the course of the subsequent quantitative research phase.

4.2.1 Introduction

When carrying out empirical social research on a qualitative level, there are basically three methods of data collection available (Bortz and Döring 1995 p. 282):

1. Qualitative interviewing
2. Qualitative observation
3. Non-reactive methods

While qualitative interviews investigate the subjective perspective of actors with respect to past events, future plans, opinions, etc., observations mainly aim at detecting external behaviour, latent motivation and meaning structures that are concluded or reconstructed indirectly. Non-reactive techniques denote methods of data collection that in the course of their execution do not exert any influence on the subjects, events or processes examined (Bortz and Döring 1995 p. 297).

The most common forms of qualitative data collection, refer to interviews and observations, in particular participant observations. With communication being the central element of social systems and their construction, interviews - besides participant observation – as direct communication between the researcher and the persons from the respective field of interest are of essential importance (Froschauer and Lueger 1998 p. 25). Also Lamnek (1989 p. 35) quotes the interview as the most important qualitative method besides participant observation. He argues that it not only enables the researcher to ask in case of ambiguities, but also provides a much more easier access to the social field than participant observation.

Against the background of these considerations, in the course of this thesis qualitative interviewing was applied to generate the required data for the exploratory phase. To this end, selected founders of university high-technology spin-offs were asked to report their personal experiences made during the founding process. In this regard, the qualitative phase furnishes a feeling for the academic spin-off formation process, in particular with respect to Austrian peculiarities.

4.2.2 The Qualitative Interview

According to Lueger and Schmitz (1984 p. 10) an open or qualitative interview can be defined as an oral questioning regarding self-experienced (real) stories, experiences, perspectives, evaluations, assessments of the environment, etc. The form of discussion is directed to a higher activity of the respondent and aims at leaving him / her the structuring and control of the conversation (Kohli 1978 adopted by; Lueger and Schmitz 1984 p. 10). Analogue defines Lamnek (1989) qualitative interviews as largely open forms of dialogue where the interviewer remains stimulatingly-passive and the interviewee assumes the active control of the conversation.

Thus, qualitative interviews enable researchers to learn about the interviewees' perspective by disclosing their structuring aspects. The respondents can show, how they understood the questions and what is relevant to them. Moreover, unknown structures can only be detected, if the persons acting in this context are allowed to talk freely (Lueger and Schmitz 1984 p. 15).

According to Patton (2001), three types of qualitative interviews can be distinguished:

(1) Informal conversational interviews, which are spontaneous and loosely structured (i.e., no interview protocol was used).

(2) Interview guide approach, which is more structured than the informal conversational interview and includes an interview guide listing open-ended questions that can be asked in any order by the interviewer. In addition, the question wording can be changed as well by the interviewer if it is deemed appropriate.

(3) Standardized open-ended interviews, which comprise open-ended questions that are asked in an exact order specified on an interview guide. The wording of the questions cannot be changed.

According to Hoepfl (1997), an interview guide is a list of questions or general topics the interviewer wants to explore during each interview. Interview guides ensure good use of limited interview time, they make interviewing multiple subjects more systematic and comprehensive, and they help to keep interactions focused (Hoepfl 1997).

While Patton's categorisation is predominantly applied within English-speaking literature on the topic, the classification of Kepper (1994 p. 39) in turn provides a good synopsis of the different forms of qualitative interviews as to be frequently found in German research areas. In her presentation Kepper focuses on forms of qualitative interviews that are particularly applicable for market research. She, too, uses a trichotomy for classification (see Figure 4-4).

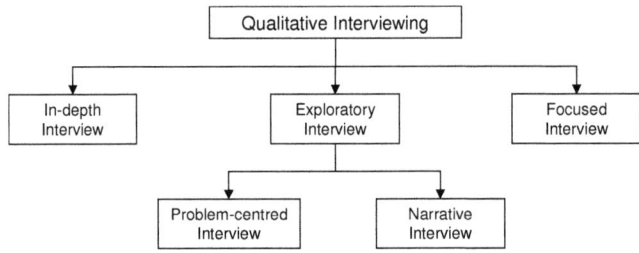

Figure 4-4: Forms of Qualitative Interviews (Kepper 1994 p. 39)

According to Kepper (1994 p. 36) exploratory interviews ask for respondents' knowledge, experiences, attitudes or know-how with special emphasis on the narrative moment. To stimulate respective stories different questioning techniques can be applied that range methodologically between to the two extreme sub-groups of narrative and problem-centred interview. The psychological in-depth interview, on the other hand, intends to reveal certain pre- or unconscious motive structures and contexts that obtain their meaning only in ex-post interpretation. Finally, it is also possible to combine qualitative interviews with the presentation of a uniform stimulus (film, radio broadcast, etc.) to govern the in principle open interviews more purposively as the case in focussed interviews as developed by Merton and Kendall (1946).

Within exploratory interviews the interviewer asks primarily self-formulated questions with free response categories. The content spectrum may range from seemingly casual conversations to problem-oriented expert talks. As opposed to in-depth interviews exploratory interviews do not intend to determine the psychological content of the generated parts of communication by means of ex post, external allocation of meaning within the boundaries of predefined theories, but concentrate on a preferably comprehensive and complete collection of subjectively relevant information and statements to the research problem (Kepper 1994 pp. 39-40).

Developed by Schütze (1976) the narrative interview has achieved great importance in the German speaking area and is considered to be the classic example of a free and open interview. In the course of narrative interviews respondents are encouraged to tell rather spontaneously and at the beginning without questions on the part of the interviewer about their personal experiences. This approach intends to reveal the structures interviewees apply as orientation for their actual behaviour (Kepper 1994 p. 41).

Witzel's (1985; 1982) problem-centred interview also makes use of the story-telling concept. Yet, it is less open than the narrative interview. By means of a flexible interview guide the researcher addresses pre-analysed aspects of the research problem (Hölzl 1994). The respondent is encouraged to talk as unstrained as possible, but the conversation centres around a certain topic or problem introduced by the interviewer and represented by the interview guide. Yet, problem-centred interviews are particularly appropriate for research questions that are more theory-guided, i.e. where problems have already been specified to a certain extent and some knowledge about the topic is available (Mayring 1990 p. 46) and where research questions are oriented to knowledge about facts or socialisation processes (Flick 1998 p. 90). For that reason this method is particularly appropriate for preparatory studies characterised by both a hypothesis-generating as well as – examining nature. In a first step, existing assumptions are confronted with empiricism while at the same time additional insights may be gained. In a mostly quantitative second step thus generated and refined hypothesis are validated (Kurz et al. 2007).

For the purpose of the first phase of this dissertation, where theory-based preliminary hypotheses shall be validated, and if necessary, adapted and amended, completely unstructured interviews, did not seem to be the appropriate method. At this point, previous theoretical knowledge shall be reviewed and further insights in a predetermined topic provided, which requires some sort of problem-focus of the interviews. While interviewees shall talk as freely as possible to capture the whole range of possible information, they should not move too far away from the actual matter of research, i.e. the academic spin-off formation process. Thus, referring to the Anglo-American research domain and following Patton (2001), an interview guide approach has been chosen. On the part of the German speaking research area, Witzel's (1985; 1982) concept of problem-centred interviews has been applied. The practical implementation is further detailed in Chapter 4.2.4.

4.2.3 Sampling

The focus of qualitative research is on process and meanings. Samples are not meant to represent large populations. While in quantitative research the dominant sampling strategy is probability sampling, which aims at the subsequent generalisation of the research findings to the population (Hoepfl 1997). By contrast, purposeful sampling, seeking information-rich cases that can be studied in depth, is the dominant strategy in qualitative research (Patton 2001). In this context, rather small samples of articulate respondents are used, because they can provide important information, not because they are representative of a larger group (Sale et al. 2002). The lack of randomness and the bias associated with the type of sampling procedure can be neglected due to the qualitative nature of this phase and its primary exploratory and descriptive purpose. No statistical evaluations are applied that would require a statistically representative set of survey participants.

When looking at the possible forms of purposive sampling an often cited (Hoepfl 1997; Miles and Huberman 1994 p. 28) overview is provided by Patton (1990 pp. 169-83), who has identified 16 types of purposeful sampling including extreme or deviant case sampling, typical case sampling, maximum variation sampling, snowball or chain sampling, confirming or disconfirming case sampling, convenience sampling, and others.

For the purpose of this dissertation, a convenience sampling strategy has been chosen. In convenience sampling, the selection of units from the population is based on easy availability and/or accessibility. The trade-off made for ease of sample obtainment is the representativeness of the sample. This may be argued by the comparably minor role the qualitative study assumes in the overall design of this research project being only of preparatory nature and by the fact that the population of investigation, i.e. Austrian founders of high-technology university spin-offs, is rather narrow to allow for a more selective strategy. The smallness of the target population also reduces the risk of lack of representativeness associated with this sampling strategy.

Based on the overall objective of the study, i.e. the identification and analysis of academic entrepreneur's social capital at the different stages of the spin-off process of Austrian high-technology ventures, aspects of the cases to be studied that directly connect to this research question were defined with respect to the given limits of time and means (Miles and Huberman 1994 p. 27). In this respect, the survey sample of the qualitative study focused on young Austrian high-technology ventures established as spin-offs out of the university context. For that purpose, selected academic entrepreneurs, who had founded a technological spin-off at last two years ago, were identified by means of Internet research and personal recommendations. To ensure a minimum of variability, heterogeneous respondents covering a broad spectrum of demographic, economic and academic properties were targeted. Unfortunately, it was not possible to enlist a woman to participate in the qualitative survey. In total a sample of 10 academic entrepreneurs was interviewed (see Table 4-3).

University	Region	Industry	Founding Year
University of Natural Resources and Applied Life Sciences	Vienna	Biopharmacy	1992
Johannes-Kepler-University Linz	Upper Austria	Mathematical Software	1996
Technical University Vienna	Vienna	Virtual Reality	1998
University of Economics and Business Administration	Vienna	Innovation Management	1998
University Vienna	Vienna	Information Technology	1999
Technical University Graz	Styria	Navigation	1999
Johannes-Kepler-University Linz	Upper Austria	Microsystems	1999
Technical University Vienna	Vienna	Embedded Systems	2000
University Innsbruck	Tyrol	Meteorology and Remote Sensing	2002
University of Natural Resources and Applied Life Sciences	Vienna	Nanotechnology	2003

Table 4-3: Composition of the Qualitative Sample

4.2.4 Data Collection

As mentioned data collection basically followed the guiding principles of Witzel's (1985) problem-centred interview and the interview guide approach according to Patton (2001). To put it in nutshell, semi-structured interviews were carried out by means of an interview guide established on the theoretical basis of a critical review of state-of-the-art research on the topic. Following the resulting core assumptions on the evolution of the academic spin-off formation process, questions were structured according to the identified three phases of new venture formation to bring the relevant topics into a reasonable order that can be easily traced and understood by the interviewee. This structure also supported to narrative string pursued by the interviewee. In addition, the interview guide served as basis for giving the interview a new turn, when the flow of conversation slowed down or moved to far away from the original topic (Flick 1998 p. 88). While the interview guide organises the prior knowledge and represents frame of orientation and mnemonic backing for the interviewer in form of thematic fields and their differentiation in buzzwords and -phrases, for the concrete interview situation the plot of the respondent is decisive (Hölzl 1994 p. 65).

Interviews lasted on average about 45 minutes and, following Witzel (1985), consisted of four main blocks or communication strategies (Lamnek 1989 p. 75). They started with a conversational entry establishing the narrative structure and the problem area of the interview via a rather general question (Hölzl 1994 p. 66). In particular, respondents were asked to narrate their experiences in the process of spinning off a new venture according to their own memory and in their own words. The major objective consisted in enabling the interviewees to tell their story on how they experienced the spin-off process starting from the initial business idea until its final realisation in form of a new venture.

Second, the core phase comprising of a general prompting, aiming at specifying individual issues and interrelations to generate further material (Kepper 1994 p. 43). Where necessary, interviewees were asked to provide concrete examples or questions asking for more details of reported information were posed. Special emphasis was put on generating input on social capital related issues to safeguard the reference to the basic underlying research topic.

In the third phase of specific prompting, different questioning techniques were applied to provide for more comprehension and improved interpretability of different interview sequences. To this end, mirroring of what had been said, i.e. summary of certain events to enable subsequent corrections by the respondent, questions of understanding, and confrontations in case of evading or contradicting responses, were used (Flick 1998 p. 89; Kepper 1994 p. 44). This approach intends to clarify the interrelation of different statements already during to interview to provide for some sort of ex ante interpretation facilitating understanding.

In case that by the time problem-relevant topics remained open or turned out to be of relevance during the interview without having been contained in the interview guide so far, direct ad-hoc questions were posed (Hölzl 1994 p. 66; Lamnek 1989 p. 76).

Yet, it was necessary to decide when to follow the narrative string by means of immanent questions derived from the said, and where to apply examinant questions to differentiate the contributions of the interviewee (Hölzl 1994 p. 65). In doing so, minor interviewer influence was an important point in order to provide respondents with the opportunity to speak as freely and unguided as possible. The interview guide focussing on social capital related issues was brought into where necessary, to safeguard the reference to the underlying research topic. Interviews were tape recorded and subsequently transcribed into an interview protocol.

4.2.5 Analysis

Qualitative analysis requires some creativity, for the challenge is to place the raw data into logical, meaningful categories; to examine them in a holistic fashion; and to find a way to communicate this interpretation to others. (Hoepfl 1997)

4.2.5.1 Theoretical Frame

According to Froschauer and Lueger (1998 pp. 25-26) the material generated by qualitative interviews basically assumes three functions:

(1) **Description** of a particular phenomenon (also prior to or following a quantitative analysis):
This is important to precisely differentiate the phenomenon and depict its diversity. Moreover, descriptive interviews are especially suitable to analyse the different perspectives to a social phenomenon and its description. Particular attention to this function is to be drawn in the course of the exploratory research phase.

(2) **Analysis** of a social phenomenon in its context or in connection to other phenomena, its development, stabilisation and change in the course of time:
Here, it is intended to capture the explication models of the members of the system investigated and compare them to the actual modes of operation. In addition, factors attributing to the understanding of the structure and the process of the phenomenon can be identified.

(3) **Reflexion** of particular themes and generation of new perspectives to them: Interviews producing such material are mostly applied in action research, consulting projects or family-therapeutic processes for intervention purposes with analysis and intervention intertwining permanently.

As Froschauer and Lueger (1998 p. 26) note, the three types of material are closely connected. While the analytic interview requires description, reflection, in turn, cannot be without analysis. The same way, the requirements towards interpretation increase. While in the first case they can be reduced to a mere summary of the manifest content in the form of a precise description (nevertheless interpretation also allows for deeper analysis), in the analytic and reflexive interview extensive analyses are necessary. In the last case, the interview also serves for producing new interpretation and external analysis is integrated in subsequent interviews.

As the purpose and position of qualitative interviews in the scope of this research imply, they assume the first function by describing the phenomenon of academic spin-off formation and providing more detailed insights into this process. As mentioned, applying descriptive interview material is especially advisable in the exploratory research phase. As a result and following Froschauer and Lueger (1998 p. 26), interpretation is effected in the form of a describing comparative summary of the transcribed interview texts. A comparing systematisation is intended by filtering typical variants with the aim to detect general patterns of action.

To this end, qualitative content analysis as proposed by Mayring (1993) was applied. This approach corresponds to the modus operandi often used when interpreting texts that result from problem-centred interviews, which is not committed to any special method of interpretation, but mostly to coding procedures and content analysis (Flick 1998 p. 90).

Mayring's qualitative content analysis enables analysing texts of all sorts with respect to certain content categories. Prior to data collection a theory or hypothesis about the research object should already be available. Examination of theories has priority over theory development (Schäfer 1995). Following Mayring (2000) this approach enables an empirical, methodologically controlled interpretation also of larger texts, where the material is examined embedded in its communication context according to content-analytic rules. He (Mayring 1993) proposes three basic methods involved in the content analysis of verbal data: summary, explication, and structuring.

According to Kepper (1994 pp. 56-57) summarising interpretation has the aim to reduce the text in a way that brings the content into a manageable, shortened form while keeping a copy of the original material. To this end, the original text is paraphrased by identifying and eliminating all non-substantial and embroidering phrases, while the rest of the text is brought onto a common level of language. Paraphrases with the same meaning are deleted and central statements kept. In a last reducing step, a bundling of similar paraphrases is sought to arrive at basic propositions.

In the course of explication, the context of the investigated passage is taken into account to enhance its understanding. In the narrower sense, context refers to the respective preceding and succeeding passage, in the broader sense, it denotes information available beyond the text (Schäfer 1995).

Within the scope of structuring, analysed passages are assigned to appropriate categories according to a category system developed in the forefront (Schäfer 1995). Figure 4-5 provides a an overview of an ideal type structuring process as depicted by Mayring (2000).

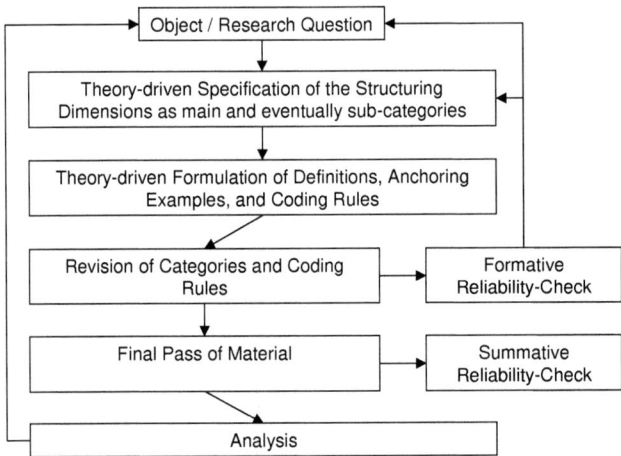

Figure 4-5: Process Model of Deductive Application of Categories (Mayring 2000)

According to Mayring (2000), ex ante determined, theoretically founded aspects of analysis are applied to the material. The qualitative step of analysis consists in allocating deductively generated categories to text passages in a methodologically secured way (Mayring 2000). At the core is the exact definition of the categories allowing for an as much as possible distinct allocation of the individual text components. The structuring dimensions that serve the development of categories are derived from the research question and the theoretical knowledge acquired a priori. Finally, concrete text passages, so called anchoring examples, are chosen for each category that assume a prototypical function and facilitate the allocation of further text components (Kepper 1994 p. 59). Categories are confirmed or modified in the course of revising the material. A modification or amendment of the category system is effected, if indicated by further cycles of data interpretation (Schäfer 1995). After a final pass of the material, the analysis is summarised and the results are written out.

Generally speaking, the qualitative content analysis represents a good approach to the systematic content-oriented analysis of qualitative data. It is appropriate for the interpretation of large amounts of text and provides for efficient across-case comparisons.

4.2.5.2 Practical Implementation

As to the practical implementation of qualitative interpretation in the course of this thesis, the transcribed interview protocols were analysed using the software package NUD*IST 4 provided by QSR (Qualitative Research Software) International[1]. NUD*IST is a common qualitative data analysis tool especially useful for analysing long interview transcripts. For the purpose of analysis, NUD*IST enables to codify text components according to evolving categories of analysis.

As regards the first summarising step foreseen in Mayring's (1993) qualitative content analysis, given the deployment of a qualitative data analysis software respective data reduction for the sake of facilitated data handling, but at the expense of limited depth and range of information, did not seem to be necessary. Above all, using an qualitative data analysis programme requires the allocation or indexing of the completely transcribed interview texts to categories or codes (Medjedovic and Witzel 2005).

Explication was obtained by drawing onto the information provided within the analysed text itself, from other interviews already analysed, and resulting from the prior knowledge acquired within the scope of the theoretical part of this research project.

Yet, main emphasis was on the structuring part of the qualitative content analysis. Basically following Mayring's (2000) scheme of deductive application of categories to interpretation (see Figure 4-5), categories were broken down according to the three stages of academic spin-off formation: (1) opportunity recognition, (2) formation, and (3) establishment.

Yet, these codes cannot be understood as categories that stand at the end of the analysis, but as thematic consolidated categories that provided support in the search of adequate data material for further analyses and related formations of theoretical concepts at each phase of the interpretation process. In this respect, analysis started with the codes related to the three identified phases as main categories, and respective sub-categories derived from the hypotheses established against the reviewed theoretical background.

Text sections of the different interview protocols were attributed to the various categories on an interview-spanning basis. In doing so, common issues as well as their weight in terms of number of nominees and text units could be identified. Additional categories were established as new information emerged in the course of the interpretation process. Table 4-4 provides an overview of the category system in its final realisation.

---

[1] http://www.qsr.com.au

Category / Code / Node	Definition
1 opportunity recognition	*reason for spinning of*
1 1 contacts	*persons relevant for opportunity recognition*
1 2 discussion	*persons with whom founder discussed is business idea*
1 3 projects	*projects on which opportunity was based, which represented the base of operations for the future spin-off*
1 4 past experience	*past experiences of the founder influencing the spin-off process*
1 5 industry	*whether there were industry projects carried out at the institute and technology transfer taking place*
1 6 family	*what was the role of the family, how did it react to the spin-off idea*
2 formation	**preparing phase**
2 1 partners	*persons participating in preparing and later on founding the spin-off*
2 2 business plan	*was a business plan established*
2 3 prototype	*was there a prototype in the formation phase and how was it funded*
2 4 public funding	*preparing activities supported by public funding programmes and initiatives*
2 5 relationship to university	*description of relationship and reaction of university and institute with respect to the spin-off idea*
2 6 information	*where was information about resources (financial, human) and about relevant potential contacts obtained*
2 7 economic know-how	*where did the economic know-how come from*
2 8 legal know-how	*where was legal know-how obtained*
2 9 infrastructure	*how was the necessary infrastructure obtained (premises, equipment, etc.)*
2 10 support	*intellectual support in any form*
2 11 advantages of university	*what advantages where derived from the university background*
3 establishment	**actual legal spinning-off and first market experiences**
3 1 start capital	*how was the initial financial capital for the realisation of the spin-off obtained*
3 2 human capital	*how was necessary human capital acquired*
3 3 recognition	*how was the visibility of the spin-off in the market realised (awareness building)*
3 4 competence	*establishment of competence signals with respect to other market participants*
3 5 emotional support	*persons who furnished emotional support*

	3 6 institutional support	institutional support in the form of a strategic partner, customer, investor, etc. serving as trust-building showcase
	3 7 business	business the spin-off is mainly engaged in
	3 8 acquisition	how are/were new projects and contracts acquired
	3 9 funding	how was the funding beyond start capital obtained
	3 10 scientific co-operation	interdisciplinary or some other form of scientific co-operation
	3 11 dual role	dual role professor / founder and related advantages
	3 12 future	outlook / intentions for the future

Table 4-4: Code System for Qualitative Data Analysis

In accordance with Lueger and Schmitz (1984 p. 200), the following questions were tackled in the scope of analysis:

- Which themes can be found with which content and which relevance? Which themes are not addressed and why?
- Which relevance is attributed to the mentioned persons or person groups and how do they differ?
- Based upon the above-stated questions, what can be said about the environment of the respondents?

Within the scope of the material review, hypotheses about the assumed relations were selected according to their redundant occurrence and related to each other. Contradicting interpretations were subjected to an in detail examination. In addition, mentioned actors were listed and described according to their roles and functions. Also the relations between them were characterised. Moreover, stated themes were listed and their attributed relevance assessed. Finally, the relationship of actors and themes was analysed looking at which actors were combined with what themes.

The summarising interpretation of all interviews and the validation of the presumed structural model of spin-off formation were effected by addressing the following questions:

- Which differences and similarities can be found between the results of the individual interviews? Which meaning do they bear and how can they be explained?
- Of which nature are the detected differences and similarities? Are they different or similar with respect to content or latent structure?
- How can the interpretation results of the individual interviews be pulled together, so that the actions of the members of this environment become understandable and explainable? Do the interpreted structures correspond to the ex ante assumed hypotheses of the research model?

The results of the analysis are presented in Chapter 5.1.

4.3 Quantitative Network Analysis

The quantitative phase represents the core of the thesis. In this phase the research model and related hypotheses, as adapted and amended in the exploratory phase, are statistically validated.

When it comes to measuring social capital, social network analysis has been proven as extremely useful, since it enables the quantitative specification of appropriate indicators and their gathering on a large statistical scale. As a result, it has been chosen as primary instrument to empirically test the hypotheses on the structure and composition of the academic entrepreneur's network at the different stages of the spin-off process as stipulated in the research model and adapted according to the results of the qualitative analysis.

4.3.1 Introduction

To provide for quantitative validation of the established research model, appropriate social capital measures are specified and subjected to hypothesis testing at each of the three identified phases of the entrepreneurial process. The main underlying assumption is that the composition of an academic entrepreneur's network is dependent on the relative stage of the spin-off process. Entrepreneurs, who have successfully passed all of the three stages, are assumed to feature typical properties of their personal networks at the different stages. It is postulated that thus provided access to different types of resources impact the likelihood of opportunity recognition and the success of subsequent spin-off formation and establishment.

This approach enables to produce transparent results that can be tested against their statistical significance thus easing argumentation as well as clear rejection or corroboration of postulated hypotheses. Nevertheless, the specific limitations of ego-centred networks, as highlighted in this chapter, have to be kept in mind and are accounted for accordingly within the course of analysis.

4.3.2 Social Network Analysis

Social network analysis is a set of methods for the systematic study of social structures that goes far beyond a mere descriptive tool for sociologists (Degenne and Forsé 1999 p. 1). Network analysis offers a new standpoint from which to look at social structures. According to Wasserman and Faust (1997 p. 21) social network analysis can be seen as distinct research perspective within the social and behavioural sciences, since it puts its main emphasis on the relationship between interacting social units instead of focussing on their attributes as customary in standard social and behavioural research. Structural or relational information is captured to study theories. As the measurement of relations among social units gives rise to new forms of data, an entire body of methods has been developed for their analysis. Yet, the social network perspective comprises theories, models, and applications that are expressed in terms of relational concepts or processes (Wasserman and Faust 1997 p. 4). Summing up, the fundamental idea of the network approach is to be able to formulate explanations for the actions / activities of individuals by describing and analysing their social embeddedness (Bührer 2002). As a result, beyond its original application to social and behavioural sciences, the concept has also found its way into economics, marketing, and industrial engineering.

Historically, the social network concept roots back to a series of different scientific disciplines that are highly interdependent and cannot always be seen in isolation.

However, the most influential ones were anthropology, sociology, the application of mathematical graph theory in social sciences, i.e. topological psychology and sociometry, and research on communication networks (Schenk 1984 p. 1). The first graphical representation of relations between individuals in small groups is the so called *sociogramme* developed by the Austrian Jacob Moreno (1954; 1934) in the USA. According to him these could be represented in diagrams similar to those of spatial geometry, with individuals represented by points and the interpersonal relations between them by lines (Scott 2000 pp. 9-10). He called the method *sociometry*, measurement of social relations. The aim was to examine how the psychological well-being is related to the structural features he referred to as social configurations. However, the sociogramme was only practicable for small groups. It primarily served the representation of structures and not as analytical tool (Jansen 2003 p. 40).

Modern network analysis, which has increasingly developed in sociology since the late 70ies, uses the term network not in a metaphoric, but in an analytic sense (Schenk 1995 p. 14). According to Wasserman and Faust (1997 p. 20) a social networks comprise of a finite set of actors and the relation or relations defined on them. In this regard, the presence of relational information is a critical and defining feature of a social network. Actors are discrete individual, corporate, or collective social units (Wasserman and Faust 1997 p. 17). They are linked to one another by social or relational ties, with the collection of ties of a specific kind among members of a group being called a relation (Wasserman and Faust 1997 p. 20).

A special advantage of social network analysis consists in the fact that networks as empiric systems can be formally represented as graphs with the nodes corresponding to the social units and the lines to the relations between the units (Schenk 1995 p. 17). The graph-theoretic foundation of network analysis introduced by Frank Harary (1969), allows describing networks by means of different structural parameters, such as size, density, closure, disparity, or reachability (Schenk 1995 p. 17).

Social networks can be approached from various perspectives. According to Pappi (1987), one distinctive feature is the type of relations to be subjected to analysis. While "total networks" are complex networks comprising all possible sorts of social relations within the total network, network research usually concentrates on only one particular social relation that represents an extraction from the total network, so called "partial network" (Schenk 1995 p. 14).

According to Marsden (2005), the broad majority of social network studies apply either "whole-network" or "egocentric" designs. These two distinct designs emerged from two distinct historical traditions (McCarty 2003). The whole or socio-centric network approach has its seeds in sociology and was significantly shaped by the work of Georg Simmel (1908). Whole-network studies investigate sets of interrelated actors that are regarded as bounded social collectives for analytical purposes (Marsden 2005). For that reason, analysing whole networks requires that the social units, between which a relationship may exist, can be attributed to a definite system context, such as the members of an organisation or the inhabitants of a small village (Schenk 1995 p. 14).

By representing relationships as numbers, many powerful mathematical and statistical analyses can be applied. The focus is on measuring the structural patterns of interactions between the members of bounded social collectives or groups, and how those patterns explain certain outcomes (McCarty 2003). Investigations of whole networks always have to face the problematic question, whether to provide propositions about individual units, e.g. persons, or about subsets or subgroups, or finally about all units and thus the whole network (Schenk 1995 p. 15).

In contrast, ego-centred or personal networks concentrate from the outset on individual actors and their direct interpersonal environment. This approach arose from anthropology and among others may be traced back to Radcliffe-Brown. Ego-centric studies concentrate on the relationships of focal actor (ego) and the objects (alters) to which it is linked. They are almost always about people rather than about groups (McCarty 2003). As opposed to the sociocentric approach, the egocentric design does not require a priori enumeration of a population of nodes (Marsden 2002). Responding actors are often sampled from a larger population (Marsden 2005). Thus, the instrument of ego-centred networks enables to maintain the usual sampling strategies of representative surveys with the add-on that also the social relations that connect the respondents with their interpersonal environment may be gathered (Schenk 1995 p. 28).

While socio-centric network analysis intends to identify structural patterns within socially defined groups that can be generalised to similar system contexts, ego-centric analysis is concerned with making generalisations about the features of personal networks to explain such things like consumer behaviour, coping with difficult life situations, economic successes or failures, etc. From this perspective, each person has their own network of relationships that cut across many groups and that influence their behaviours and attitudes (McCarty 2003).

Against the background of the previous considerations and in light of the objectives of this research that relate to the economic success of individual entrepreneurs, ego-centric network analysis is the perspective adopted within this research project. With regard to the research question posed, academic entrepreneurs cannot be seen as members of a homogenous group primarily consisting of other entrepreneurs, as the case for example when studying high-technology clusters, but rather their various contacts cutting across different stakeholder groups, such as university peers, partners from industry, family members, etc. are of interest. Yet, it is the individual spin-off founder and his or her personal network that is in the focus of research. Furthermore, the ego-centred design allows to apply standard survey sampling techniques that enable statistical examinations with subsequent generalisations.

4.3.3 Ego-centred Networks as Unit of Analysis

As the focus of this research is on studying and measuring the social capital of academic entrepreneurs at different stages of the spin-off process by means of a quantitative survey, data has to be collected at the level of the individual (Stone 2001). In this regard, the use of ego-centred networks is a common and widely applied approach (Burt 1984; Burt 1992; Marsden 1990; McCallister and Fischer 1978) to obtain information on individual social capital. Furthermore, this form of analysis is especially appropriate for collecting network data from a target population that is a small percentage of a population, and whose relations are not concentrated in a single social structure such as the case with entrepreneurs (Greve and Salaff 2003).

As mentioned, ego-centred networks consist of a focal actor (ego) and a set of alters, who have ties to the focal actor, as well of the ties among these alters (Wasserman and Faust 1997 p. 42). Ego-centred network data assemble information about the local environments surrounding an actor. To obtain data about ego, the alters, and the relations among them, ego is the primary source of information. They are the respondents asked to report on their relationships to alters (first order star) and state their assumptions on the relationships among the alters (first order zone). Further indirect or multi-level relations (2nd, 3rd order see Figure 4-6) usually remain out of consideration due to research-technical reasons (Schenk 1995 p. 15). In addition, absolute attributes of ego and its respective alters are also collected.

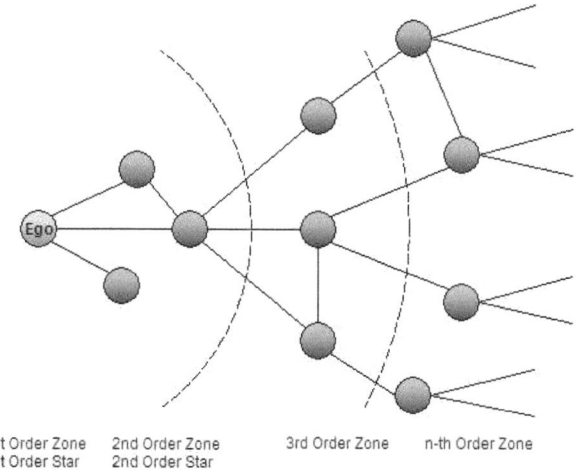

| 1st Order Zone | 2nd Order Zone | 3rd Order Zone | n-th Order Zone |
| 1st Order Star | 2nd Order Star | | |

Figure 4-6: Personal Network (Schenk 1995 p. 16)

The unit of analysis of this research comprises the responding academic entrepreneur, as focal actor, and the information they provide on their own relationships to alters and on the presumed linkages between alters. Moreover, proxy reports about alters are requested. Thus, the analysis focuses on the elicitation and examination of academic entrepreneurs' first order zones. Respondent's alters are not interviewed, due to practical, i.e. ensured anonymity of quoted alters, and resource considerations.

Against the background of the numerous possible relations that can put into the focus of analysis, the main problem of ego-centred network analysis consists in the choice of appropriate types of relations to be subjected to investigation. As to the information on relations questioned, following Jansen (2003 p. 59) they can be characterised in terms of their content, their strength, and their form.

With respect to the content of the interactions, in correspondence with Bührer (2002) three types of sub-networks were investigated:

1) resource networks
2) information networks
3) support networks

While resource networks provide the entrepreneur with required material assets such as property, capital and credit, information networks consist of people the entrepreneur addresses to discuss the business opportunity and aspects of establishing and running the spin-off. From the support networks, in turn, entrepreneurs receive backing and encouragement by means of emotional and institutional support.

Tie strengths are usually determined according to the frequency and importance of the relations, and the extent of resource transfer (Jansen 2003 p. 59). In case of the underlying research, reference is made to the frequency of interaction, the context role of the alter, as well as its weight in terms of multiplexity.

The form of the relation refers to the direction of the ties. It has to be determined whether there is a difference, if the tie goes from ego to alter or vice versa, or not.

For example, in case ego quotes an alter as important discussion partner, this alter does not necessarily need to quote ego as important discussion partner as well. As the alters are not questioned on their evaluation of the contacts, following common procedures (Jansen 2003 p. 82) relations were treated as non-directed and symmetric.

In contrast, to other network-analytic research traditions, ego-centred network analysis is strongly concerned with the question of multiplexity, since it may analyse several relations simultaneously. A relationship between ego and alter is multiplex, if it not only exists in one relational dimension, but in more (Jansen 2003 p. 80). In this regard, multiplexity denotes the degree of overlap between the different sub-networks of a respondent's network.

Yet, ego-centred networks deliver only a section of the social networks of a population. They do not yield an overall description of its social structure. Moreover, the method strongly depends on egos' ability and readiness to correctly describe alters and the relations between them (Jansen 2003 p. 85). As with most survey methods this entails recall problems and systematic biases that have to be treated accordingly. However, the approach provides representative samples of the social environments surrounding particular elements and is compatible with conventional statistical methods of generalisation to large populations (Marsden 1990). Moreover, due to the fact that ego-network data can be gathered by means of standardised questionnaires thus facilitated, large-scale data collection, contributes to its increasingly wide-spread application in different fields of research.

Beyond its increased range of methodological applicability the concept of ego-centred networks also encompasses total networks with respect to its broader content of analysis when it comes to the selection of social interactions and environments. Based on one focal person different forms of interaction within different social environments or systems of action can be captured (Pfenning 1996 p. 22).

4.3.4 Network Measures of Social Capital

As Lin (1999) notes, the premise behind the notion of social capital is rather straightforward: investment in social relations with expected returns. Thus, social capital, as a concept, is rooted in social relations and social networks, and should be measured relative to this root (Lin 1999).

Summarising and compiling the different notions on relevant network measures of social capital (Bührer 2002; Burt 2000; Hall and Wellman 1985; Jansen 2003; Jansen 2001; Lin 1999), a basic distinction can be made between relational, i.e. content characteristics of the dyadic relationships, and structural aspects relating to personal networks as a whole (see Table 4-5).

Characteristics of Relationships	Characteristics of Networks
(1) content	(1) size or scope of the network
(2) intensity	(2) density
(3) frequency of interaction	(3) centrality
(4) multiplexity	(4) heterogeneity
(5) intimacy	
(6) duration of the relationship	
(7) symmetry and reciprocity of exchange	

Table 4-5: Network Measures of Social Capital

Depending on the type of network, i.e. the nature of the underlying interaction that shall be promoted by the respective structural and relational characteristics of the network, different realisations of the measures are deemed to provide positive contributions to the entrepreneurs' social capital. In case of supplying information, the larger the network, the more alters it contains, the less these alters are connected, and the more they differ in their attributes, the better the performance of the network as source of information. When it comes to solidarity and support, in the form of both material and emotional backing, the same measures apply, but - besides network size - with inverse signs. In this case, dense networks with many alters that know each other from several contexts characterised by a high level of multiplexity and homogeneity are of benefit (Jansen 2003 pp. 106-07). In addition, it is assumed, as elaborated in Chapter 3.4, that information about opportunity structures is transferred more efficiently via weak ties and sparse networks, while consulting services and in particular costly resources like financial capital require strong ties, whose resilience and trustworthiness increase with their social embeddedness.

4.3.4.1 Relational Measures

As mentioned relational measures capture the characteristics of the relationships between the network actors. With respect to the hypotheses to be tested in the scope of this research the following three relational measures are deemed to be most appropriate to grasp the relevant essentials transported: *content*, *tie strength*, and *multiplexity*.

Content

The content of relationships refers to the meaning of the links between ego and its respective alters. As mentioned in Chapter 4.3.3, in this research three major types of exchanged content are distinguished:

(1) information,

(2) resources, and

(3) support.

They are measured in terms of categorical variables presented to the respondent. Yet, the content transferred through a tie is less a direct measure of an entrepreneur's social capital, but rather a contingency factor determining which realisation of the other measures has a positive or negative effect with respect to its contribution to social capital. Taking the next measure as example, strong ties are assumed to be favourable for the exchange of resources and enable the transfer of emotional support, while novel, non-redundant information is more probably accessed via weak ties to distant contacts.

Tie Strength

The umbrella term "strength of a tie" can be applied to compile several relational dimensions. According to Granovetter (1973), the strength of ties entails a combination of amount of *time*, emotional *intensity*, *intimacy* (mutual confiding) and the *reciprocal services* that characterise the tie. Hanneman (2001 p. 12) acts on this notion by stipulating that the strength of a tie comprises a variety of things. According to him (Hanneman 2001 p. 12) one dimension is the *frequency* of interaction; another dimension is "intensity" reflecting the degree of emotional arousal associated with the relationship; and a third dimension is reciprocity with ties being stronger to the extent that they are reciprocated.

Thus, to determine the strength of a tie, reference is made to the role of the alters quoted by the responding academic entrepreneur. In this regard strong ties are associated with contacts to family members and close friends that are assumed to entail more emotional arousal and consequently relational intensity than rather distant acquaintances such as colleagues or contacts from industry. An other indicator is the frequency of interaction. It expresses how regularly a link is activated. That may be daily, weekly, monthly, and several times a year, or even less, indicating decreasing tie strength. Again, this is determined by means of categorical questions directly asking the respondent to choose the applicable frequency category for the relationship currently under investigation.

As mentioned (see Chapter 4.3.3), only symmetric relationships are collected, thus, measures involving directional ties, such as mutual confiding or reciprocity, are neglected.

Moreover, high levels of multiplexity are also deemed as indicator of strong relations (Degenne and Forsé 1999; Jansen 2001), since the provision of different types of content by one alter suggests a close relationship to ego.

Multiplexity

Multiplexity refers to the extent to which ego's network contacts deliver more than one form of relational content. For determining the multiplexity of relations in ego-networks, reference is made to the relationships between ego and its alters. Following Jansen (2003 p. 109), linkages between alters are neglected, due to the lack of related information. In a sub-matrix for each alter mentioned by ego the number of different types of relation, i.e. provision of information, resources, and support, quoted between the respective dyads is recorded (Jansen 2003 p. 109). This matrix is dichotomised at a pre-defined threshold m = 2, assigning cases ≥ 2 the value 1 (= multiplex) and the others the value 0.

Consequently, the measure of multiplexity is the number of ego-alter-relations x_{ij} meeting the threshold m related to the total number of all ego-alter-relations (Jansen 2003 p. 109).

(4.1) $\quad$ *Multiplexity of Actor i* $\quad M_i^{'} = \dfrac{\sum_{j=1}^{n} x_{ij(m)}}{(n-1)} \quad$ *for* $i \neq j$

As regards the *duration* of the relationship, this measure is not elicited. Given the temporal character of the questionnaire requiring respondents to resemble network data from three different, subsequent phases of the spin-off process, they cannot be expected in addition to provide information about the duration of their relationships to stated alters relative to the respective stage of the process. For that reason, this dimension of social capital has been neglected.

As argued in the previous Chapter 4.3.3, due to high cognitive pressure imposed on respondents, when requiring them to assess the direction and reciprocal nature of their alters relations among each other, respective questions are omitted and thus data on *symmetry and reciprocity* of exchange are not available. However, given the underlying focus of research respective information is negligible, in particular in the light of the associated high danger of loss of reliability.

4.3.4.2 Structural Measures

Burt (1983) consolidates the different properties of ego-centred networks under the term "range": the interpersonal environment of an actor has a range to the extent in which it connects the actor with totally different sets of other actors. Burt (1983) proves that the range possesses several different dimensions with the structural parameters of *size, density*, and *constraint* or *effective size* as measure for *structural holes* being the most important.

Network Size

The size of an ego-centred network N is determined by the number g of alters it contains.

(4.2) $\quad N = \{n_1, n_2, ..., n_g\}$

Density and Closure

Density represents the degree of connectedness, "mesh" (Schenk 1995 p. 17). It is defined by the ratio of the number of relations existing in a network to the number of possible relations. Due to the form of data collection asking ego to name its alters, the relations of ego to its alters are quasi an artefact necessarily conditional on the method of data collection. Consequently, density does no longer range between the theoretically possible values zero and one, but is biased towards dense networks. To this end, it is common to exclude ego-alter-relations from density calculation. Hence, density can be calculated as follows, with the sigma sign starting with i=2 and j=2, if ego is denoted as first actor, and given symmetric relationships between alters:

(4.3) $\quad Density\Delta = \dfrac{\sum\limits_{i=2}^{n}\sum\limits_{j=2}^{n} x_{ij}}{(n-1)\times(n-2)} \quad$ for $i \neq j$ and $i/j \neq 1$

Following Coleman's (1990) main argument that closed networks are characterised by a high degree of interconnectedness of its actors, network density is usually deemed as one measure of network closure. Beyond this assumption this thesis applies an additional measure of network closure taking into account the beneficial effect network closure is deemed to have according to Coleman on the generation of trust in the sense of bonding social capital. As the author states network closure directly affects outcomes, in particular higher closure generates more trust between friends facilitating cooperation.

Trusting environments are however not only generated by a high degree of connectedness, but also by the quality of the relations ego possesses. In particular, a high amount of close relations or in other words strong ties measured by the total amount of relations ego has, provides for high interaction with a friendly social environment that can be trusted and vice versa is also more likely to trust ego. Against the background of Allcott et alii's (2007) idea of trust flow, which measures the degree to which agents in a network can cooperate, the proportion of intense pairwise relationships between ego and alters is taken as measure of the closure of the network around ego. High closure means high pairwise trust with a small number of agents, while lower closure means having low pairwise trust with a large number of agents. In line with Kalish and Robins strong-tie network closure (2006) for ego i is calculated as proportion of strong ties s_{ij} ego has to its alters j of the total amount of alters N:

(4.4) $$Closure = \frac{\sum_{j=1}^{n} s_{ij}}{N} \text{ for } i \neq j$$

Structural Holes measured by Effective Size

According to Burt (1992 p. 51) structural holes refer to network measures indicating to what extent individual actors have non-redundant contacts and to what extent actors are constrained by the structure of relationships of their direct contacts. Depending on the underlying research question, either effective size or constraint, or both measures may be applied when analysing structural holes in the context of social capital. In the scope of this research, structural holes as sources of non-redundant information are in the focus of interest. Yet, Burt's redundancy-related measure of effective size is applied.

As to effective size, Burt developed a measure, which principally calculates the number of non-redundant contacts in the personal network of an actor. To this end, Burt takes the size of an actor's ego-network and deducts a redundancy factor to result in the effective size.

The line of argumentation rendered is as follows: Given an actor p_i has direct relationships with actors p_j and p_q that in turn are also directly connected. As a result, the time and energy p_i invests in the relationship with p_j is redundant, since the information or resources that p_j has to offer can be also accessed by p_i through the connection between p_j and p_q (Böhner 2007 p. 132). Burt (1992 p. 52) gives this measure for node *i* as:

(4.5) $$E_i = \sum_j \left[1 - \sum_q p_{iq} m_{jq} \right] \text{ for } q \neq i,,j$$

where p_{iq} reflects node p_i's time and effort invested in a relationship with p_q (interaction with p_q divided by the sum of p_i's relations), while m_{jq} denotes the strength of p_j's relation with contact p_q. The larger the summation inside the brackets the more redundant is the contact with p_j for p_i. Redundant contacts reduce the effective size of an egocentric network (Marsden 2002). When aggregating the product $p_{iq}m_{jq}$ across all contacts q, the portion of p_i's relationship with p_j that is redundant to other relations of p_i is measured. Subtracting $p_{iq}m_{jq}$ from 1 delivers the non-redundant portion of the relationship (Böhner 2007).

Essentially, network effective size is the number of alters a focal individual has minus the average number of ties its alters have within the ego network, while the ties to ego are not counted (Rowley and Baum 2004). To put it in other words, effective size measures how many different or non-redundant other actors (potential sources of information or other resources) the focal actor can access in his ego-network.

Heterogeneity

Heterogeneity refers to the composition of the network and expresses the dissimilitude of alters in ego's network. Usually, heterogeneity is measured in socio-demographic terms or social strata. With increasing heterogeneity also the range of a network increases, since, as a result, ego is integrated in different social spheres and groups, thus obtaining access to different resources of information, communication, and influence (Schenk 1995 p. 17).

According to van der Gaag (2005), a wide-spread measure for heterogeneity is the Index of Qualitative Variation (IQV). It is a standardised version of Simpson's (1949) index of diversity, which is represented by the following formula:

(4.6) $$D = 1 - \sum_{i=1}^{k} p_i^2$$

Where $p_{i'}$ is the proportion of observations in class i of k classes. D can be defined as the probability of two individuals in a random sample being the same category.

D takes the value 0, when all observations fall in one class, and has a maximum value of 1 − 1/k, when all observations are evenly distributed over all k classes. With increasing number of classes, this value asymptotically approaches 1. Since D is consequently dependent on the number of considered classes, multiple comparisons with observations in different numbers of classes are difficult. To this end, Agresti and Agresti (1978) suggested to employ the standardised Index of Qualitative Variation developed by Mueller and Schuessler (1961):

(4.7) $$IQV = \frac{\left(1 - \sum_{i=1}^{k} p_i^2\right)}{\left(1 - \frac{1}{K}\right)}$$

The IQV summarizes the amount of diversity in one category. It ranges between 0 and 1, representing the state of no diversity or of maximum diversity, respectively.

Centrality

According to Freeman (1979), three different centrality measures exist for sociocentric or whole-networks: (1) degree centrality, (2) closeness centrality, and (3) betweenness centrality. Degree centrality is measured as the number of direct ties that involve a given node. Closeness centrality, in turn, is measured as a function of geodesic distances, i.e. the shortest path connecting two actors, with decreasing closeness centrality as the geodesics increase in length (Marsden 2002). And finally, betweenness centrality reflects the intermediary location of a node along indirect relationships linking other nodes.

When transferring these measures to ego-centred networks, in principle only one measure remains that delivers meaningful information. While the closeness centrality measure is uninformative when based on the egocentric data, since all geodesic distances from ego to other nodes in the first-order zone are 1 by definition (Marsden 2002), degree centrality is basically equal to the size of the network due to the definition of ego-centred networks featuring the respondent or ego as focal actor. This leaves us with betweenness centrality as main centrality indicator for egocentric networks.

Freeman (1979) gives a conceptual interpretation for betweenness measures in coordination/control terms: a node with high betweenness has a capacity to facilitate or limit interaction between the nodes it links.

Freeman's (1979) betweenness measure for sociocentric data is

(4.8) $$C_B(p_i) = \sum_{j=1}^{N} \sum_{k=1}^{j-1} b_{jk}(p_i)$$

where b_{jk} (p_i) (defined as g_{jk} $(p_i)/$ g_{jk}, where g_{jk} is the total number of geodesic paths linking p_j and p_k, and g_{jk} (p_i) is the number of those geodesic paths that include p_i) reflects the extent to which node p_i lies between nodes p_j and p_k.

The corresponding egocentric betweenness measure refers to the number of pairs of actors (not involving ego) in the egocentric network that are not directly linked to each other and, therefore, are indirectly connected via a geodesic through ego. This measure represents, in essence, the reverse of egocentric network density (Marsden 2002):

(4.9) $$C_{B_e}(p_i) = \sum_{j \neq i}^{N} \sum_{k \neq i}^{j-1} [1 - a(p_j, p_k)]$$

The egocentric analogue of Freeman's betweenness measure additionally takes account of the fact that some pairs (p_j, p_k) may be connected not only via ego, but also through other nodes in the egocentric network that thereby share the intermediary position for that relationship with ego (Marsden 2002).

4.3.5 Sampling

In statistical sense, a sample is a subset of the population of interest with manageable size. Statistics are calculated from the samples to make inferences or extrapolations from the sample to the population. This process of collecting information from a sample is referred to as sampling (Wikipedia 2007a).

According to Diekmann (2002 p. 328) main groups of sampling techniques can be distinguished:

(1) Probability sampling (random sampling)

(2) Purposive sampling

(3) Arbitrary sampling

In case of a simple random sampling, elements of the population are treated equally and feature the same probability of entering the sample. Where the population consist of a number of distinct categories, segmentation may be carried out and elements then selected randomly form the identified segments, such as the case with stratified or cluster sampling.

In the second case, which is also often equated with quota sampling, the population is also first segmented into mutually exclusive sub-groups. Then judgement is applied to select the subjects from each segment based on a specified proportion. This second step makes the technique one of non-probability sampling.

Convenience sampling or opportunity sampling is the method of choosing items arbitrarily and in an unstructured manner. The sampling procedure is not controlled. Though almost impossible to treat rigorously, it is the method most commonly employed in many practical situations. As the name implies, the selection of units from the population is based on easy availability and/or accessibility.

This also applies to this research project. Due to the predominant objective of the dissertation project of analysing successful academic entrepreneurs' social capital at different stages of the high-technology venture creation process, a couple of stringent requirements had to be put on the range of possible informants.

First of all, target persons had to have founded a high-technology spin-off according to the definitions of Chapter 2.1.2. This means that, taking Viardot's (2004 p. 6-7) characterisation of high tech products, high-technology ventures are companies that incorporate sophisticated technology to produce products with a short product life cycle that are new to the market.

Secondly, this founding process had to have a direct link to their university activity, without any interruptions in their professional career. Next, to be a successful entrepreneur or new venture all stages of the start-up process had to be completed at least two years ago. Consequently, the selection was limited to entrepreneurial researchers, who had founded their high-tech spin-off company at least two years ago. This approach should safeguard that only those academic entrepreneurs with substantial research opportunities were considered for analysis. Thus mere "mayflies" being present just at the time the research was carried out and disappearing shortly after form the marketplace were excluded. The comparatively short period of two years was chosen for pragmatic-methodological reasons. Although some sources cite a longer period of time required to assess to substantiality of a new venture, longer time restrictions would have significantly impaired the validity of data reported due to respondents' recall problems.

Finally, of exclusive interest was the initiating entrepreneur as the person, who had the idea for the venture and/or established the new business. While this definition neither excludes partnerships or other collective action nor the existence and importance of supporting entrepreneurs, the focus remained primarily on the original initiator/s (Christensen et al. 2000).

However, the above stipulated characteristics only refer to a rather small proportion of the Austrian population of entrepreneurs. To this problem attributes the fact that available databases were rather limited in their ability to identify the interested subgroup of academic spin-off founders. Most of the time information was only provided on new businesses in general without the possibility to systematically address the sub-group of start-ups that exclusively spun-off from a university.

To reach as many members of the target population as possible, the websites of the nine incubators founded within the scope of the Austrian spin-off support programme Academia plus Business[1], the Life Science start-up support programme LISA as well as technology centres and technology transfer offices of universities were addressed. Sampling was realised such that all apt academic entrepreneurs, i.e. high-technology university spin-offs of more than two years age, thus identified were approached by e-mail and requested to complete the online questionnaire.

The method for sampling respondents greatly depends on the type of study. A difficult and critical task is to achieve the a balance between the number of respondents, the number of alters they will be asked about, and the amount of information about each alter elicited (McCarty 2003). This is an important aspect, which has to be considered when it comes to the method of data collection.

4.3.5.1 Population

Until recently systematic collection of data on academic spin-offs in Austria was rather moderate. In particular, it was not possible to make valid statements on their actual number apart from information on new high technology businesses in general. However, with the finalisation of two empirical studies on academic entrepreneurship in Austria (Egeln et al. 2006; Egeln et al. 2003) carried out by the Centre for European Economic Research[2] (ZEW) in Mannheim in cooperation with the

---

[1] http://www.ffg.at/aplusb
[2] http://www.zew.de/

Viennese Institute of Technology and Regional Policy of Joanneum Research[1], systematic information on this subject is available for the first time.

The main results of these studies have been concisely summarised in the Austrian Research and Technology Report 2007 (Federal Ministry of Science and Research et al. 2007). As pointed out here (Federal Ministry of Science and Research et al. 2007 p. 130), the authors of the two studies, Egeln et al. (2006; 2003) distinguish between so called "exploitation spin-offs" and "competence spin-offs". While the first are start-ups requiring the use of new research findings or scientific processes obtained from public research, in which at least one of the founders has been involved, the latter refer to start-ups essentially based on the utilisation of special skills acquired by the founders through their work in science or from their studies. The authors found that 2003/2004 out of the almost 20,000 new companies founded annually in Austria, 4,730 occurred in research- and knowledge-intensive segments. Of these, 1,990 or 42% were set up by academics. Using the above definitions, spin-offs made for 560 formations, of which 250 were exploitation spin-offs and 310 were competence spin-offs.

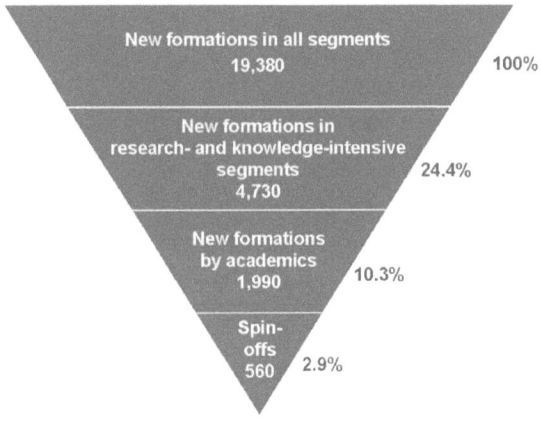

Figure 4-7: Spin-off Population in Austria (Federal Ministry of Science and Research et al. 2007)

As Figure 4-7 indicates the share of academic spin-offs in the Austrian population of new business foundations is rather low, i.e. less than 3%. Moreover, it is difficult to estimate how many of them can be assigned to the high-technology sector. As Eglen et al. (2006; 2003) further present in their work, a large number – about 40% - of the spin-offs founded 2003/2004 originated from the tertiary sector, such as consulting and trade. Consequently, only about 336 companies remain as basis for calculating the population of academic high-tech start-ups relevant to this research project. Yet, this last question remains unanswered. As to the actual amount of spin-offs belonging to the field of high technology, only vague assumptions can be made. Furthermore, the above presented data have to be taken with a pinch of salt, since they only reflect the number of newly founded businesses within a given period and not the total number of companies actually in existence at this time. Nevertheless, given the prevalent lack of information on this subject, the information provided can be taken as good point of reference.

---

[1] http://www.joanneum.at/en/fb5/rtg.html

4.3.5.2 Sample Characteristics

The dimensions described in the previous chapter have to be taken as comparative value for assessing size and properties of the sample actually chosen. In total 199 spin-off founders were sampled and contacted by e-mail. The e-mail contained a brief introduction explaining the motivation and background of the survey and a link to the online questionnaire.

As depicted in Figure 4-8 the majority of new business founders identified were male. Only 5.6% of the total sample could be ascribed to female academics. As to the age of the companies set up by the sampled entrepreneurs most of them were founded between 2004 and 2006 with another peak of almost 10% at 2002 (see Figure 4-9).

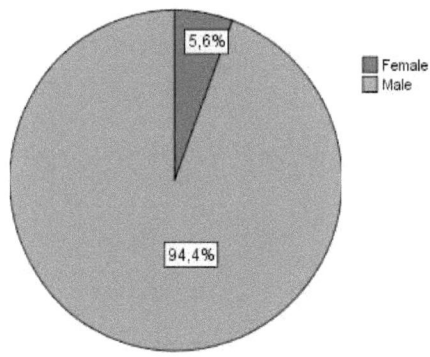

Figure 4-8: Gender Distribution of Sample

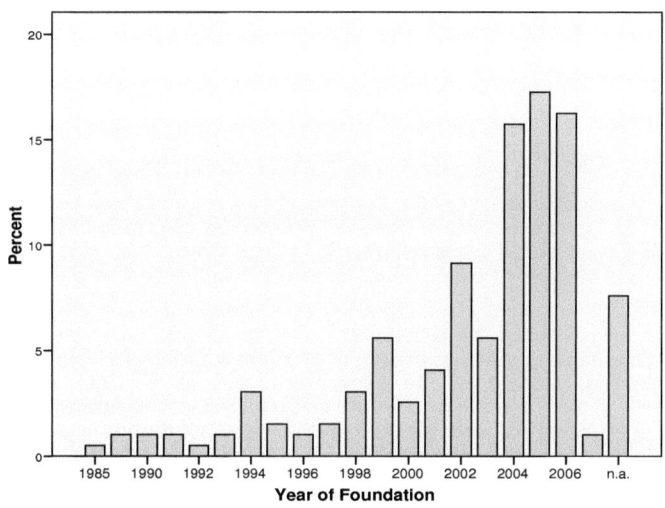

Figure 4-9: Year of Foundation of Sampled Companies

Regarding the distribution of incubating academic institutions as revealed in Figure 4-10 the most of the sampled entrepreneurial researchers, i.e. almost 30%, originated from the Technical University of Graz in Styria. At some distance follows the University of Innsbruck in Tyrol with almost 15% academic business founders. Still a rather larger proportion of identified spin-off initiators can be attributed to both the Montanuniversity of Leoben also in Styria and the Technical University of Vienna. "n.a." stands for "not applicable" and refers to companies the year of foundation of which could not be determined.

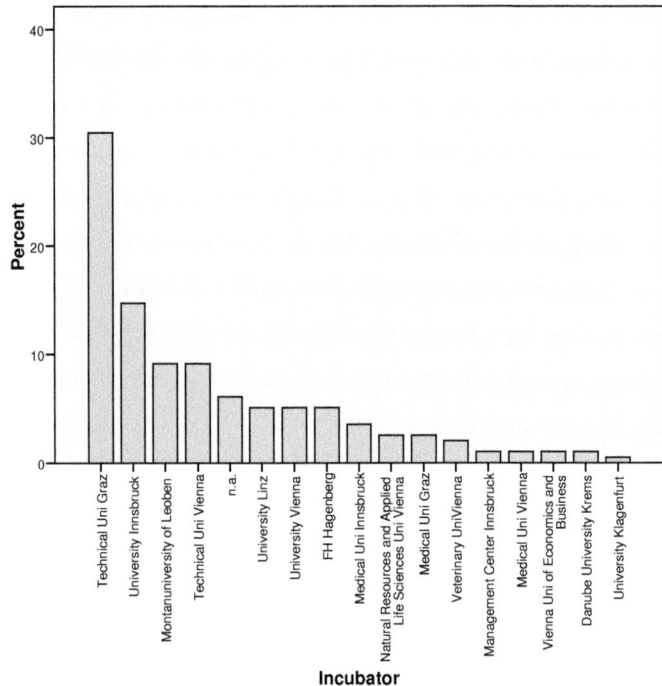

Figure 4-10: Distribution of Incubating Organisations

Coming to the regional allocation of the young companies of the sampled academics as illustrated by Figure 4-11, most of them, i.e. 40%, can be found in Styria followed by Tyrol and Vienna with 21%, respectively. Still a large share of 12% is located in Upper Austria, while Lower Austria, Carinthia, and Salzburg only range a one-digit percentage area.

And finally having a look at the industries the spin-offs operate in as presented in Figure 4-12, one can see a clear dominance of electronic and information-processing lines of business, followed by life technologies such as bio- and medical technologies. The third rank belongs to industries associated with manufacturing, i.e. production technologies and materials. All in all the majority of sectors, in which the spin-offs operate, can be ascribed to technological areas, while natural and social sciences are hardly represented.

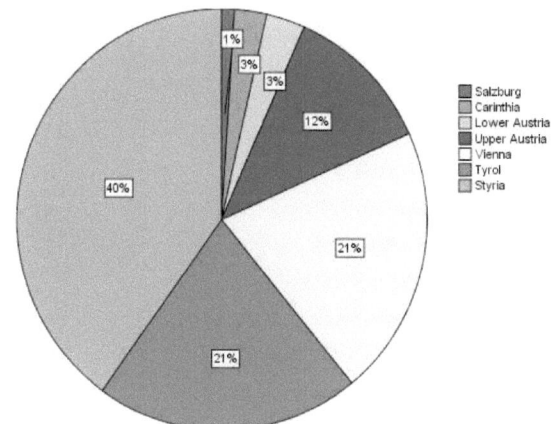

Figure 4-11: Regional Distribution of Spin-offs

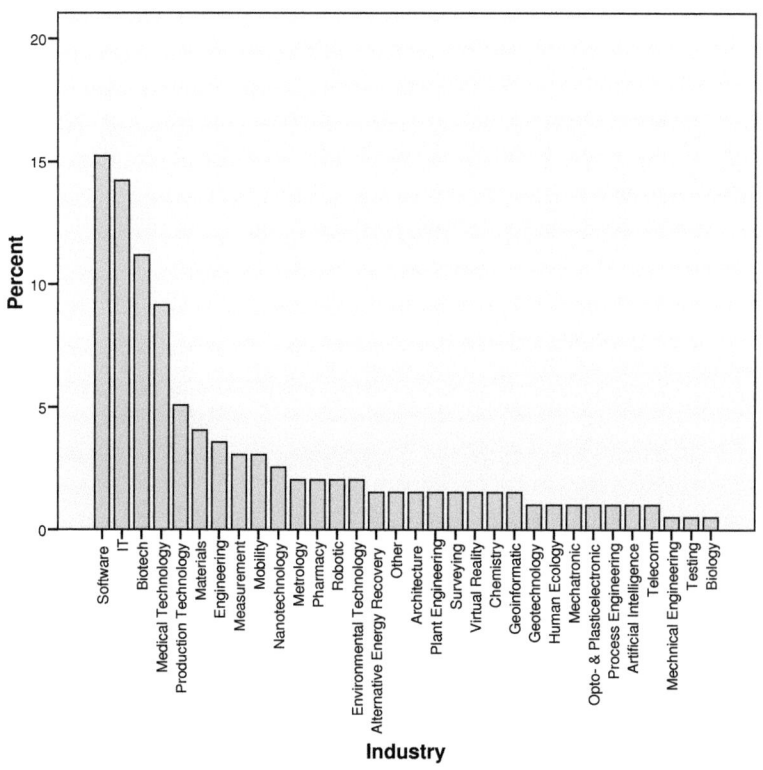

Figure 4-12: Distribution of Industries

4.3.6 Data Collection

According to Wasserman and Faust (1997 p. 45), there are a various ways in which social network data can be gathered. These techniques are:
- Questionnaires
- Interviews
- Observations
- Archival records
- Experiments

The most commonly used data collection method refers to surveys and questionnaires. Surveys enable researchers to decide on relationships to measure and on actors to be approached for data. In case no archival records are available, questionnaires are often the most practical alternative, since they are not so demanding on participants than, for example, diary methods ore observations. Yet, surveys introduce artificiality and the results are heavily dependent on the presumed validity of self-reports (Marsden 2005).

As regards the collection of ego-centred network data, the most widespread measurement method is the name generator / interpreter approach originally developed by McCallister and Fischer (1978). The purpose of the name generator consists in producing an extensive list of alters belonging to ego's network. The name interpreters in turn furnish additional information about the alters and the relations between ego and alters and between the alters. The name generator determines the boundaries of the network. It consists of one or more type(s) of relationship with respect to which ego has to nominate its relevant network persons. The selection of types of relationship relevant for constructing the network is grounded in the underlying research question and / or the theoretical concepts (Jansen 2003 p. 80).

4.3.6.1 Questionnaire

Following the above described method, data for this research project was generated on the basis of a standardised network questionnaire in line with the name generator / name interpreter approach. The questionnaire was administered online to facilitate the correct passage through the large number of filtering questions and relational specifications shaped by the amount of network members quoted beforehand. In line with the research model established and presented in Chapter 3.4, the questionnaire was structured along the three phases of new venture creation (1) opportunity recognition, (2) formation, and (3) start-up and establishment on the market.

4.3.6.1.1 Introduction

To prepare the ground, the questionnaire started with an introductory text explaining the motivation of the research project and the underlying concepts of entrepreneurial phases and social capital applied. Respondents were informed about the expected duration of the survey and of course assured anonymity. In addition, they were also offered a summary of the final work as incentive to complete the questionnaire. To ease orientation, each phase was coded with an individual colour that was followed throughout the questionnaire (see Appendix 8.3). A bar indicated the progress of the enquiry both graphically in the colour of the respective phase and in percent. In case of ambiguous questions or terms a question mark in a blue spot provided explanatory information in the form of mouse-over pop-up window. Similarly, each entrepreneurial phase was shortly explained again when the respondent brought his / her mouse over the respective arrow of the illustration of phases on top of each page (Figure 8-1 of Appendix 8.3).

4.3.6.1.2 Social Network Questions

As expected, the core of the questionnaire consisted of social capital related questions. These were realised by McCallister's and Fischer's name generator / operator approach. To be able to observe the development of entrepreneurs' social networks in the course of the spin-off creation process, the respective battery of questions remained the same for all three phases and thus had to be completed trice. Respondents were informed beforehand that they were free to nominate both new and already quoted contact persons, and in the latter case to either stick to the person's original role and content of the interaction or to change them depending on their respective appreciation of the past situation.

Name Generator

The central question in this context refers to the method of alter determination. Basically, two types of name generators can be distinguished: *global questions* addressing entire categories of persons, such as "best friends", and *socio-metric questions* that collect persons directly (Bastians 2001).

The main problem of global name generators consists in their selectivity. Only one segment of the network is detected, possibly at the cost of other groups of persons, e.g. when asking for "best friends" kin are hardly quoted. Other problems refer to the meaning associated with the terms – for example "friend" – applied, which may differ from (sub-)culture to (sub-)culture. Moreover, effects of social desirability may influence the answers; out of the fear of being unpopular many respondents tend to exaggerate the number of their friends. Socio-metric questions as name generators help to overcome these problems by asking for persons, with whom particular interactions have taken place. Thus, certain groups of persons are not excluded a priori und the network is focussed on the subject or interaction areas concerned (Bastians 2001).

From there, socio-metric name generators representing the respective social interactions under investigation were applied. For each of the three phases respondents were asked to name a maximum of 10 persons, with whom they were in contact at this time (see Figure 8-2 of Appendix 8.3).

Name Interpreter

At this stage information about the characteristics of the network members obtained as result of the name generator question was collected. The only general demographic question asked in this context referred to the gender of the quoted network persons. Due to the focus on network-related information, more room was given to the characteristics of the ties between ego and his / her alters. These included relational characteristics, such as the content (information, resource or support) and frequency of contact, and relational types, i.e. whether the alter relates to ego's private (family or friend), academic (same or other discipline) or business (industry or funding agency) network (Müller et al. 1999). Respondents were also given the opportunity to specify an additional category under the heading 'Others' (see Figure 8-3 of Appendix 8.3).

When it comes to determining the content of the interactions of ego with its network partners, the generic categories of information, resource and support networks were further specified. As to the information exchanged a distinction was drawn between legal information, economic knowledge, information about of potential customers, and information about and brokering of personnel. In the case of resources provided, material, capital, infrastructure and personnel were taken into account. In accordance with the considerations of Chapter 3.4 support networks were investigated along the categories of emotional and institutional support.

In both cases, i.e. the question on relational types and the question on the content of relations, respondents could choose several categories, but had to indicate at least one of them. Otherwise, the questionnaire produced an error message asking to tick at least one box, before allowing moving on to the next question.

To measure the structure within each respondent's network, respondents not only had to report on their relationship with each alter, but also on the relationships of all pairs of alter. As McCarty (2003) highlights, it is common practice among researchers to ask only about the existence a of a tie and to avoid questions about asymmetric ties, since these are unlikely to be understood correctly by respondents. In line with this experience responding entrepreneurs were only asked to report on their assumption whether pairs of network members knew each other or not.

However, even when limiting reports to symmetric ties, as in this case, the number of tie evaluation grows geometrically as alters are added (McCarty 2003). This is the reason why the name generator was limited to a maximum of ten persons taking into account that information may be lost in case of larger entrepreneurial networks. As empiricism showed ex post, this was not the case, as the number of network members specified did not exceed 9 and amounted to about 4 persons in average.

In this context, it has to be admitted that even ten persons is a large number of alters to be treated cognitively. Given that a respondent nominates the maximum of ten contacts for a phase, he or she has to report for a total of 45 pairs ($n(n-1)/2 = 10*9/2$) whether they know each other or not. To account for associated cognitive boundaries responding entrepreneurs could not only choose between the two values 'Know each other' and 'Do not know each other', but also tick 'I do not know'.

In this context, the interactive nature of the online questionnaire proved to be extremely helpful. This was also one important factor in favour of the original decision for this mode of enquiry. The dyadic matrices could be constructed dynamically such that only those contacts actually cited at the beginning of the phase were included (see Figure 8-5 of Appendix 8.3).

4.3.6.1.3 Phase-specific Questions

In addition to the general social network questions that were basically kept the same throughout all three phases of the spin-off process, specific questions were posed at each phase accounting for its particular peculiarities.

Phase 1 started with an open question on the underlying idea or motivation that had led to the subsequent spin-off creation and whether one specific person had served as main impetus to spin-off. After these initiating questions respondents were directly led to the name generator of the first phase.

Additional questions for *Phase 2* dealt with the general reaction of the entrepreneur's professional academic environment to his or her articulated intentions to spin-off a firm. A distinction was made between the responses of the institute the entrepreneurial research worked for and the university as such. This differentiation accounted for the results of the qualitative exploratory phase, which revealed in part different climates of support on the level of the department and on the level of the faculty or university. Respondents could specify whether the reaction of their department had been negative, neutral or positive. The same was asked for the university as a whole. Moreover, interviewees could indicate whether they were allowed to use the infrastructure (e.g. laboratories, IT) of the university at this stage of the process. The second battery of questions referred to the economic and financial issues of this preparing phase of venture formation. Two binary questions (yes/no) asked for a business plan and for the availment of venture capital. An open question enabled respondents to describe possible sources of seed capital.

Regarding the *third* and last *phase* consisting of the actual legal formation of the new business and its first steps on the market, informants were asked about the existence and number of partnering co-founders. Depending on the number of founding partners quoted, the questionnaire interactively popped up additional fields for each partner asking for his or her name. Respondents were offered the possibility to choose among persons already stated or to nominate a new contact (see Figure 8-6 of Appendix 8.3).

Furthermore, to determine the primary source of first staff, respondents could chose between 'department/university', 'open market', 'circle of acquaintances' or 'others'. The last questions associated with this phase asked for the respondent keeping his or her position at the university after having founded the company and for promotive factors resulting from his/her university background when entering the market. Out of five categories entrepreneurs could choose, whether it was their 'academic career', 'existing industry contacts', 'connections from the circle of friends', the 'family background', or 'other' factors that had helped them to build up a reputation on the market.

4.3.6.1.4 General Questions

To conclude some phase-independent demographic questions on the person of the entrepreneur and his / her firm were posed.

As to the person-related questions, they comprised of sex and age of the respondent and his or her university background such as discipline, academic position at the time of business foundation and the name of the incubator university. While discipline and university could be freely described, the academic position had to be chosen from a drop-down menu listing 'professor', 'associate professor', 'assistant professor', 'postdoctoral research fellow' or 'student'.

Concerning company-related data, founding year and industry was asked for. The respondents had to choose among a pre-defined set of nine industries derived on the basis of the Technology Sectors[1] specified by the former Bureau for International Research and Technology Co-operation (BIT). The tenths industry category was reserved for an open specification 'others'. Interviewees also had to state, whether they had founded a company before and whether they had received some sort of public support. An accompanying description field allowed them to specify the source of public funding.

4.3.6.1.5 Overview

The subsequent Table 4-6 provides a recapitulatory overview of the questions asked and their classification with respect to phase and network actors addressed.

Following the central assumption of the research model that network composition and structure change in the course of the spin-off process, the same network-related questions are repeated at each of the three entrepreneurial phases. To provide for a comprehensive picture of the academic start-up process, additional socio-economic items are prompted at the different stages that will also contribute to generating structural network parameters in the course of the statistical network analysis.

---

[1] http://www.bit.or.at/irca/220.htm

	General	Phase 1 - Opportunity Recognition	Phase 2 - Formation	Phase 3 - Establishment
Ego / Spin-off	Sex	Underlying idea / motivation	Reaction of University Environment	Number of Co-founders
	Age		Business Plan	Sources of Personnel
	University Background		Venture Capital	Keeping Position at University
	Serial Entrepreneur		Seed Capital	Reputation on the Market
	Public Support		Shared Infrastructure	
	Founding Year			
	Industry			
Network / Alters		Contact persons at respective phase (max. 10)		
		Gender		
		Social Role		
		Relational Content		
		Relations between Alters		
				Co-Founders

Table 4-6: Overview of Questions

4.3.6.2 Implementation and Data Handling

As mentioned the questionnaire was administered online in the form of computer-assisted self-administered interviewing (CASI). This means that respondents had to complete the questionnaire on their own without any direct intervention, but also without any support from the interviewer. To access the sampled population an e-mail containing the link to the start page of the questionnaire was send to the respondents identified in Chapter 4.3.5.

Response Rate

Out of the 199 entrepreneurs addressed 54 at least started to complete the questionnaire. However, only 34 of them accomplished all three phases thus delivering useful data records. Talking in relative terms, a response rate of 27% was achieved, out of which an exhaustion rate of 63% useful data sets could be realised. These 34 respondents in turn delivered information about 246 alters.

Data Bases

Against the background of the intended data analysis with the statistical software package SPSS, data were gathered and recorded in Excel. Following the suggestion of Mueller et al. (1999) as to the treatment of ego-centred network data in SPSS, data were organised according to three levels of analysis comprising of:

(1) *Ego* consisting of attributes of the entrepreneurs and their firms,

(2) *Alters* comprising of the network members quoted and their relations to ego,

(3) *Alters-Alters* indicating the relationship between egos' network members.

These three levels correspond to three Excel sheets in one Excel workbook labelled *'Ego'*, *'Alter'* and *'Alter-Alter'* (to be found in Appendix 8.2). For the purpose of analysis the datasets in these sheets will be subjected to different aggregation and matching procedures in the course of the statistical analysis of the next chapter. A crucial issue in this context is the consistent identification of all datasets in all sheets with the associated ego represented by its Ego ID.

4.3.7 Variables

In the following the variables subject to the quantitative analysis are presented in more detail. A general distinction is made between "*original variables*", which were collected in the present form and are applied to statistical analysis without any further modifications, "*aggregated variables*" derived by merging answers about and between alters on the level of ego, and finally "*compound variables*" being the result of computations of several original or aggregate variables.

4.3.7.1 Original Variables

As mentioned above, *original variables* were collected as presented and need no further adaptations to be applied to analysis. In principle, they correspond to general, non-network questions about ego, his/her business and other environmental factors possibly influencing the start-up process.

Table 2-1 enlists all original variables processed as well as their corresponding SPSS labels, general labels, scales and values. A short description informs about the content of the variable. Following the usual procedure when using SPSS for analysis, multiple choice questions were split into binary sub-questions coded 1 for "Yes" when ticked and 0 for "No" when left empty. In the course of further analysis these variables were reassembled as one multiple response set with the intended SPSS feature to be treated as one variable.

SPSS Label	Variable Label	Scale	Values	Description
main	Person Giving Main Impetus for Business Foundation	Nominal	"Yes" = 1 "No" = 0	Denoting whether ego was stimulated by one influential person to spin off.
busp	Business Plan	Nominal	"Yes" = 1 "No" = 0	Indicating whether a business plan was written in the course of spinning off.
vent	Venture Capital	Nominal	"Yes" = 1 "No" = 0	Indicating whether venture capital was used to start the business.
univ	Reaction University	Nominal	„Positive" = 1 „Negative" = 2 „Neutral" = 3	Denoting the reaction of the university – other than the department of the entrepreneur – to his / her articulated intention to start an own business.
inst	Reaction Department	Nominal	„Positive" = 1 „Negative" = 2 „Neutral" = 3	Denoting the reaction of the department the entrepreneur worked at to his / her articulated intention to start an own business.

SPSS Label	Variable Label	Scale	Values	Description
infra	Joint usage of University Infrastructure	Nominal	"Yes" = 1 "No" = 0	Indicating whether the academic entrepreneur could use infrastructure of the university for his / her business.
part	Foundation with Partners	Nominal	"Yes" = 1 "No" = 0	Indicating whether the entrepreneur had co-founders or not.
nrpart	Number of Partners	Metric	Integer	Informing about the number of co-founders – if any.
$emp	Personnel from ...	Nominal	Multiple Choice	Revealing the sources where the founder recruited his / her personnel.
empuni	... University / Department	Nominal	"Yes" = 1 "No" = 0	
empmkt	... Market	Nominal	"Yes" = 1 "No" = 0	
empfrien	... Circle of Friends	Nominal	"Yes" = 1 "No" = 0	
empoth	... Other Sources	Nominal	"Yes" = 1 "No" = 0	
keepfunc	Keeping Function at University	Nominal	"Yes" = 1 "No" = 0	Indicating whether the entrepreneur kept his / her academic position after foundation.
$rep	Reputation on the Market Due to ...	Nominal	Multiple Choice	Revealing from where the responding entrepreneur and his / her company derived their reputation.
repacad	... Academic Career		"Yes" = 1 "No" = 0	
repind	... Industry Contacts		"Yes" = 1 "No" = 0	
repfrien	... Friends		"Yes" = 1 "No" = 0	
repfam	... Family		"Yes" = 1 "No" = 0	
repoth	... Others		"Yes" = 1 "No" = 0	
specoth	Specification of Others	String	Text	
sexego	Sex of Ego	Nominal	„Female" = 1 / „Male" = 0	Indicating respondent's sex.

SPSS Label	Variable Label	Scale	Values	Description
age	Age of Ego	Metric	Years	Indicating respondent's age.
scidisc	Scientific Discipline of Ego	String	Text	Allowing the responding researcher to specify his / her scientific discipline.
inc	University of Origin	String	Text	Enabling the respondent to specify the incubating university.
unipos	Ego's Position at University	Nominal	"Professor" = 1 "Lecturer" = 2 "Assistant" = 3 "Research Associate" = 4 "Student" = 5 "Other" = 6	Indicating respondent's position at university at the time of spin-off.
year	Year of Foundation	Metric	Year	Denoting the year the spin-off was legally set up.
bran	Industry of Spin-off	Nominal	"Micro/IT/Telco" = 1 "Manufacturing/ Transport" = 2 "Industrial Technologies" = 3 "Energy" = 4 "Sciences" = 5 "Medicine/Bio/Bio-tech" = 6 "Micro- & Nanotech" = 7 "Agriculture & Forestry" = 8 "Food" = 9 "Others" = 10	Denoting the industry the spin-off operates in.
serent	Serial Entrepreneur	Nominal	"Yes" = 1 "No" = 0	Indicating whether the respondent has already founded a company.
pubfin	Public Finance	Nominal	"Yes" = 1 "No" = 0	Indicating whether the spin-off received support in the form of public funding.
fundspec	Specify	String	Text	Allowing the respondent to comment on the public funding.

Table 4-7: Original Variables

4.3.7.2 Aggregated Variables

Aggregated variables were generated by applying the aggregate procedure of SPSS, which refers to the process of consolidating cases belonging to a certain break variable by applying a particular mathematical function. In the present case, the variables of the *Alter* or *Alter-Alter* Excel sheet containing data about the alters quoted by the respondents were summarised using the associated distinct Ego ID as break variable. The resulting output refers to variables on the level of ego to be included in the Ego Excel sheet as one aggregated value per responding ego. At large these variables refer to *relational network parameters* like type of content, role of alters, frequency of interaction or tie strength as presented in Chapter 4.3.4.1.

Table 4-8 provides an overview of the respective variables, their values and the underlying aggregation function. # serves as replacement character for the respective phase of the spin-off process addressed by the variable to avoid lengthy redundant descriptions. Following the theoretical considerations of the previous chapters, three phases were distinguished. Consequently, when for example talking about the SPSS label *netsiz_#* in realita three variables *netsiz_1*, *netsiz_2*, and *netsiz_3* standing for the respective network size in Phases 1, 2 and 3 were analysed. Variables without underscore "_" refer to the raw data from the *Alter* or *Alter-Alter* worksheets subjected to aggregation. Underscores denote that the variables have been brought to the ego level and may be treated accordingly. More information about Alter and Alter-Alter variables can be found in Appendix 8.2.

SPSS Label	Variable Label	Scale	Values	Aggregation Function	Description
netsiz_#	Network Size in Phase #	Metric	0 to 10 (discrete)	= N(sexa#) by Ego ID [a]	Indicating the total size of ego's network, i.e. number of all contacts, in Phase #.
sexa_#	N° of Females - Phase #	Metric	0 to 10 (discrete)	= Σ(sexa#) by Ego ID [b]	Indicating the number of women ego had contact with in Phase #.
fam_#	N° of Family members - Phase #	Metric	0 to 10 (discrete)	= Σ(fam#) by Ego ID [b]	Indicating the number of family members ego had contact to in Phase #.
friend_#	N° of Friends - Phase #	Metric	0 to 10 (discrete)	= Σ(friend#) by Ego ID [b]	Indicating the number of friends ego had contact to in Phase #.
acad_#	N° of Academics - Phase #	Metric	0 to 10 (discrete)	= Σ(acad#) by Ego ID [b]	Indicating the number of academics ego had contact to in Phase #.
disc_#	Same Discipline - Phase #	Metric	0 to 10 (discrete)	= Σ(disc#) by Ego ID [b]	Indicating the number of academics from the same discipline as ego for Phase #.
inst_#	Same Department - Phase #	Metric	0 to 10 (discrete)	= Σ(inst#) by Ego ID [b]	Indicating the number of academics from the same department as ego for Phase #.
ind_#	N° of Industry	Metric	0 to 10	= Σ(ind#) by	Indicating the number of

SPSS Label	Variable Label	Scale	Values	Aggregation Function	Description
	Partners - Phase #		(discrete)	Ego ID [b]	industry representatives ego had contact to in Phase #.
funds_#	N° of Funding Partners - Phase #	Metric	0 to 10 (discrete)	= Σ(funds#) by Ego ID [b]	Indicating the number of representatives of funding institutions ego had contact to in Phase #.
other_#	N° of Others - Phase #	Metric	0 to 10 (discrete)	= Σ(other#) by Ego ID [b]	Indicating the number of other contacts of ego in Phase #.
info_#	Informational Content - Phase #	Metric	0 to 10 (discrete)	= Σ(info#) by Ego ID [b]	Denoting the number of contacts providing information in Phase #
infola_#	Legal Information - Phase #	Metric	0 to 10 (discrete)	= Σ(infola#) by Ego ID [b]	Denoting the number of contacts providing legal information in Phase #
infoec_#	Economic Information - Phase #	Metric	0 to 10 (discrete)	= Σ(infoec#) by Ego ID [b]	Denoting the number of contacts providing economic information in Phase #
infocu_#	Information about Potential Customers - Phase #	Metric	0 to 10 (discrete)	= Σ(infocu#) by Ego ID [b]	Denoting the number of contacts providing information about customers in Phase #
infoem_#	Information about Potential Employees - Phase #	Metric	0 to 10 (discrete)	= Σ(infoem#) by Ego ID [b]	Denoting the number of contacts providing information about Employees in Phase #
infoot_#	Other Information - Phase #	Metric	0 to 10 (discrete)	= Σ(infoot#) by Ego ID [b]	Denoting the number of contacts providing other information in Phase #
res_#	Provision of Resources - Phase #	Metric	0 to 10 (discrete)	= Σ(res#) by Ego ID [b]	Denoting the number of contacts providing resources in Phase #
resmat_#	Material Resources - Phase #	Metric	0 to 10 (discrete)	= Σ(resmat#) by Ego ID [b]	Denoting the number of contacts providing material resources in Phase #
rescap_#	Financial Resources - Phase #	Metric	0 to 10 (discrete)	= Σ(rescap#) by Ego ID [b]	Denoting the number of contacts providing financial resources in Phase #
resin_#	Provision of Infrastructure - Phase #	Metric	0 to 10 (discrete)	= Σ(resin#) by Ego ID [b]	Denoting the number of contacts providing infrastructure in Phase #
resemp_#	Provision of Personnel - Phase #	Metric	0 to 10 (discrete)	= Σ(resemp#) by Ego ID [b]	Denoting the number of contacts providing personnel in Phase #
resoth_#	Other Resources -	Metric	0 to 10 (discrete)	= Σ(resoth#) by Ego ID [b]	Denoting the number of contacts providing other

SPSS Label	Variable Label	Scale	Values	Aggregation Function	Description
	Phase #				resources in Phase #
sup_#	Support - Phase #	Metric	0 to 10 (discrete)	= Σ(sup#) by Ego ID [b]	Denoting the number of contacts providing support in Phase #
supem_#	Emotional Support - Phase #	Metric	0 to 10 (discrete)	= Σ(supem#) by Ego ID [b]	Denoting the number of contacts providing emotional support in Phase #
supins_#	Institutional Support - Phase #	Metric	0 to 10 (discrete)	= Σ(supins#) by Ego ID [b]	Denoting the number of contacts providing institutional support in Phase #
oth_#	Other Relational Content - Phase #	Metric	0 to 10 (discrete)	= Σ(oth#) by Ego ID [b]	Denoting the number of contacts providing other things in Phase #
strong_#	N° of Strong Ties – Phase #	Metric	0 to 10 (discrete)	= Σ(strong#) by Ego ID [b] with strong# = 1 if frequ# > 3 and multi# = 1	Denoting the number of strong ties in ego's network in Phase #; with contacts defined as strong when featuring high frequency and multiplexity.
weak_#	N° of Weak Ties – Phase #	Metric	0 to 10 (discrete)	= Σ(weak#) by Ego ID [b] with weak# = 1 if frequ# < 3	Denoting the number of weak ties in ego's network in Phase #; with contacts defined as weak when featuring low frequency
strhole_#	Structural Holes - Phase #	Metric	0 to 10 (discrete)	= Σ(iso#) by Ego ID [b]	Indicating the number of structural holes (iso# = alters with no tie to other alters) of ego's network in Phase #.
know_#	Ties between alters - Phase #	Metric	0 to 45*	= Σ(know#) by Ego ID [b]	Indicating the number of ties between alters of ego's personal network in Phase #.
multi_#	Degree of Multiplexity – Phase #	Metric	0 to 1 (continuous)	= Σ(multi#) / N(multi#) by Ego ID [a,b] with multi# = 1 if Σ(info#, res#, sup#) > 1 else 0	Being a measure of the degree of multiplexity of ego's network in Phase #; with contacts (multi#) defined as multiplex when providing more than one relational content.
frequ_#	Frequency of Interactions - Phase #	Metric	0 to 4 (continuous)	= Σ(frequ#) / N(frequ#) by Ego ID [a,b]	Being a measure of the mean frequency of interaction ego had with his / her network members in Phase #.

[a] N = total number of cases of the aggregated alter variable
[b] Σ = sum of the values of the aggregated alter variable

Table 4-8: Aggregated Variables

While the other tie qualities such as relational role and content were directly prompted in the course of the online survey, the strength of ties is a derived parameter generated by combining frequency and multiplexity of interaction. The associated binary variable *strong#* takes the value 1, when *frequ#* > 2, i.e. ego sees alter either frequently or permanently, and *multi#* is 1, i.e. the relationship is multiplex transferring more than one type of content. *strong#* then was aggregated by Ego ID resulting in the ego-level variable *strong_#* denoting the number of strong ties in ego's network in phase #. On the contrary, ties were regarded as weak and thus taking the value 1, when *frequ#* < 3, i.e. ego meets alter either sporadic or occasionally. Again to obtain *weak_#* *weak#* was summed by Ego ID.

As presented in Chapter 4.3.4.2, structural holes (*strhole_#*) are measured in terms of the effective size of the respondent's network expressed by the number of non-redundant contacts, i.e. alters that, apart from ego, are not connected to any other alter.

Apart from the last three variables, values of the aggregated variables may range from 0 to 10 reflecting the limitation of the name generator to 10 nominations. Thus, the maximum network size for each phase is 10 and also the maximum possible realisation of different relational roles such as family member or friend is restricted to 10. The same refers to relational contents like information or support and other qualities of ego's ties to alters like strength and structural holes, which can only be a function of the number of underlying network contacts.

The variable *know_#* however relates to the relations of alters among each other, thus featuring a proportionally higher number of possible combinations and consequently a higher maximum value. The number of ties between alters is derived from the *Alter-Alter* worksheet indicating how many of ego's contacts knew each other in the respective phase. Given a maximum of 10 possible alter nominations (n = 10), *know_#* might range from 0 to 45 with N = (n * (n-1))/2 = 45 maximal possible ties between n alters.

Multiplexity, like the variables prior to *know_#*, comes back to ego's relations to his / her alters. A tie between ego and alter is defined as multiplex, when it transfers more than one type of relational content expressed in terms of information, resources or support. In that case the underlying binary variable *multi#* assumes the value 1 otherwise it is 0. *multi#* is than aggregated as sum of all multiplex ties in Phase # related to ego's total network size at this time. The resulting variable *multi_#* stands for the degree of multiplexity the personal network of an entrepreneur exhibits in the respective phase. Consequently, the variable *multi_#* may assume all possible rational values between 0 and 1, with 1 being the maximal possible realisation of multiplexity.

As to the last variable regarding frequency of interaction, it represents the average frequency of ego's interactions with his / her network members in the respective Phase # expressed as arithmetic mean of *frequ#* from the *Alter* worksheet by respective Ego ID. While all other original alter variables were dichotomised with possible values 0/1 thus discretely ranging from 0 to 10, frequency of interaction could assume four different realisations: 1 = "Once/Sporadic", 2 = "Occasionally", 3 = "Frequently" or 4 = "Permanently". Consequently, mean frequency of interaction ranges from 0 to 4 on a continuous scale.

4.3.7.3 Compound Variables

The term compound refers to variables that are derived from original or aggregated variables by means of different computational procedures. As Table 4-9 illustrates, they predominantly consist of *structural social capital measures* such as density, centrality, and effective network size. The variables presented have in common that they were in principle generated by relating aggregated tie specific information to the total size of ego's network at the phase concerned. Such they allow for cross-phase comparisons of the individual parameters being independent from the underlying network size.

Beyond these structural network measures, relative variables relating to the three major types of content transported between ego and alter, i.e. information, resources and support, were also calculated to be put in the position to compare them across the three spin-off phases. This approach particularly accounts for the phase-related hypotheses specified in Chapter 3.4.

Due to the relative nature of the compound variables in essence expressing a proportion or percentage of a certain characteristic of the total personal network size, their values continuously range between 0 and 1.

SPSS Label	Variable Label	Scale	Values	Computational Function	Description
infopc_#	Proportion of informational relations - Phase #	Metric	0 – 1 (continuous)	= (info_#) / netsiz_#	Indicating the proportion of informational contacts of ego's personal network in Phase #.
respc_#	Proportion of resource relations - Phase #	Metric	0 – 1 (continuous)	= (res_#) / netsiz_#	Indicating the proportion of resource providing contacts of ego's personal network in Phase #.
suppc_#	Proportion of support relations - Phase #	Metric	0 – 1 (continuous)	= (sup_#) / netsiz_#	Indicating the proportion of supporting contacts of ego's personal network in Phase #.
spars_#	Sparsity - Phase #	Metric	0 – 1 (continuous)	= (strhole_#) / netsiz_#	Indicating the proportion of structural holes in ego's personal network in Phase #.
closu_#	Closure - Phase #	Metric	0 – 1 (continuous)	= strong_# / netsiz_#	Indicating the proportion of strong ties in ego's personal network in Phase #.
dens_#	Density - Phase #	Metric	0 – 1 (continuous)	= know_# / ((netzsiz_# * (netzsiz_# - 1)) / 2)	Describing the proportion of all possible ties actually realised in ego's personal network in Phase #.
cent_#	Centrality - Phase #	Metric	0 – 1 (continuous)	= (((netzsiz_# * (netzsiz_# - 1)) / 2) – know_#) / ((netzsiz_# * (netzsiz_# - 1)) / 2)	Referring to the number of pairs of alters not directly linked to each other and, therefore, indirectly connected via a geodesic through ego in Phase #.

Table 4-9: Compound Variables

Sparsity (*spars_#*) refers to the proportion of non-redundant contacts ego can access in his / her network. It is deducted by relating the number of structural holes (*strhol_#*) to the total size of the respondent's network in phase #. A similar procedure was applied in the case of closure (*closu_#*) being calculated as amount of strong ties (*strong_#*) of the total network size at the phase concerned.

The last two parameters, density and centrality, both relate to connections between alters as opposed to sparsity and closure that deal with the number of ties between ego and alters. Density (*dens_#*) in essence renders the proportion of connections between alters actually realised related to totality of all possible connections.

Centrality or egocentric betweenness refers to the number of pairs of actors (not involving ego) in the egocentric network that are not directly linked to each other and, consequently, are indirectly connected via a geodesic, i.e. shortest path, through ego. In essence, this measure corresponds to the reverse of egocentric network density (Marsden 2002). *cent_#* was generated by deducting the number of connections between pairs of alters (*know_#*) from the possible number of ties between alters (*(netzsiz_# * (netzsiz_# - 1)) / 2*) thus obtaining the number of pairs of alters featuring no connection between each other. This number in turn is related to the possible number of ties (*(netzsiz_# * (netzsiz_# - 1)) / 2*).

4.3.8 Descriptive Statistics

In the scope of descriptive analysis the distribution of variables is looked at. For interval scaled variables parameters presented include arithmetic mean and associated standard deviation as well as minimum and maximum values. In the case of nominal scaled categorical variables only frequency tables are depicted, since other measures would not allow for meaningful conclusions.

For the purpose of descriptive statistical depiction the variables introduced above are split into four major groups: first, personal variables providing information about the underlying research units of the academic business founder and his / her spin-off; second, environmental variables, referring to external factors influencing the start-up process other than entrepreneurs' personal networks, and finally the social-network attributes distinguished in relational and structural parameters.

4.3.8.1 Personal Variables

Personal variables refer to attributes of the entrepreneur and his / her newly founded company. In the following, distribution parameters, i.e. mean and standard deviation, for metrically scaled personal variables are given. Furthermore, frequency tables are provided for categorical personal variables. Where appropriate additional graphical depictions shall illustrate the figures presented.

Descriptive Statistics for Metric Variables

As Table 4-10 and Figure 4-13 illustrate, respondent's age ranged from 26 to 61 years with an average age of about 38 years. Most entrepreneurs participating in the survey were between 35 and 40 years old.

	N	Minimum	Maximum	Mean	Std. Deviation
Age of Ego	34	26	61	38,76	7,932
Year of Foundation	34	1992	2006	2003,21	3,391
Number of Partners	34	0	5	1,62	1,181
Valid N (listwise)	34				

Table 4-10: Descriptives – Personal Variables

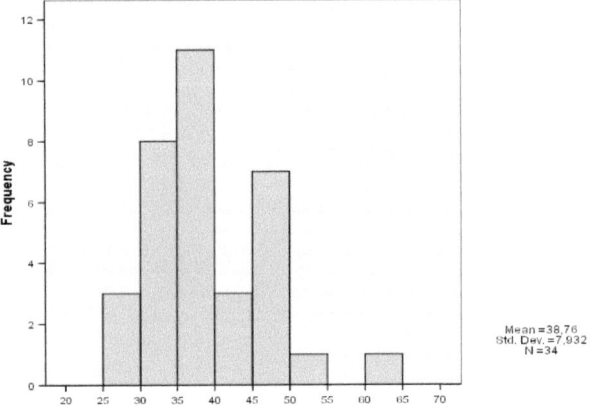

Figure 4-13: Histogram – Respondent's Age

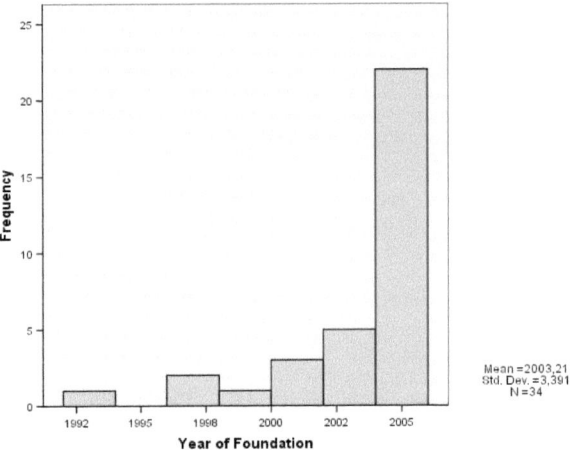

Figure 4-14: Histogram – Year of Foundation

Looking at companies' year of foundation, a bias towards younger businesses founded around 2005 can be observed. This potentially reflects a tendency towards increased academic spin-off creations over the last years. A development that may be partly attributed to strengthened public efforts and related support for this type of new venture over the last decade. A major impetus may be derived in particular from the Academia plus Business initiative (see also Appendix 8.1.1), whose first incubating centres were initiated back in 2002.

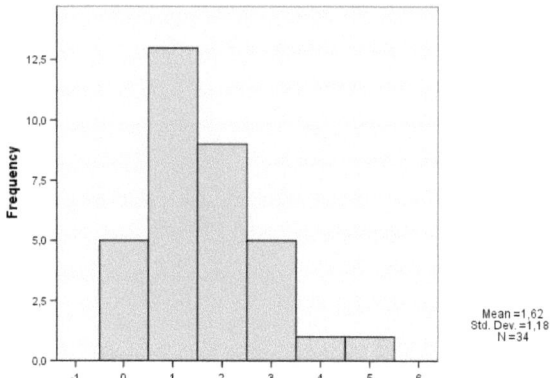

Figure 4-15: Histogram – Number of Co-Founders

Regarding the number of co-founders, most respondents quoted one or two partners together with whom they had started the business. In one case even five partners were specified. In total five respondents or 14.7% had acted as solo entrepreneurs without any partners.

Frequencies

While the previous section referred to metrically scaled personal variables and their distributions that can be described by arithmetic mean and standard deviation, the tables and associated figures given below intend to provide a picture of the realisations of categorical personal variables.

Table 4-11 starts with an overview of frequencies and percentages of the binary coded variables: sex of ego, nomination of one specific person as decisive factor for the spin-off decision, development of a business plan, foundation with partners, prior start-up experience, and whether respondent kept his / her position at university despite spinning off.

Variable	Frequency	Percent	Valid Percent	Cumulative Percent
Sex of Ego				
Valid Male	27	79,4	79,4	79,4
Female	7	20,6	20,6	100,0
Total	34	100,0	100,0	
Person as Impetus				
Valid Yes	34	100,0	100,0	100,0
Business Plan				
Valid No	2	5,9	5,9	5,9
Yes	32	94,1	94,1	100,0
Total	34	100,0	100,0	

Variable		Frequency	Percent	Valid Percent	Cumulative Percent
Serial Entrepreneur					
Valid	No	25	73,5	73,5	73,5
	Yes	9	26,5	26,5	100,0
	Total	34	100,0	100,0	
Keeping Function at University					
Valid	No	20	58,8	58,8	58,8
	Yes	14	41,2	41,2	100,0
	Total	34	100,0	100,0	

Table 4-11: Frequencies of Binary Personal Variables

As Table 4-11 shows about 20% of the respondents were female business founders, 80% of them men. This corresponds to the rather low distribution of women in the addressed sample indicating rather unbiased response behaviour as regards this characteristic of the target population.

Interestingly all 34 respondents stated that there had been a specific person acting as main trigger of the eventual decision to start an own business, which strongly confirms the findings of the exploratory qualitative phase.

The opposite holds true for the answer to the question on whether a business plan had been written in the course of preparing the spin-off. While the entrepreneurs interviewed in the preparatory study mostly negated related activities, more than 90% of the participants in the online survey have prepared a business plan. The associated assumption of whether this may be attributed to the fact that the majority of the businesses of the qualitative phase were founded comparatively early back in the late 90ies and early 2000 is subject of the subsequent quantitative analysis (Chapter 5.1.2).

Surprisingly, more than one fourth or about 26% of the academic entrepreneurs participating in the survey had already started an own business before. More than 73% however were absolute novices in the field.

As for the position entrepreneurs held at the university at the time of spinning-off, about 41% kept their function after having started their company, while more than 41% were completely taken up with the new task and quitted their original job.

University of Origin		Frequency	Percent	Valid Percent	Cumulative Percent
Valid	Uni Vienna	5	14,7	14,7	14,7
	TU Vienna	2	5,9	5,9	20,6
	Uni Linz	3	8,8	8,8	29,4
	TU Graz	2	5,9	5,9	35,3
	MUL	6	17,6	17,6	52,9
	Uni Innsbruck	10	29,4	29,4	82,4
	FH	2	5,9	5,9	88,2
	Other	4	11,8	11,8	100,0
	Total	34	100,0	100,0	

Table 4-12: Frequencies of Egos' Universities of Origin

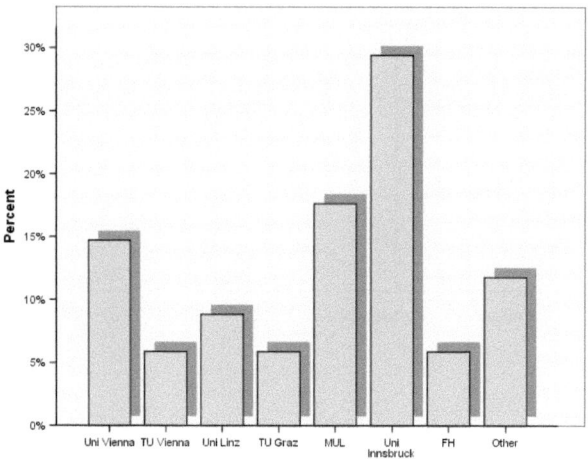

Figure 4-16: Distribution of Universities of Origin

At first sight the regional distribution of the spin-offs' parent universities as illustrated in Figure 4-16 seems rather unusual. While Eastern Austria and here in particular the region of Vienna features the highest accumulation of universities in Austria, still most spin-offs (about 30%) participating in the survey originated from the West of Austria, i.e. the University of Innsbruck in Tyrol, and also from the centre of the country, more precisely from Styria and here from Graz and its surrounding area, represented by the Technical University (TU) of Graz and the Montanuniversity of Leoben (MUL) amounting together to about 24%. The area of Vienna consisting of the Technical University and the University of Vienna altogether only produced 20% of the spin-offs. Yet, this is more or less in line with the distribution of the sampled population (see Figure 4-10 of Chapter 4.2.3). An open point however is the category *Others*, which has to be analysed in more detail. Looking at the four universities actually quoted under this heading, three of them are smaller Viennese institutions, which attribute another 8.8% to the present 20% adding up to almost 29% for the Vienna region. While this relativates the initial findings suggesting a dominance of Western Austria, the large proportion of spin-offs form the University of Innsbruck still remains rather impressive. Summing up, one can state that parent universities are equally distributed among the East and West of Austria, about 30% being situated in each part of the country respectively, with a yet large central anchoring, close to 24%, in Styria. Another 8.8% of start-ups were founded at the University of Linz. *Fachhochschulen (FH)* or colleges of higher education with 5.9% made up a rather small proportion of incubators. Yet, they already represent a not neglectable factor with increasing tendency.

As to the position held at the university at time of spinning off the venture, most entrepreneurial researchers indicated that they were either research associates (26.5%) or university assistants (23.5%). Almost 15% decided to start their business in the course of their studies, while the same percentage amounts to the category *Others*, which subsumes sabbaticals, maternity leaves and positions other than the stated not closer specified. 11.8% of the entrepreneurs were lecturers at their department and only 8.8% held a chair. This strongly opposes to the participants in the qualitative preparatory phase, most of whom were professors. Again this may be ascribed to the earlier years of venture foundation compared to the founding years of the quantitative sample.

Ego's Position at University		Frequency	Percent	Valid Percent	Cumulative Percent
Valid	Professor	3	8,8	8,8	8,8
	Lecturer	4	11,8	11,8	20,6
	Assistant	8	23,5	23,5	44,1
	Research Associate	9	26,5	26,5	70,6
	Student	5	14,7	14,7	85,3
	Other	5	14,7	14,7	100,0
	Total	34	100,0	100,0	

Table 4-13: Frequencies of Egos' Position at University

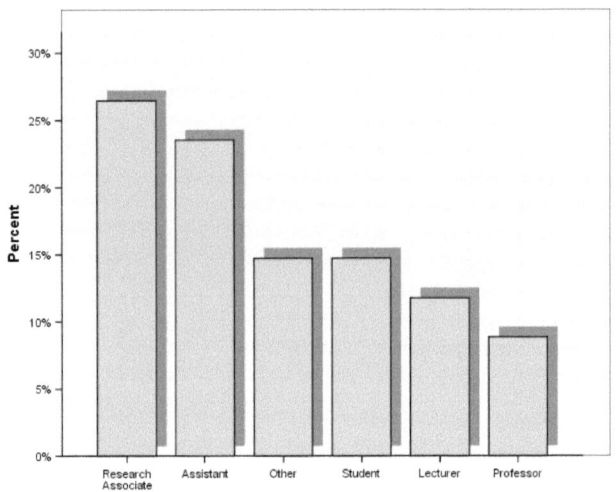

Figure 4-17: Distribution of Egos' Position at University

Asked for the line of business their company operates in, 35.3% of the respondents denoted that they were engaged in micro- and information technologies or the telecom business, followed by 20% in medicine, biology and biotechnology. Thus, more than 55% of the companies can be assigned to only two sectors of industry. This tendency reflects the strong emphasis of Austrian funding programmes on IT, telecommunication and biotechnology since the end of the 90ies. The category *Others* basically refers to the service sector including consulting, measurement technique, which unfortunately was not included in the pre-specified industry categories, and architecture. With 11.8% they feature the same relative frequency as sciences that consist of basic sciences such as mathematic, physics, chemistry, etc. Ex aequo follow manufacturing and transport and industrial technologies with 8.8%. Only one respondent declared that he operated in the energy sector, which ranks at the bottom of the table with 2.9%.

Industry of Spin-off		Frequency	Percent	Valid Percent	Cumulative Percent
Valid	Micro/IT/Telco	12	35,3	35,3	35,3
	Manufacturing/Transport	3	8,8	8,8	44,1
	Industrial Technologies	3	8,8	8,8	52,9
	Energy	1	2,9	2,9	55,9
	Sciences	4	11,8	11,8	67,6
	Medicine/Bio/Biotech	7	20,6	20,6	88,2
	Others	4	11,8	11,8	100,0
	Total	34	100,0	100,0	

Table 4-14: Frequencies of the Spin-offs' Industries

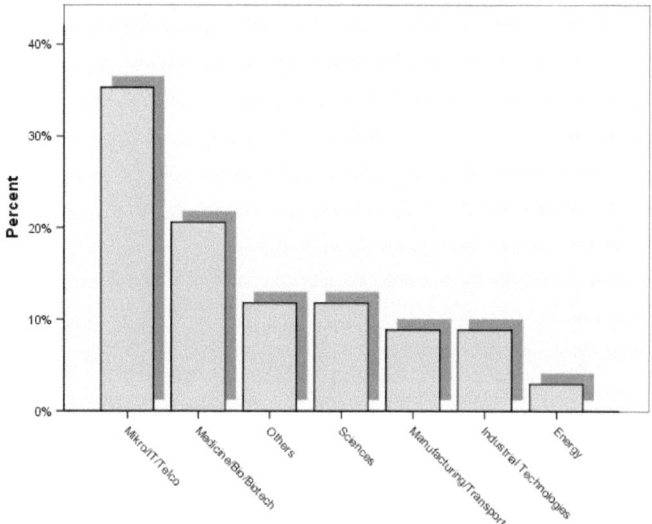

Figure 4-18: Distribution of Spin-offs' Industries

To obtain information about entrepreneurial researchers' scientific discipline, respondents were offered the possibility to indicate their academic background in their own words unrestricted by any predefined category. As no categories with corresponding values were specified, it is not possible to present a frequency table at this point. However, counting the total number of same nominations, it is possible to derive Figure 4-19 providing a snapshot of the distribution of academic disciplines among egos.

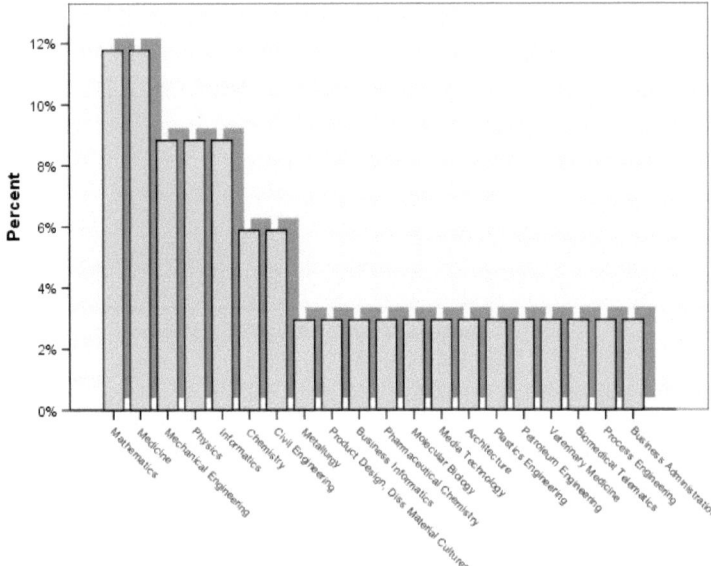

Figure 4-19: Distribution of Scientific Disciplines among Egos

The distribution of disciplines basically reflects the resulting lines of business as presented in Figure 4-18 with mathematics, physics and informatics representing important prerequisites of IT and micro technologies, and medicine and chemistry of the medical sector. Measurement technique and architecture, hidden in the industry category *Others*, are in turn primarily supported by mechanical and civil engineering. The remaining scientific disciplines quoted are equally distributed at a rather low level.

4.3.8.2 Environmental Variables

Environmental variables deal with external factors of assumed influence on the process of spin-off formation beyond particular network-related aspects. They either refer to specific forms of finance, such as venture capital and public funding, or to special conditions resulting from the distinctive feature of spinning off from a university background. Additional consideration is given to sources of personnel acquisition and reputation on the market. The same manner as in the previous chapter separate frequency tables for binary categorical variables and those with more than two categories are provided respectively. Special provision is made for the variables capturing sources of reputation and personnel, since they were constructed as multiple response sets.

Variable		Frequency	Percent	Valid Percent	Cumulative Percent
Venture Capital					
Valid	No	31	91,2	91,2	91,2
	Yes	3	8,8	8,8	100,0
	Total	34	100,0	100,0	

Joint Usage of University Infrastructure					
Valid	No	17	50,0	50,0	50,0
	Yes	17	50,0	50,0	100,0
	Total	34	100,0	100,0	
Public Finance					
Valid	No	4	11,8	11,8	11,8
	Yes	30	88,2	88,2	100,0
	Total	34	100,0	100,0	

Table 4-15: Frequencies of Binary Environmental Variables

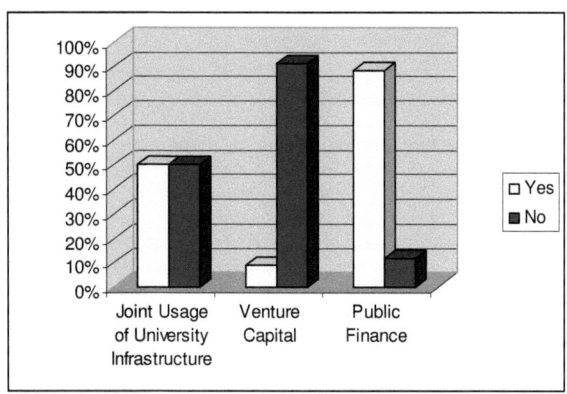

Figure 4-20: Distribution of Binary Environmental Variables

Looking at the different forms of external finance, only three respondents or 8.8% made use of venture capital to start their business. In contrast, more than 95% of the academic entrepreneurs had access to public funding. University infrastructure, such as equipment, software, space, and laboratories, could be shared by the newly founded enterprise in 50% of the cases.

Variable		Frequency	Percent	Valid Percent	Cumulative Percent
Reaction University					
Valid	Positive	15	44,1	44,1	44,1
	Negative	1	2,9	2,9	47,1
	Neutral	18	52,9	52,9	100,0
	Total	34	100,0	100,0	
Reaction Department		Frequency	Percent	Valid Percent	Cumulative Percent
Valid	Positive	19	55,9	55,9	55,9
	Negative	2	5,9	5,9	61,8
	Neutral	13	38,2	38,2	100,0
	Total	34	100,0	100,0	

Table 4-16: Frequencies of Reaction University and Department

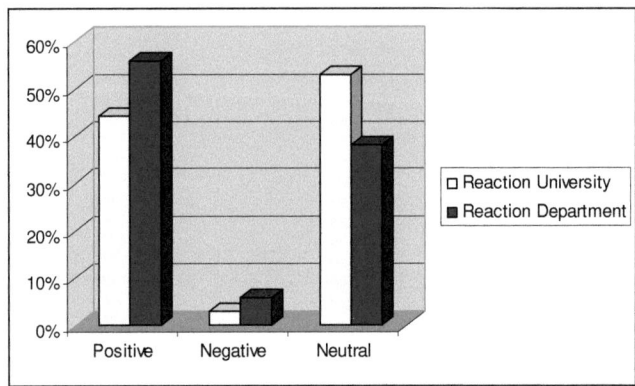

Figure 4-21: Distribution Reaction of University and Department

Regarding the reaction of both the entrepreneurial researcher's parent university and the department he or she worked for to his/her articulated intention to spin off, related attitudes are similarly distributed. Only in exceptional cases, i.e. one university and two departments or 2.9 and 5.9%, the response was clearly negative. However, when it comes to a positive or neutral attitude researchers' departments were more polarised tending to a supportive behaviour, while universities remained distant and adopted a more neutral position. This may be explained by the fact that in particular departments directly benefited from the start-up creation of their researchers by common research projects, practical experience for their graduates and doctoral candidates, etc. thus naturally considering the entrepreneurial venture more positively.

As for the questions about the channels through which the entrepreneurial researcher and his / her venture gained the required reputation on the market and the necessary human resources, both of them were conceived as multiple choice allowing for more than one possible answer. For the purpose of data handling each realisation was treated as one dichotomous variable taking the value 1 when ticked and 0 when left empty. To provide for an appropriate representation of the captured information in terms of descriptive parameters, these variables then had to be reassembled as one multiple response set applying the respective SPSS command. As depicted in Table 4-17 and Table 4-18 two different types of frequencies can be derived: one expressing the choices for one category as percentage of the total number of choices made, '*Percent*'; and one relating the selections of one category to the total number of respondents, '*Percent of Cases*'.

$rep Frequencies		Responses		Percent of Cases
		N	Percent	
Reputation [a]	Reputation on the Market Due to Academic Career	22	31,0%	64,7%
	Reputation on the Market Due to Industry Contacts	27	38,0%	79,4%
	Reputation on the Market Due to Friends	15	21,1%	44,1%
	Reputation on the Market Due to Family	1	1,4%	2,9%
	Reputation on the Market Due to Others	6	8,5%	17,6%
Total		71	100,0%	208,8%

[a] Dichotomy group tabulated at value 1.

Table 4-17: Frequencies of Sources of Reputation

Interpreting the parameter *'Percent of Cases'* of Table 4-17, it can be said that the majority of respondents (79.4%) confirmed that they had ascertained their reputation on the market through their contacts to industry closely followed by the academic career, which was selected by almost 65% of the participating entrepreneurs. While friends were also important mouthpieces for a good reputation, other contacts and most of all family ties contributed only marginally.

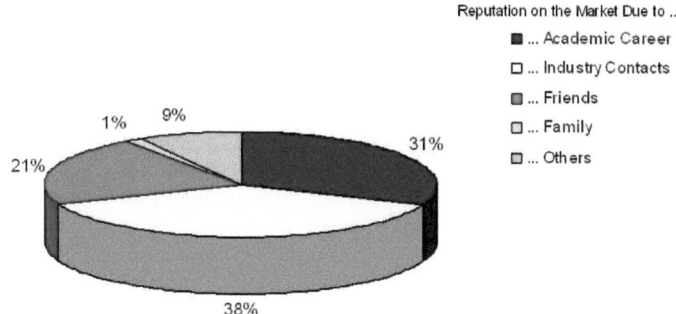

Figure 4-22: Distribution of Sources of Reputation

Looking at the relation of the different selections made to the total number of responses, as illustrated by Figure 4-22, 38% of the sources quoted refer to industry contacts, 31% to the entrepreneur's academic career. 21% of the answers indicated friends as important reputational assets; far behind rank others and family members with 9 and 1% respectively.

Regarding the utilisation of different sources of personnel as presented below (see Table 4-18 and Figure 4-23), 72.7% of the respondents stated, or to put it differently 47.1% of all answers given related to the university or the entrepreneurial researcher's department as main supplier of manpower. With some distance followed the circle of friends with nearly 40% of respondents or 25.5% of answers provided. The labour market and other sources played only a minor role equally amounting to 21.2% of respondents or 13.7% of responses.

$pers Frequencies		Responses		Percent of Cases
		N	Percent	
Personnel [a]	Personnel from University/Department	24	47,1%	72,7%
	Personnel from Market	7	13,7%	21,2%
	Personnel from Circle of Friends	13	25,5%	39,4%
	Personnel from Other Sources	7	13,7%	21,2%
Total		51	100,0%	154,5%

[a] Dichotomy group tabulated at value 1.

Table 4-18: Frequencies of Sources of Personnel

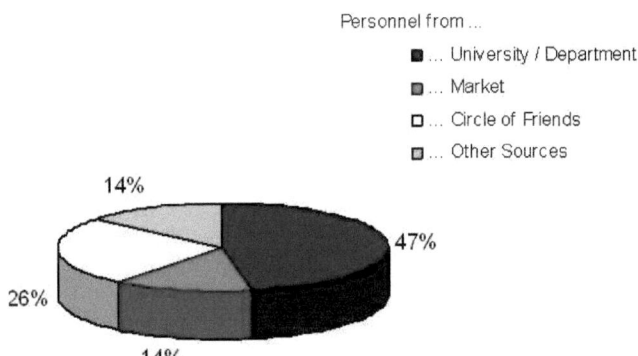

Figure 4-23: Distribution of Sources of Personnel

4.3.8.3 Relational Variables

As mentioned relational measures capture the characteristics of the relationships between the network actors. As specified in Chapter 4.3.4.1, *relational roles*, *content*, *frequency*, *tie strength*, and *multiplexity* are deemed to be the most relevant relational measures in view of the underlying objective of this research. Being interval-scaled the distribution of relational parameters may be described by their arithmetic mean, the associated standard deviation, as well as minimum und maximum values. Against the background of the assumed phase-dependence of relational characteristics, underlying distribution parameters are given for each phase of the spin-off process possibly displaying different values from one phase to another.

Relational Roles

Relational roles basically refer to the functions that alters assume in their relation to ego. As such alters can be friends, family members, academic peers, industry and funding partners or a combination of them. An exceptional position inheres to the sex of the contact, which in the narrower sense does not incorporate a certain function, but is nevertheless an important relational characteristic to be looked at in more detail.

As Table 4-19 indicates the smallest average networks can be found for the role of family members with an even decreasing tendency over the three phases, i.e. a mean of 0.4118 for Phase 1, 0.3235 for Phase 2, and 0.2941 for Phase 3. This empirical evidence supports the assumption that the number of family members an entrepreneur can draw upon during the start-up process is rather limited. At the beginning, i.e. in the first phase, family support seems to be more important, when concrete actions are not yet required, but the general decision to spin-off has to be discussed and the family's consent to be secured.

Ties to other academics and contacts from industry exhibit an equal average size. However, the development over time seems to take place in the opposite direction. While the average number of scientific contacts is the largest in the second phase with a mean of 0.7941, the number of business relations is the smallest with a mean of 0.4706. Nevertheless, this is no statistical evidence indicating any significance of this trend.

As to be expected, while there is already a lot of financial acquisition going on in the first phase of the spin-off process (mean: 0.9412), the network of funding contacts is the largest in the second phase (mean 1.0588), when the new venture is actually spun off and most of the financial support is required. This observation is further supported by the sudden drop in average network size (0.7941) in the last phase.

The network of friends is the largest for all three phases, i.e. with an average number of contacts of 1.2059 for Phase 1, 1.3235 for Phase 2 and 1.2647 for the last phase, as compared to the other functional roles. This accounts for the fact that the online questionnaire allowed for multiple nominations regarding the relational role of alters and that other types of contact were often cited together with this category. As to the development over phases, a slight increase in the average number of network ties may be observed for the second phase, however not really considerable.

A more distinct dynamics features the number of other contacts with a steady increase over time, i.e. average values of 0.4706, 0.6176, and 1 respectively. An attempt of an explanation can be made by having a look at respondents' actual specifications regarding this category and its development over time. While in the first phase *Others* primarily referred to consultants and employees, in the second phase it also included contacts providing legal advice such as lawyers, notaries and accountants, as well as funding relations, i.e. banks, and partners for sales and distribution. In the last phase this accumulated list is further augmented by strategic co-operation partners and associates.

	N	Minimum	Maximum	Mean	Std. Deviation
N° of Females - Phase 1	34	,00	4,00	,7647	,88963
N° of Females - Phase 2	34	,00	4,00	,9118	1,05508
N° of Females - Phase 3	34	,00	4,00	,8824	1,17460
N° of Family Members - Phase 1	34	,00	1,00	,4118	,49955
N° of Family Members - Phase 2	34	,00	3,00	,3235	,63821
N° of Family Members - Phase 3	34	,00	2,00	,2941	,52394
N° of Friends - Phase 1	34	,00	4,00	1,2059	1,29754
N° of Friends - Phase 2	34	,00	6,00	1,3235	1,70063
N° of Friends - Phase 3	34	,00	4,00	1,2647	1,21378
N° of Academics - Phase 1	34	,00	5,00	,8824	1,14851
N° of Academics - Phase 2	34	,00	5,00	,7941	1,34343
N° of Academics - Phase 3	34	,00	5,00	,7647	1,20752
N° of Industry Partners - Phase 1	34	,00	6,00	,7353	1,33278
N° of Industry Partners - Phase 2	34	,00	5,00	,4706	1,02204
N° of Industry Partners - Phase 3	34	,00	4,00	1,0000	1,34840
N° of Funding Partners - Phase 1	34	,00	3,00	,9412	,98292
N° of Funding Partners - Phase 2	34	,00	3,00	1,0588	,88561
N° of Funding Partners - Phase 3	34	,00	4,00	,7941	,84493
N° of Others - Phase 1	34	,00	3,00	,4706	,82518
N° of Others - Phase 2	34	,00	3,00	,6176	,81704
N° of Others - Phase 3	34	,00	7,00	1,0000	1,53741
Valid N (listwise)	34				

Table 4-19: Descriptive Statistics – Relational Roles

Coming to the special case of distribution of sexes among egos' alters, the highest average number of female contacts can be found in the second phase with an arithmetic mean of 0.9118. This represents a visible increase in average female network size compared to a mean of 0.7647 in the first phase. A slight decrease may be observed for the last phase (mean 0.8824).

	N	Minimum	Maximum	Mean	Std. Deviation
Same Discipline - Phase 1	34	,00	5,00	,7353	1,05339
Same Discipline - Phase 2	34	,00	4,00	,5882	1,13131
Same Discipline - Phase 3	34	,00	4,00	,5588	,99060
Same Department - Phase 1	34	,00	4,00	,5294	,92884
Same Department - Phase 2	34	,00	4,00	,4118	,95719
Same Department - Phase 3	34	,00	4,00	,4412	,82356
Valid N (listwise)	34				

Table 4-20: Descriptive Statistics – Academic Contacts

Table 4-20 enables a more detailed examination of the composition of egos' academic networks. It represents the outcome of extra questions dealing with the discipline and the institutional setting of academic alters. For a consistent picture, figures of Table 4-20 have to be compared to the overall specifications on academic contacts of the previous Table 4-19. As can be learned there, average academic network size decreases over time. This is also reflected by the development of the average number of contacts departing from the same discipline as ego. Yet, when comparing it to the figures of Table 4-19 the decrease is even more dramatic. In relative terms this signifies a diminution of the proportion of colleagues from the same scientific background of the overall academic network, which is surprisingly large in the first phase, i.e. 0.7353 out of 0.8824 average number of academic contacts. While in the first phase the average academic network size is 0.8824 compared to 0.7353 of academics form the same discipline, it drops to 0.7941 associated with an even stronger reduction to 0.5882 for contacts from the same specialty in the second phase, and 0.7647 or 0.5588 respectively for the last phase. Maybe this tendency reflects the successive disentanglement of the entrepreneurial researchers from their parent institution in the course of the spin-off process, and an increased outward focus on other fields of discipline required for realising a market-ready product, i.e. complementary technical know-how, economic inputs, etc.

As to the average number of contacts originating from the same university department as respondents, it remains rather stable, but also on a comparably lower level as the number of contacts from the same discipline. Still also in this case a slight decrease of average network size can be observed, and may be ascribed to the same reasons as explained above.

Content

The content of relationships refers to the meaning of the links between egos and their alters. As mentioned in Chapter 4.3.3, three major types of content were distinguished for the purpose of this research: (1) information, (2) resources, and (3) support. These three categories were in turn split into several sub-categories for a more fine-tuned examination of the composition of the contents transferred. In the following descriptive statistics are provided for each type of content and its related sub-categories.

The first type of content to be presented regards to the different forms of information gathered by egos in the course of the spin-off process. As Table 4-21 reveals, is the average number of informational ties approximately the same for the first and the last phase, while a decrease can be observed for the second phase. Possibly at this time of actually spinning off, more concrete resources other than information are required.

Coming to the detailed composition of egos' information networks, a closer look has to be taken at the different types of information exchanged. As to be expected, in the first phase, where the idea to spin of has just emerged and information exchange is characterised by a more exploratory character, legal information plays a rather inferior role with an average number of contacts of 0.9412. As the start-up process advances, legal information requirements also increase with a mean of 1 for the second and 1.3235 for the last phase.

Economic information is already quite important in the first phase. Still the amount of contacts providing commercial input slightly increases over time with means 1.5294 for the first, 1.6471 for the second, and 1.6765 for the last phase, reflecting advanced firm establishment. This tendency is even more distinct for the collection of information about potential customers. While already in the first phases some contacts are established in this direction (Phase 1: 0.7647, Phase 2: 0.8824), yet, it is in the last phase of establishment on the market, where most interactions, i.e. 1.2059 ties on average, refer to the acquisition of future customers.

The average number of interactions dealing with recruiting staff for the new venture is comparably low. After some activities in this regard in the course of the first phase (mean: 0.2647), the number of related contacts (0.2353) drops for the second phase indicating a reduced need of new personnel. Still the largest average amount of interactions in this regard can be observed for the last phase with a mean of 0.5882, possibly as a result of intensified business expansion.

	N	Minimum	Maximum	Mean	Std. Deviation
Informational Content - Phase 1	34	,00	8,00	2,9706	1,93038
Informational Content - Phase 2	34	,00	7,00	2,6176	2,01532
Informational Content - Phase 3	34	,00	7,00	3,0000	2,10339
Legal Information - Phase 1	34	,00	4,00	,9412	,95159
Legal Information - Phase 2	34	,00	5,00	1,0000	1,15470
Legal Information - Phase 3	34	,00	5,00	1,3235	1,29616
Economic Information - Phase 1	34	,00	5,00	1,5294	1,30814
Economic Information - Phase 2	34	,00	7,00	1,6471	1,79025
Economic information - Phase 3	34	,00	7,00	1,6765	1,57096
Information about Potential Customers - Phase 1	34	,00	3,00	,7647	,92307
Information about Potential Customers - Phase 2	34	,00	6,00	,8824	1,34310
Information about Potential Customers - Phase 3	34	,00	4,00	1,2059	1,24996
Information about Potential Employees - Phase 1	34	,00	2,00	,2647	,61835
Information about Potential Employees - Phase 2	34	,00	2,00	,2353	,55371
Information about Potential Employees - Phase 3	34	,00	4,00	,5882	1,04787
Other Information - Phase 1	34	,00	6,00	1,0882	1,33411
Other Information - Phase 2	34	,00	4,00	,8824	1,27362
Other Information - Phase 3	34	,00	4,00	1,0294	1,44569
Valid N (listwise)	34				

Table 4-21: Descriptive Statistics – Informational Content

The category referring to other types of information, offered respondents the possibility to specify in a more tailored manner, which information other than legal, economic, and about customers and employees, they had exchanged in the course of starting their business. In the first phase which already displays a rather high amount of average interaction (1.0882) in this context, most specifications related to the exchange of technical and scientific know-how. In the second phase of actually spinning off, this type of information, while still present, took a back seat, with references to funding agencies coming to the fore. However, as the average amount of interactions discloses, the degree of interaction in this respect decreased to an average of 0.8824 to rise again in the last phase to a mean value of 1.0294. When establishing on the market, it was again professional and scientific knowledge, but with more focus on production processes, as well as feedback and reflexion, and information about funding agencies that dominated the interaction.

	N	Minimum	Maximum	Mean	Std. Deviation
Provision of Resources - Phase 1	34	,00	7,00	1,1765	1,50697
Provision of Resources - Phase 2	34	,00	4,00	1,8529	1,30575
Provision of Resources - Phase 3	34	,00	5,00	2,0000	1,74078
Material Resources - Phase 1	34	,00	2,00	,2059	,47860
Material Resources - Phase 2	34	,00	3,00	,4706	,86112
Material Resources - Phase 3	34	,00	5,00	,6765	1,12062
Financial Resources - Phase 1	34	,00	3,00	,6176	,95393
Financial Resources - Phase 2	34	,00	4,00	1,1176	1,20012
Financial Resources - Phase 3	34	,00	5,00	,9706	1,33678
Provision of Infrastructure - Phase 1	34	,00	2,00	,4412	,70458
Provision of Infrastructure - Phase 2	34	,00	4,00	,5588	,85957
Provision of Infrastructure - Phase 3	34	,00	4,00	,7647	1,04617
Provision of Personnel - Phase 1	34	,00	3,00	,4118	,74336
Provision of Personnel - Phase 2	34	,00	3,00	,7647	1,07475
Provision of Personnel - Phase 3	34	,00	4,00	,9412	1,22947
Other Resources - Phase 1	34	,00	1,00	,0294	,17150
Other Resources - Phase 2	34	,00	1,00	,0294	,17150
Other Resources - Phase 3	34	,00	1,00	,0588	,23883
Valid N (listwise)	34				

Table 4-22: Descriptive Statistics – Resources

As Table 4-22 shows, the average number of contacts providing resources of all types strongly increases over time with corresponding mean values of 1.1765, 1.8529, and 2. Compared to the average number of information-related interactions, resource networks on average are smaller. For the distribution of different types of resources obtained, as to be expected, the largest proportion refers to financial support. While rather small in the first phase (0.6176), the average number of respective contacts rises considerably in the second phase (1.1176), where the venture needs most of its money to realise the actual spin-off. This need slightly abates to an average amount of 0.9706 in the last phase.

Ranking second is the provision of personnel, with the average number of relations rising from a mean value of 0.4118 in first phase over 0.7647 in the second phase to 0.9412 in the last phase, reflecting the growing operative character of business. It is closely followed by the provision of infrastructure, such as office space or research facilities. Again demand for respective resources increases over time with a mean value of 0.4412 in Phase 1, 0.5588 in Phase 2, and 0.7647 in Phase 3.

The same holds true for material resources like components, raw material, or equipment, with related interactions featuring heightened prevalence in the last two start-up phases. Interactions supplying other resources than the stipulated remain on a rather low level and, furthermore, were not closer specified by respondents.

The last content category refers to the support provided to respondents in the course of the venture creation process as depicted in Table 4-23. Comparing respective results for the overall support network with the ones given for resource-related interactions in Table 4-22, it is interesting to learn that support seemed to play a more important role for the participants in the survey than resources obtained. Being rather stable during the first two phases with corresponding mean values of 2.0882 and 2.0294, the average number of contacts, i.e. 2.4706, rendering support encounters a slight increase for the last phase.

Opposing the two support types prompted, institutional support defined by contacts to alters transferring credibility and trust on ego, or helping the entrepreneur to increase his / her visibility and reputation on the market, often in the sense of strategic partners, was even more prevalent than emotional support. Yet, both types feature a slight slump in the second phase, with institutional support more strongly recovering in the last phase. This can be seen as indicator for the findings of the exploratory phase claiming that existence of a strategic partner providing institutional support considerably contributed to spin-offs' market success.

	N	Minimum	Maximum	Mean	Std. Deviation
Support - Phase 1	34	,00	8,00	2,0882	1,95971
Support - Phase 2	34	,00	7,00	2,0294	1,67848
Support - Phase 3	34	,00	8,00	2,4706	2,14949
Emotional Support - Phase 1	34	,00	5,00	1,2353	1,37199
Emotional Support - Phase 2	34	,00	5,00	1,1471	1,32876
Emotional Support - Phase 3	34	,00	5,00	1,5000	1,46163
Institutional Support - Phase 1	34	,00	5,00	1,4412	1,63654
Institutional Support - Phase 2	34	,00	5,00	1,3824	1,37101
Institutional Support - Phase 3	34	,00	8,00	1,7941	1,85504
Valid N (listwise)	34				

Table 4-23: Descriptive Statistics – Support

Tie Strength

According to Granovetter (1973) strong ties can be seen as some combination of intimacy or socially-close, multiplexity, i.e. interacting in a variety of contexts, and frequent contacts. Following this assumption, strong ties were operationalised, as explained in Chapter 4.3.7.2, by combining frequency and multiplexity of interaction. In this context, ties were deemed as strong when respondents had frequent or permanent interactions with alters and the underlying relationship was multiplex in the sense that it transferred more than one type of content.

Table 4-24 enables to observe the development of the average number of strong ties in respondents' networks over time. As the respective mean values, i.e. 1.3529 for Phase 1, 1.4118 for Phase 2, and 1.7941 for Phase 3, indicate, personal networks feature an increased tendency towards stronger interactions. A characteristic that may be attributed to deepening relationships as result of the lasting co-operation in the course of the spin-off process.

Interestingly, the average number of weak ties is also the lowest for the first phase of the start-up process, yet, with a mean value of 1.5 still higher than the average number of strong ties at this time. The same holds true for the second phase, which encounters a slight increase in weak ties up to an average of 1.6765. However, while the average number of strong ties continues to increase, in the third phase the number of weak ties declines to an average of 1.5588, possibly being an effect of the deepening relationships as assumed above.

	N	Minimum	Maximum	Mean	Std. Deviation
Strong Ties - Phase 1	34	,00	4,00	1,3529	1,25245
Strong Ties - Phase 2	34	,00	5,00	1,4118	1,59768
Strong Ties - Phase 3	34	,00	6,00	1,7941	1,82208
Weak Ties - Phase 1	34	,00	6,00	1,5000	1,63763
Weak Ties - Phase 2	34	,00	6,00	1,6765	1,60908
Weak Ties - Phase 3	34	,00	8,00	1,5588	1,92576
Valid N (listwise)	34				

Table 4-24: Descriptive Statistics – Tie Strength

Multiplexity

Multiplexity denotes the degree to which egos' networks are characterised by multiplex ties; with ties being regarded as multiplex when transferring more than one type of content. In literature the assumption prevails that multiplex relations are particularly intensive and trusting ties. Uniplex ties, on the other hand, are used more for instrumental and material aid (Jackson et al. 1977).

Being aggregated as mean values in egos' networks, multiplexity ranges between 0 and 1 as reflected in subsequent Table 4-25. Interestingly other than strength of ties multiplexity slightly decreases in the second phase from a mean value of 0.4606 in first phase to 0.4579. This may be seen as indication of the increased need for material aid during the actual start-up of the business, which takes place in the second phase. Subsequent establishment on the market, as captured by the third phase, on the other hand, requires trust and commitment, which may be the reason for the increase in multiplexity during the last phase.

	N	Minimum	Maximum	Mean	Std. Deviation
Multiplexity - Phase 1	34	,00	1,00	,4606	,37426
Multiplexity - Phase 2	34	,00	1,00	,4579	,40063
Multiplexity - Phase 3	34	,00	1,00	,4889	,43069
Valid N (listwise)	34				

Table 4-25: Descriptive Statistics – Multiplexity

4.3.8.4 Structural Variables

As described in Chapter 4.3.4.2 the structural parameters applied are *size, number of structural holes (sparsity), density* and *centrality* of the personal networks. As with most ego-centred network analyses, the structural characteristics are summarised to the respondent level and used as independent or dependent variable (McCarty 2003).

Network Size

Network size captures the total number of alters entrepreneurial ego has in his or her network at the respective phase independent from their functional role. As Table 4-26 reveals, the average network size of academic spin-off founders is the largest in the last phase with a mean of 4.26 alters. Starting with an average value of 3.91 in the first phase, average size slightly drops in the second phase down to 3.88 contacts. Whether this represents a significant development has to be examined in the course of the statistical analysis of the subsequent chapters.

	N	Minimum	Maximum	Mean	Std. Deviation
Network Size in Phase 1	34	1	8	3,91	2,065
Network Size in Phase 2	34	1	8	3,88	1,919
Network Size in Phase 3	34	1	9	4,26	2,233
Valid N (listwise)	34				

Table 4-26: Descriptive Statistics – Network Size

Sparsity

Sparsity is an indicator of the unconnectedness of the alters in ego's network. Yet, it expresses the amount of non-redundant or isolated contacts the entrepreneurial researcher can draw upon. As presented in Chapter 4.3.7.3 sparsity is derived by relating the structural holes of ego's network to the total network size at the respective phase. As Table 4-27 indicates, the largest degree of sparsity can be found in the second phase with an average value of 0.0956. This is also reflected by the maximum value of 0.75, which exceeds the 0.5 of the other two phases. Yet, the average sparsity of networks in all three phases is relatively low, with possible values between 0 and 1 and actually realised mean values of not even 0.1.

	N	Minimum	Maximum	Mean	Std. Deviation
Sparsity - Phase 1	34	,00	,50	,0681	,14162
Sparsity - Phase 2	34	,00	,75	,0956	,18927
Sparsity - Phase 3	34	,00	,50	,0547	,12149
Valid N (listwise)	34				

Table 4-27: Descriptive Statistics – Sparsity

Density

Network density is a variable ranging from 0 to 1 describing the proportion of all possible ties actually realised. There is no standard definition as to when a network becomes densely-knit, but most researchers apply this term to a network with a density of at least 0.67 (two-thirds of all possible ties actually exist). Consequently, the networks of the surveyed spin-off founders in average can be regarded as dense for all three phases of the start-up process with mean values consistently beyond 0.67. Opposed to sparsity, density increases over time from an average of about 0.75 in the first phase to 0.84 in the last phase. This signifies that also egos' alters get known to each other and become more and more connected over time.

	N	Minimum	Maximum	Mean	Std. Deviation
Density - Phase 1	31	,00	1,00	,7524	,25695
Density - Phase 2	33	,20	1,00	,8048	,24920
Density - Phase 3	31	,33	1,00	,8415	,19922
Valid N (listwise)	28				

Table 4-28: Descriptive Statistics – Density

Centrality

As explained in Chapter 4.3.7.3, centrality or egocentric betweenness refers to the number of pairs of alters in ego's network that are not directly linked to each other and, consequently, are indirectly connected through ego. In essence, this measure corresponds to the reverse of egocentric network density (Marsden 2002). As density increases the number of ties between alters increases. Table 4-29 clearly documents this fact with decreasing average values of centrality in the course the three phases of the spin-off process analogous to the increase in network density of Table 4-28. In other words, one can say that entrepreneurs loose the central position in their network due to the heightened connectedness of their alters. However, while being at the expense of the novelty of information obtained, this development provides for strengthened embeddedness and consequently more substantial support due to improved trust resulting from mutual recognition.

	N	Minimum	Maximum	Mean	Std. Deviation
Centrality - Phase 1	30	,00	1,00	,2392	,25697
Centrality - Phase 2	32	,00	,80	,1856	,24701
Centrality - Phase 3	30	,00	,67	,1471	,19210
Valid N (listwise)	27				

Table 4-29: Descriptive Statistics – Centrality

Closure

While density expresses the extent of interrelation between the contacts of the academic founder and thus relates to a property between ego's alters, closure reflects a certain quality of ego's relation to its alters by measuring the amount of strong ties in entrepreneurs' personal networks at the different spin-off phases. As Table 4-30 shows, relationships between ego and alters intensify over time and thus average network closure increases from a mean value of 0.3389 in the first to 0.3832 in the last phase. This may be attributed to the increased life-time of the underlying relationships entailing stronger interactions of the actors involved.

	N	Minimum	Maximum	Mean	Std. Deviation
Closure - Phase 1	34	,00	1,00	,3398	,31290
Closure - Phase 2	34	,00	1,00	,3485	,37542
Closure - Phase 3	34	,00	1,00	,3832	,35800
Valid N (listwise)	34				

Table 4-30: Descriptive Statistics – Closure

4.3.9 Statistical Power

At large the data at hand is non-parametric. Due to the small sample size, even interval-scaled variables are most of the time not normally distributed and thus may not be subjected to the usual parametric tests. Consequently, non-parametric (distribution-free) tests are applied. As they process rankings and not the actual values themselves, they do not require normally distributed data and can also be used in cases, where no interval but only ordinal scale levels exist.

Yet, they can also be applied to normally distributed data with a lower efficiency of 95% of respective parametric tests. Thus, they are suitable for the comparison of means of differently distributed variables as to be analysed in the course of this research project that intends to compare both normally and not normally distributed, as well as interval and ordinal scaled variables across the different phases of the spin-off process.

When comparing samples by means of statistical tests, a distinction between dependent and independent samples has to be made.

In case of *independent or unrelated samples* the general probability of a research unit of one sample to feature a certain value is the same as the conditional probability that just this unit features the same value on condition that a research unit of the other sample features a certain value. In other words, there is no connection between the two samples; the individual probabilities of the different samples are not correlated.

Dependent or related samples are obtained, when several samples are collected from the same group of research subjects or from two groups of research subjects matched in pairs. In other words, formally different variables are captured that measure the same content, but at different points in time or in different contexts, i.e. under different conditions. Samples are denoted as dependent, if the probability that one research unit features a certain value significantly influences the probability of a certain value in the other sample. This is always the case when causal relations between samples exist.

In principle, there exists at least one non-parametric equivalent for each parametric general type of test. The following categories can be distinguished:

- Tests of differences between groups (independent samples);
- Tests of differences between variables (dependent samples);
- Tests of relationships between variables.

Table 4-31 provides an overview of the different non-parametric tests according to their intended fields of application.

Given two samples to be compared concerning their mean value for some variable of interest, non-parametric alternatives for the usual parametric t-test for independent samples test are the Wald-Wolfowitz runs test, the Mann-Whitney U test, and the Kolmogorov-Smirnov two-sample test. The Moses extreme reactions test refers to experimental settings. In case of multiple groups, the Kruskal-Wallis analysis of ranks and the Median test are applied instead of the parametric analysis of variance.

If two variables measured in the same sample are to be compared, instead of the custom parametric t-test for dependent samples, non-parametric alternatives to this test are the Sign test and Wilcoxon's matched pairs test. For dichotomous variables McNemar's Chi-square test is appropriate. In case of more than two variables in the same sample, non-parametric alternatives to parametric repeated measures ANOVA are Friedman's two-way analysis of variance and Cochran Q test. Cochran Q is particularly useful for measuring changes in frequencies (proportions) across time.

Relationships between variables are usually expressed in terms of correlation coefficients. Nonparametric equivalents to the standard correlation coefficient are Spearman R, Kendall Tau, and coefficient Gamma. In case of categorical variables, appropriate non-parametric statistics for testing the relationship between them are the Chi-square test, the Phi coefficient, and the Fisher exact test.

Object	Test
Independent Samples	
Two Samples	Mann-Whitney-U
	Wald-Wolfowitz
	Kolmogorov-Smirnov
	Moses Extreme Reactions
K Samples	Kruskal-Wallis
	Median
Dependent Samples	
Two Samples	Wilcoxon
	Sign Test for Paired Data
	McNemar's Chi-Square
K Samples	Friedman
	Kendall-W
	Cochran-Q
Relationship between Variables	
Ordinal	Spearman R
	Kendall Tau
	Coefficient Gamma
Nominal	Chi-Square
	Phi Coefficient
	Fisher Exact

Table 4-31: Non-parametric Tests

4.3.10 Methodological Limitations

In the course of ego-centred social network analysis a series of methodologically induced problems occur that are basically caused by the fact that respondents act as informants about their relationships. Thus, the obtained set of alters is ultimately based on the subjective perception of the informant. Consequently, the method strongly depends on egos' ability and readiness to correctly describe alters and the relations between them (Jansen 2003 p. 85). As with most survey methods this entails recall problems and systematic biases that have to be treated accordingly. Examples include the pursuit of social popularity entailing a tendency to quote more support contacts than actually present or inter-individually differing interpretations, e.g. of who is deemed as important contact. Providing information of people other than they themselves imposes extreme cognitive demands on respondents. This also entails a bias toward the inclusion of stronger ties (Lin 1999), since they are more easy to remember. Moreover, ego-networks display a tendency towards a larger density, since a relationship between ego and their alters is compulsory by definition. Only if also the relationships between alters are elicited, ego-networks may be structurally described (Jansen 2001).

Research results about the reliability and validity of ego-centred network analysis may be summarised as follows (Haß 2002):

4.3.10.1 Reliability

Ego-centred network analysis is usually based on a cross-sectional design, i.e. persons are asked about their social network only once and normally retrospectively. This means that a specific and at a given point in time presumably unique constellation of persons is captured. If surveyed again ego may quote more or less different persons. In addition, as a result of interim changes in the quality of the relation, even information about the same person may differ (Marsden 1993).

The ZUMA Institute in Mannheim, Germany has published several fundamental studies about the usage of ego-centred networks in mass surveys indicating a moderate reliability of the approach (Hoffmeyer-Zlotnik 1987; Pfennig et al. 1991). Test-retest studies yielded in average more than 70% identical nominations of alters, with the reliability increasing by the number of name generators applied. In this context, differences according to the type of network could be observed. As to role networks a test-retest reliability of 88% was achieved, regarding exchange networks reliability ranged from 63 to 74%, and for affective networks it amounted to 78% (Marsden 1993). Regarding the reliability of respondents' information about alters' demographics and attitudes, the reliability of socio-demographic specifications proved to be significantly higher than for attitude measurements. Observable attitudes, in turn, were reproduced more reliable than latent attitudes, such as religiosity (Pfennig et al. 1991).

The size of the network also plays an important role. The larger the networks the more homogeneous they appear in the correspondence of attitudes. Looking at the concept of relational characteristics describing the relationship between respondent and network persons, their reliability decreases with the size of the social network.

As to the reliability of individual network parameters regarding interactional properties quantitative measures such as frequency and duration of contact proved to be more reliable than qualitative. Structural network parameters were also subjected to reliability checks. However, since they represent aggregated items, particular estimation methods were applied. Relatively high reliability could be observed for network composition and density (Marsden 1993). Yet, it has to be noted that respective reliability measures are strongly dependent from the homogeneity of the networks (Pfennig et al. 1991).

4.3.10.2 Validity

The validity of ego-centred networks touches the question of bias and error perception in ego's statements. In this context, it has to be examined to what extend the specifications of the focal actor correctly reflect the characteristics and attitudes of alters. Different methodological experiments have tested the validity of ego-network surveys by means of control interviews of egos' alters asking for their relation to ego and for socio-demographic and attitude variables. A problematic aspect in this regard is the fact that validity is measured by corresponding choices of ego and alter, i.e. the symmetry of their relation. Friendship choices for example need not be symmetric. Furthermore they depend on the extroversion of the actors involved. It may be perfectly the case that a rather reclusive ego actor nominates among his or her rather few friends a rather extroverted person with many friends. This alter in turn may not choose ego among his or her many friends as important contact (Marsden 1990).

Although not always consistent, results of existing studies at large report a satisfactory validity, i.e. respondents' quotations about their social network may be generally trusted. Rather hard facts such as contact frequency, duration and kinship status of a relation or alters' socio-demographics, turned out to be more consistent than soft facts, such as personality characteristics or attitudes of alters (Hammer 1985; Pappi and Wolf 1984). Egos' specifications about these items tend to feature a latent tendency towards their own preferences (Pfennig et al. 1991). Validity thus is also a problem of how environmental aspects are perceived. Unspecific questions about friendship relations deliver only low degrees of accordance. Specific questions regarding emotional support, guidance and material aid on contrary feature accordances between 70 and 90%.

Systematic biases are also caused by respondents' sex: Women for example apply more stringent criteria about the quality of a relation to be attributed to their personal network (Hammer 1985). Moreover, women tend to perceive emotional affection, confirmation or instrumental support to a lesser extent than men.

Further biases result from situative factors on the part of the respondent influencing the perception of his or her network or from the interview situation. The relatively monotonous prompting of network members' attributes may entail fatigue of respondents in the course of the interview, thus also reducing the validity of results.

As to the validity of retrospectively gathered data results are rather heterogeneous. Bernard et alit (1984) discovered that respondents recalled in particular non-routinized interactions rather poor. Hammer (1985) in addition pointed to systematic selection criteria in mentioning relations with high contact frequency, up-to-datedness of last contact and high intensity more frequently. Furthermore, contacts of multiplex character are also more likely to be stated.

However, as many authors state (Hammer 1985; Jansen 2003; Pfenning 1995) from a social network perspective a detailed description of single interactions is not necessarily required. More of an interest is the stable long-term composition of the social structure. However, this and not single sequences of interactions are remembered and reported by respondents in the course of interviews. And it is also this stable social environment that determines respondents' room for action and thus their actual behaviour.

Against the background of the results of existing social studies, data on network members at large can be regarded as sufficiently reliable and valid (Bernard et al.

1982; Pfennig et al. 1991). In his review about the measurement of network data, Marsden (1990) comes to the conclusion that in particular concrete and specific relational types and properties can be reliably and validly gathered by means of ego-centred networks. All in all, specific name generators as the Fischer instrument outmatch unspecific generators with respect to reliability (Pfenning et al. 1991).

Yet, compared to other survey methods the low number of studies on reliability and validity is still regarded as considerable deficit of ego-centred network analysis (Schenk 1995).

5 Research Results and Interpretation

Starting with the results of the qualitative exploratory phase of the research project, yet the main focus will be on the core of the thesis, which is the quantitative part grounded in social capital theory and realised by means of social network analysis. Subsequent to a more or less neutral presentation of the main results of the social network survey, these results will be subjected to an interpretative analysis as basis for concluding remarks.

First, qualitative results are taken as reference to reappraise the initial hypotheses established in Chapter 3. Second, the results of the quantitative network analysis are interpreted with respect to their ability to confirm the thus adapted and amended list of hypotheses.

5.1 Qualitative Exploratory Study

The results of the qualitative exploration phase are intended to throw more light on the academic spin-off process as perceived by the Austrian spin-off founders interviewed. Emphasis was put on a most comprehensive picture of the spin-off history with respect to the founder's social environment. For that purpose extensive personal interviews were carried out in a narrative style enabling the interviewees to fully elaborate on their founding experiences at the different stages of the entrepreneurial process.

5.1.1 Presentation of Results

Results were analysed with respect to the three different phases of the spin-off process specified in the research model. Consequently, they are presented in line with this stage-model as depicted below.

5.1.1.1 Opportunity Recognition

Apart from one academic entrepreneur none of the interviewees could state a particular research result as the basis or core of the subsequent spin-off. Rather various coincidences were stated that in their combination led to the idea of spinning off a firm. In this context an important factor were industry projects carried out by the researcher's university institute that at certain point exceeded the institute's capacity and structural setting. Yet, in most cases the driving factor was the market and its increasing demand for particular technology transfer projects. Often it was one large industry client that provided the initial impetus. Moreover, the upcoming introduction of the autonomy of Austrian universities and the related necessity for third-party funding also contributed to the entrepreneurial intentions. Two academic entrepreneurs also stated legal issues in particular problems related to university institutes' inability or lack of willingness to assume liability for the results of industry commissioned research and development projects as one important reason to chose the way of legal independence in form of a new venture.

A majority of founders also reported that individual contacts, primarily out of the scientific community, significantly contributed to the genesis of the spin-off idea. If not from the scientific community, the contacts emerged from business activities within technology transfer projects.

For most entrepreneurs the actual decision was a jump into the cold water without any considerable considerations. If any discussions were held they were directed to colleagues of the scientific community and to some extent to the own family. However, most respondents restricted the notion of family to their spouse and children, where present, whereas only the spouse was explicitly conferred regarding the intended company formation. Basically, the role of the family was considered as minor. In two cases, family members (wife, sister) assumed administrative functions later on in the newly founded enterprise.

5.1.1.2 Formation

As stipulated in Chapter 3.1 the formation phase consists of activities targeted to establish the necessary basis for the subsequent spin-off. In this context it has to be noted that from this stage on most of the academic entrepreneurs were no longer alone but had pooled together with one or more other founding partners. Only one respondent founded alone. In many cases respective partners were colleagues from the institute to whom the entrepreneur was bound by a long common research history characterised by mutual understanding and trust. Partners, who came from outside of the university background were usually obtained via contacts to the scientific community either at conferences or by other academic peers known by prior research. In some cases also the industry partner of the industry project underlying the entrepreneurial opportunity decided to have a stake in the planned new venture.

At this stage the acquisition of seed capital played a vital role. In most cases academic entrepreneurs obtained this initial funding within the scope of the first industry project(s) constituting the launching pad for the entrepreneurial opportunity. One respondent quoted the selling of patents filed in the course of his academic career at the university institute as important source of founding capital. Moreover, only one of the respondents had made use of debt capital granted by banks to finance the subsequent start of his business. In no case public funding could be made available to pre-finance the business formation.

Regarding the establishment of a business plan interestingly only two of the respondents had written a respective plan. The majority reported themselves as enthusiastic layman that had just taken the step of business creation without any significant planning in this respect. In those cases where a business plan had been established respective preparation was either supported by one of partners, who had an economic background, or by an external accountant. Since the decision to found a venture usually was not based on the wish to commercialise a particular research result but to carry out industry projects in a favourable legal framework outside the structural constraints of the institutional background of the university, in the rare cases a prototype was build it was done within the scope of the respective project thus financed accordingly.

Concerning the information network providing for the respective knowledge on where to obtain financial funding, required infrastructure and important contacts, entrepreneurs mainly referred to the network of their scientific colleagues. Of course also partners' networks and related know-how were consulted. Yet, the academic network at the university and the network of the respective scientific discipline the respondent was member of were stated as most vital in this regard. Often also the individual contacts that had served as initial impetus in the opportunity recognition phase provided access to important sources of information. When it comes to legal information in particular with regard to the establishment and legal form of the subsequent venture, external accountants or advocates most of the time served as the main source of information. Contacts were usually established as a result of recommendations of colleagues. In one case it was an old schoolmate of the entrepreneur.

In the formation phase also the relationship between the intended spin-off and the universitarian mother institution comes into play. It is at this phase that the entrepreneur usually openly articulated his / her founding intentions for the first time. In this context it has to be distinguished between the university institute the research belongs to and the university in general. Reactions of the respective institutes to the spin-off plans of their members were predominantly positive. No direct competition was assumed, since the spin-off was going to undertake tasks and projects the institute was not really appropriate for. Moreover, in most instances relationship between the two institutions remained rather close with the academic founder frequently adopting a dual role keeping some of his teaching duties. In this connection the spin-off further provided for a practical framework for dissertations and master theses in interesting industry projects. Also the former colleagues from the institute benefited by being put in the position to escape to some extent the constraining framework of peer-review procedures and institutional pressures enabling them to play around and evolving their creativity and innovativeness. On the other hand the spin-off derived advantages by having access to basic research results, laboratory infrastructure and high-quality personnel. Reactions from the university itself were neutral to positive. Subsequent spin-offs were often used for promotional purposes serving as show cases of the institution's innovative potential and societal relevance.

5.1.1.3 Establishment

The establishment phase is dominated by the actual legal set-up of the new venture and its first steps on the market. When standing its ground in the market an important factor is the recognition and reputation of the new company, which in the beginning is closely related to that of its founders. In this context the academic background played an important role. Most interviewees stated their research careers and prior industry projects significantly contributed to their reputation and renowance in the market. Potential customers addressed them, because they trusted in their abilities and competencies due to past successes in research and generally accepted publications. In particular being a well-known head of a reputed institute or working under such a person considerably accounted for awareness and competence building. Former students and industry contacts build up valuable acquisition networks. Another important factor in this context was the establishment of strategic partnerships in the form a big industry player either as direct associate in the newly founded firm or as show case customer within the scope of long-term contracts.

Both factors: publicity resulting from the academic career, and strategic partners, also granted access to additional funding channels such as bank loans and investors. Yet, the main emphasis of funding was put on industrial project contracts. Partly also public funding in the form of nationally and internationally co-funded research projects was engaged. Still in most cases it was only deemed as nice "background music". However, all of the respondents strongly opposed the use of venture capital as financial source entailing to much external influence.

Human capital required for the operation of the new venture was primarily recruited from the entrepreneurial researcher's university institute. This recruitment channel was considered as extremely valuable offering access to reliable personnel with a known and guaranteed qualification standard. It provided the entrepreneur with the unique possibility to assess possible job candidates on the basis of their performance and engagement within the scope of master theses and dissertations and to partly influence their educational focus. Former collaboration of students in projects of the spin-off offered both sides the chance to judge the applicant's suitability for the job and his or her fitting in the team. By nature this held only true for technological personnel operationally working for the firm. On the administrative side other channels were drawn on, basically recommendations of the academic network of former colleagues and of already employed staff. As mentioned in two cases members of the family, i.e. sister and wife, assumed administrative tasks.

Regarding the economic know-how complementing the founder's technological expertise, in principle two approaches were chosen: economic know-how embodied in the person of one of the associates or employment of an external accountant. The former usually came in the form of a financial executive officer deputed by the strategic partner.

Emotional support in difficult times was basically obtained from co-founders and to some extent from family members, in particular from spouses. Another form of support stated by one respondent consisted in the additional security provided by his university teaching position. In the event of a failure of the spin-off, he would have disposed of a financial fallback position. In addition, it also enabled him to waive his management wages in the early phases of the spin-off to ease its financial burden.

5.1.1.4 Summary

Summing up, it can be stated that according to the results of the qualitative interviews the social network of academic founders in the *opportunity recognition phase* predominantly consists of industry contacts and scientific community members. Moreover, the main reason for the intention to spin off can be found in the structural conditions at the university hindering the commercial realisation of industry research projects.

At the *formation phase* the academic network is of predominant importance. On the one hand it serves as important pool for winning founding partners, on the other hand it is an important source of information with regard to funding channels and important legal and economic know-how carriers. When it comes to seed financing the business network of prior industry contacts also played a vital role.

Concluding, also in the last phase of *establishing* the technological venture on the market the entrepreneur's academic network is of particular importance both when it comes to market acceptance and accessing financial resources and human capital. Moreover, formerly established business networks further contribute to the success of the venture. In this regard the existence of a strategic partner melds most of the advantages cited by offering increased visibility, trust building, access to economic know-how and an interesting business network.

5.1.2 Reappraisal of Hypotheses

Against the background of the research results of the qualitative study, as previously presented in Chapter 5.1.1, the hypotheses preliminarily established in the course of Chapter 3 can be examined as to their supportability by empiricism. Table 5-1 reflects this process by opposing the original hypotheses to their modified counterparts resulting from introducing the results of the exploratory phase. Modifications are indicated at the original hypotheses. Where assumptions turned out to be not tenable, initial hypotheses were omitted at all. New findings arising from the qualitative interviews were documented in the form of new additional hypotheses.

Phase	Hypo-thesis	Original	Final
Opportunity Recognition	H1.A	Predominance of informational content	Predominance of informational content
	H1.B	Sparse networks rich in structural holes	Sparse networks rich in structural holes
	H1.C	Weak ties > strong ties	Weak ties > strong ties
	H1.D	Information from academic actors ~~of different research areas~~	Information network: informational content from academic actors and industry contacts
	H1.E	~~Informational content from family / friends~~	One person as main impetus for spin-off
Formation	H2.A	Informational and resource content	Informational and resource content
	H2.B	Information networks sparse rich in structural holes	Information networks sparse rich in structural holes
	H2.C	Resource networks dense	Resource networks dense
	H2.D	Information network: weak ties > strong ties	Information network: weak ties > strong ties
	H2.E	Resource network: strong ties > weak ties	Resource network: strong ties > weak ties
	H2.F		Information network: informational content from academic actors
	H2.G		Resource network: dominant exchange of financial resources
	H2.H		Resources: shared usage of university infrastructure
	H2.I		Reaction department -> positive Reaction university -> neutral to positive
Establish-ment	H3.A	Informational, resource and support content	Informational, resource and support content
	H3.B	Information network dense	Information network dense
	H3.C	Information network: strong ties > weak ties	Information network: strong ties > weak ties
	H3.D	Resource network dense	Resource network dense
	H3.E	Resource network: strong ties > weak ties	Resource network: strong ties > weak ties
	H3.F	Support network: strong ties > weak ties	Support network: strong ties > weak ties
	H3.G	Support network: emotional support from ~~family / friends~~	Support network: emotional support from academics / co-founders

Phase	Hypo-thesis	Original	Final
	H3.H	Support network: institutional support from business contacts	Support network: institutional support from business contacts
	H3.I		Resource network: financial resources from academic contacts
	H3.J		Resource network: human capital from academic contacts
	H3.K		Reputation on the market due to academic career and industry contacts

Table 5-1: Original Hypotheses

5.1.2.1 Opportunity Recognition

As the in-depths interviews with the academic entrepreneurs revealed, information in deed is an important content to be gathered when considering starting your own business. Colleagues out of the scientific community and contacts to industry served as main sources of the required information. Regarding the assumed predominant access to information via colleagues from other scientific disciplines, this assumption could not be confirmed. Consequently, while hypothesis H1.A is retained unchanged, hypothesis H1.D is modified in the sense that the appendix "of different research areas" is omitted. The importance of private contacts at this stage of the process could not be corroborated as well and led to discarding hypothesis H1.E.

An important finding of the qualitative phase, however, was the fact that most interviewees cited one person as having provided the main impetus to spin-off a firm. This is accounted for by including a respective hypothesis in the place of H1.E.

As to the structural network characteristics, such as sparsity and strength of ties as addressed in hypotheses H1.B and H1.C, these can not be judged via qualitative interviews and have to be validated in the course of the subsequent quantitative analysis.

5.1.2.2 Formation

Interview partners supported the assumed relevance of both informational and resource contents for the second phase of the spin-off process. Thus, hypothesis H2.A can be regarded as confirmed by the qualitative results.

As to the validation of the structural assumptions of hypotheses H2.B to H2.E, again they can only be verified against the background of the results of the subsequent quantitative analysis.

Having a closer look at the characteristics of the information networks, interviewees emphasised the significance of the academic contacts as dominant source of information. Regarding necessary resources, the most important type to be gathered at this stage according to interviewees was capital, i.e. financial resources, which strongly dominated the exchange structure. The two newly introduced hypotheses H2.F and H2.G capture entrepreneurs' statements of qualitative enquiry.

Coming to the reaction of entrepreneurial researchers' home departments, it was quoted consistently positive, while universities' reaction to some extent was judged to feature a more neutral notion. Moreover, shared infrastructure was stipulated as important asset provided by the institute or university. This additional input of the exploratory phase finds its equivalent in hypotheses H2.H and H2.I.

5.1.2.3 Establishment

Coming to the last phase of the spin-off process, the importance of all threes forms of network content, i.e. information, resources and support both emotional and in the form of a strategic partner, was supported by the qualitative interviews. Consequently, hypothesis H3.A can be retained.

However, emotional support was stated to come more from colleagues and co-founders rather than family members and friends. Yet, hypothesis H3.G was modified accordingly. Furthermore, the qualitative analysis confirmed the importance of institutional support of business contacts, in particular in the form of strategic partners. Thus, hypotheses H3.H was kept as it is.

In addition, three further hypotheses were included: two highlighting the importance of the academic network as access to financial resources and human capital, i.e. H3.I and H3.J; the third accounting for the increased relevance interview partners attributed to their academic background and their industry contacts as essential source of reputation and recognition on the market, i.e. hypothesis H3.K.

As to hypothesis testing of structural parameters, H3.B to H3.F, also at this point reference is made to the quantitative phase of the research project.

5.1.2.4 Synthesis

Figure 5-1 captures the main final hypotheses along the research model established in Chapter 3. The different assumed realisations of the social capital parameters, i.e. structure, content, tie strength and social role (affiliation) of network partners, are structured according to the three main network types identified: information, resource, and support networks. Such the development and dynamics of the characteristics of the respective network type can be well retraced over time, or in other words along the different phases of the spin-off process. More detailed information on the final hypotheses is then provided in the written compilation of all hypotheses as depicted in the subsequent boxes.

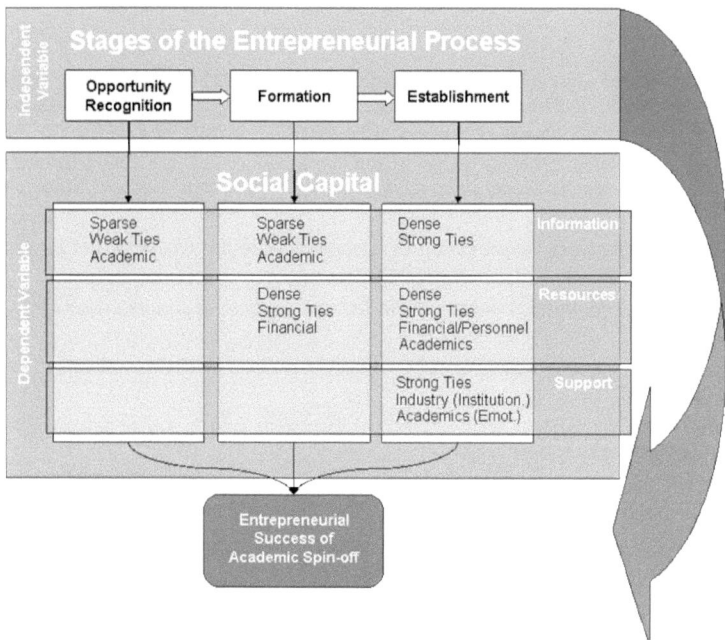

Figure 5-1: Social Capital Measures along the Spin-off Process

As already mentioned the following boxes offer an overview of the final hypotheses that will be subjected to statistical hypothesis testing in the course of the quantitative analysis in their complete wording and grouped by the three phases of the spin-off process.

Opportunity Recognition

H1.A: The opportunity recognition phase of the entrepreneurial process is characterised by information networks, i.e. the personal social network of the academic entrepreneur consists predominantly of information-related interactions.

H1.B: The opportunity recognition phase of the entrepreneurial process is characterised by sparse information networks of academic entrepreneurs that are rich in structural holes.

H1.C: The opportunity recognition phase of the entrepreneurial process is characterised by information networks dominated by weak ties.

H1.D: During the opportunity recognition phase of the spin-off process the information networks of academic entrepreneurs are dominated by contacts to academic actors and industry.

H1.E: It is usually one special contact that provides the main impetus for opportunity recognition.

Formation

H2.A: The formation phase of the entrepreneurial process is characterised by both information and resource networks, i.e. the personal network of the academic entrepreneur contains information- and resource-related interactions.

H2.B: The information networks of the formation phase are characterised by sparse social networks of academic entrepreneurs that are rich in structural holes.

H2.C: The resource networks of the formation phase are characterised by dense social networks of academic entrepreneurs.

H2.D: The information network of the formation phase is characterised by weak ties.

H2.E: The resource networks of the formation phase are characterised by strong ties.

H2.F: The information networks of the formation phase are characterised by contacts to academic actors.

H2.G: The resource networks of the formation phase are dominated by exchanging financial resources.

H2.H: In the formation phase infrastructural resources are provided by the university in the form of shared assets.

H2.I: In the formation phase the home department features a positive attitude; while the home university features a more neutral to positive attitude towards the spin-off intention.

Establishment

H3.A: The establishment phase of the entrepreneurial process is characterised by information, resource and support networks, i.e. the personal network of the academic entrepreneur comprises of interactions transferring information as well as material and symbolic support.

H3.B: The academic information networks of the establishment phase are characterised by a dense and interconnected structure.

H3.C: The information networks of the establishment phase are characterised by strong ties.

H3.D: The resource networks of the establishment phase are characterised by a dense and interconnected structure.

H3.E: The resource networks of the establishment phase are characterised by strong ties.

H3.F: The support networks of the establishment phase are characterised by strong ties.

H3.G: Emotional support is predominantly obtained from members of the entrepreneur's academic network.

H3.H: Institutional support is predominantly obtained from members of the entrepreneur's business network.

H3.I: The resource networks of the establishment phase are characterised by the provision of financial resources from academic contacts.

H3.J: The resource networks of the establishment phase are characterised by the provision of personnel resources from academic contacts.

H3.K: During the establishment phase reputation on the market is gained from the academic career and from industry contacts.

5.2 Quantitative Analysis

In the following, the data gathered in the course of the online survey of Austrian spin-off founders as presented in Chapter 4.3 is analysed against the background of the reappraised hypotheses of Chapter 5.1.2. To this end a closer look is taken at the relations between the different variables on the basis of various statistical procedures with a particular focus on changes between the phases of the start-up process. Respective results are opposed to the preliminary assumptions established in the hypotheses to examine their supportability. Additional findings not considered so far, are also documented and presented in an individual chapter.

5.2.1 Comparison of Means

To start with, the main structural parameters as presented in Chapter 4.3.8.4 are analysed with respect to their development over time, i.e. across the different phases of the spin-off process. The following parameters are included: network size, sparsity, density, centrality, multiplexity, closure, and tie strength. This shall account for the assumption that the structural composition of entrepreneurs' personal networks features a generalisable dynamics over time and thus as stipulated is dependent on the different phases of the entrepreneurial process.

To obtain statistically significant evidence, a comparison of means of the respective variables for the different spin-off phases is carried out. As presented in Chapter 4.3.9, dealing with the statistical power of the gathered data, the majority of parameters is not normally distributed, which implies the usage of non-parametric tests. Since the means of the same variables are to be compared over different points in time, dependent samples are involved, more precisely three dependent samples as three points of time are considered. Consequently, statistical tests for k-dependent samples are addressed. As mentioned, in principle Friedman, Kendall-W and Cochran-Q are possible tests. Given interval-scaled, but not normally distributed variables Friedman is the most appropriate test to be applied in this case. It is based on ranks that are calculated case-wise for the values of the variables involved, and thus needs no assumptions about the underlying distribution of the data.

As Table 5-2 shows and was already presented in the course of descriptive statistics of Chapter 4.3.8, respective means indicate different variations of structural network characteristics in time. While density, closure, and the number of strong ties of entrepreneurs' personal networks seem to increase in the course of spin-off process, centrality decreases. Networks size and multiplexity feature a slight drop in the second phase. Sparsity and also the number of weak ties in turn are characterised by a peak in Phase 2.

	Means							
	Network Size	Sparsity	Density	Centrality	Multiplexity	Closure	Strong Ties	Weak Ties
Phase 1	3,9118	,0681	,7524	,2392	,4606	,3398	1,3529	1,5000
Phase 2	3,8824	,0956	,8048	,1856	,4579	,3485	1,4118	1,6765
Phase 3	4,2647	,0547	,8415	,1471	,4889	,3832	1,7941	1,5588

Table 5-2: Means of Structural Network Parameters

However, these developments are only indications. To determine whether these trends are not purely random, but follow a real development that can be generalised to the underlying population, statistical testing is required. As described above, the appropriate statistical test to be applied in this context is the Friedman test. Respective results are presented in Table 5-3.

Friedman Test	Network Size	Sparsity	Density	Centrality	Multiplexity	Strong Ties	Weak Ties	Closure
N	34	34	28	27	34	34	34	34
Chi-Square	5,832	1,240	,925	,925	,022	,214	,867	,283
df	2	2	2	2	2	2	2	2
Asymp. Sig.	,054	,538	,630	,630	,989	,898	,648	,868

Table 5-3: Comparison of Means (Friedman Test)

As the asymptotic significance values in Table 5-3 reveal, none of the differences in the mean values of the analysed variables between the three phases is significant, i.e. all p > 0,05. Consequently, the observed developments of structural network characteristics in time cannot be regarded as systematic, but are possibly only random.

5.2.2 Exploring the Data

In the following chapter the various network variables generated are analysed to reveal possible interrelations between them at the different stages of the spin-off process. Main focus is put on the different contents transferred and their relation to other relational, i.e. alter roles, tie strength etc., and structural characteristics of entrepreneurs' personal networks.

This exploration part perpends the subsequent regression analysis, which is intended to statistically validate the presumed relations of this chapter. As mentioned, sample size is relatively small and consequently data not normally distributed. Thus, interpretations of regression results may be misleading due to distorted parameter estimations. To safeguard against false conclusions, at this point non-parametric correlation coefficients between the variables are computed, which are insensitive against the lack of normal distribution. Only relationships supported by significant non-parametric correlations are then subjected to regression analysis.

The standard correlation coefficient for normally distributed variables is Pearson's product-moment correlation coefficient. Commonly applied non-parametric equivalents to this coefficient are Spearman Rho, Kendall Tau, and Gamma coefficients.

Spearman Rho assesses how well an arbitrary monotonic function could describe the relationship between two variables, without making any assumptions about the frequency distribution of the variables. In principle, it is simply a special case of the Pearson product-moment coefficient, in which the data are converted to rankings before calculating the coefficient (Siegel and Castellan Jr. 1988).

Kendall Tau can be used to determine and test correlations between non-intervally scaled ordinal variables. While Spearman's coefficient corresponds to the Pearson product-moment coefficient, but is computed from ranks, Kendall Tau is calculated as difference between the probability that in the observed data the two variables are in the same order and the probability that the two variables are in different orders.

Gamma can be regarded as equivalent to Kendall tau for data that contain many tied observations. In short, it is also a probability like Kendall Tau, but specifically computed as the difference between the probability that the rank ordering of the two variables agree minus the probability that they disagree, divided by 1 minus the probability of ties (Siegel and Castellan Jr. 1988).

As the data analysed is not normally distributed, but yet interval-scaled, Spearman Rho is applied to uncover possible relationships between the different network variables in terms of their correlation coefficients. These first explorative observations will then be subjected to multivariate regression analysis in Chapter 5.2.3.

5.2.2.1 Relational Content in General

At first the three main types of content, i.e. information, resources, and support, to be exchanged in entrepreneurs' ego networks are analysed at the different phases of the spin-off process. As stipulated in Chapter 3.2, looking at the content of the interaction the personal network of an entrepreneur can be decomposed in different subsets. Yet, this approach allows structuring the analysis with respect to the three major subsets identified in the course of establishing the initial hypotheses of Chapter 3.4: information networks, resource networks and support networks.

While the first round of analysis, as carried out in the following chapter, focuses on the three general content categories 'Information', 'Resources', and 'Support', and a parallel comparison of their correlations and their development in time, the subsequent second round, as presented in Chapters 5.2.2.2, 5.2.2.3, and 5.2.2.4, takes a closer look at the different subcategories of each content type and their correlations across the different start-up stages.

Phase 1 – Opportunity Recognition

Table 5-4 contains the Spearman correlations for the three main content types - information, resources, and support - and all other relevant network variables for the first phase of the spin-off process. Significant correlations are marked ** for a significance level of 0.01 and * for the lower significance level of 0.05. As Pearson's product-moment correlation coefficient, the Spearman coefficient may take values ranging from -1 to 1. At a value of +1 or –1 a complete positive or negative relation between the observed properties exists. In case of a correlation coefficient of 0, the two variables are not interdependent at all. It has to be noted that significant correlations are only indicators of the strength and direction of the relationship between two variables, yet, they provide no information about the dependency of the variables, i.e. which variable influences which.

Looking at the *size* of entrepreneurs' personal networks, in the first phase several significant interdependencies to other variables can be detected. With respect to the magnitude of the interrelation, the most influential relationship can be observed between network size and the number of contacts providing informational content with a correlation coefficient of 0.816. This correlation is followed by the number of weak ties that feature a correlation coefficient of 0.705. Significant, while lower correlations can be further determined for the number of friends, the number of funding partners (0.457), amount of support provided (0.474) and the number of strong ties in the network (0.458). Furthermore, there are significant interrelations between network size and the number of academics (0.362) and the number of industry partners (0.382) contacted at the first stage of the spin-off process; yet, at a lower significance level of 0.05.

Regarding the number of *female contacts* in the personal network, the strongest (0.457) and most significant (significance level 0.01) correlation can be found with support provided. Almost equally strong - coefficient of 0.433, yet at a lower level of significance is the correlation to the number of friends in the network. Significant correlations also exist with the degree of multiplexity (0.365) and closure (0.346) of egos' networks. One possible interpretation for these correlations could be that women stand for closer relations transferring supportive behaviour.

As to the influence of family members in the first stage of the spin-off process, no significant correlations to other variables can be observed.

As mentioned, the number of *friends* besides funding partners features a significant effect on network size. Moreover, it is also significantly correlated with the amount of information provided (0.475) and the number of strong ties (0.487) in the network. At a lower significance level of 0.01, friends are also positively correlated to the amount of support rendered displaying a correlation coefficient of 0.419.

Academics in turn stand for sparsity (0.513) and an increasing number of structural holes (0.499). At a lower significance level they are also positively correlated (0.429) to the number of informational ties. And as a result of the underlying logic of the concept being closely connected to the number of structural holes, also a positive correlation of 0.403 with the degree of centrality of the networks and their number of weak ties (0.363) exists.

When looking at the correlation coefficients *industry contacts* definitely are one of the most important alter types regarding the provision of information. The correlation coefficient between number of industry contacts and number of informational ties amounts to 0.475. Thus, industry contacts together with friends that feature the same correlation coefficient to the amount of information related interactions are those networks roles that mostly interact with the provision of information in the first phase of the spin-off process. In addition, industry contacts are also positively related (0.401) to the number of strong ties, however, at the lower level of significance.

Funding partners as to be expected strongly and significantly correlate with the provision of resources (0.572) and support (0.473). The concrete type of resources and the actual type of support has to be determined later on in Chapters 5.2.2.3 and 5.2.2.4. In addition, an increasing number of funding partners signifies increased multiplexity in the network (0.566) and more strong ties (0.5). Yet, almost equally high also an increase in the number of weak ties (0.482). And finally, funding partners also provide information correlating to the number of informational ties at a lower significance level with a coefficient of 0.359.

Other types of alters only correlate with a coefficient of 0.36 at a lower level of significance of 0.05 to the number of relations involving the provision of resources. Moreover, a crowding-out effect with respect to the number of industry contacts can be observed. The two variables are negatively correlated with a correlation coefficient of -0.344, which means that an increased number of industry relations reduce the amount of unspecified contacts with alters of other functions than the predefined categories.

At the level of the content transferred *information* as depicted above strongly correlates with the number of friends and industry partners and to some extent while at a lower level of significance also with the number of academics in the network. Additional strong correlations can be observed with the number of strong and weak ties, yet, at an almost equal amount of 0.579 and 0.549, respectively. The provision of information and support is also somehow connected with a correlation coefficient of 0.368 at a lower level of significance.

Resources in turn are dominantly related to funding partners. Moreover, they are also positively correlated (0.441) with the number of strong ties, and at a lower level of significance to multiplexity (0.379) and closure (0.341) of the personal networks. This seems to support the assumption that in particular the provision of financial resources is effected via strong and multiplex ties in a trusting environment. However, respective indications still require the confirmation of further statistical analyses, as carried out in the Chapter 5.2.3.

Support is mainly related to female contacts and funding partners. In this context, it may be expected that in each case the respective other type of support is addressed. However, this will be determined later on when looking in more details at the correlations of the supportive content exchanged (see Chapter 5.2.2.4). Moreover, even much stronger than resources support is transported by multiplex and strong relations in a closed network. This is indicated by the respective correlation coefficients with multiplexity of 0.71, number of strong ties of 0.693, and closure of 0.593; all of them at the more stringent level of significance.

As mentioned above the number of *structural holes* significantly correlates to the number of academic contacts in the network. At a lower level of significance the number of structural holes is also to some extent correlated with the number of weak ties (0.359). One could possibly assume that weaker contacts of ego also to tend to not knowing each other. As to the other significant correlation coefficients regarding sparsity, density and centrality, they are the result of the underlying computation logics of these three parameters, which, same as the concept of structural holes, are a function of the lack of connections between the alters in ego's network, in case of density a reverse function as the negative correlation coefficient indicates.

The same holds true for the correlation coefficients of *sparsity, density* and *centrality*, and since number of academics and structural holes correlate, so do sparsity (0.499) and centrality (0.403) with the number of academics. The same accounts for the correlation with the number of weak ties and the correlation coefficient of strong ties and network closure of 0.881.

As presented above, *multiplexity* relates to female and industrial contacts, and the provision of information and resources. However, most strongly the degree of network multiplexity is positively correlated with the number of strong ties (0.71) and closure (0.797).

Spearman's Rho		Network Size in Phase 1	N° of Females - Phase 1	N° of Family Members - Phase 1	N° of Friends - Phase 1	N° of Academics - Phase 1	N° of Industry Partners - Phase 1	N° of Funding Partners - Phase 1	N° of Others - Phase 1	Informational Content - Phase 1	Provision of Resources - Phase 1	Support - Phase 1	Other Relational Content - Phase 1	N° of Structural Holes - Phase 1	Sparsity - Phase 1	Density - Phase 1	Centrality - Phase 1	Multi-plexity - Phase 1	Strong Ties - Phase 1	Closure - Phase 1	Weak Ties - Phase 1
Network Size in Phase 1	Corr. Coeff.	1,000	0,186	0,081	,470**	,362*	,382*	,457**	0,206	,816**	0,285	,474**	0,194	0,167	0,157	-0,240	0,262	0,123	,458**	0,092	,705**
	Sig.	.	0,293	0,651	0,005	0,035	0,026	0,007	0,243	0,000	0,103	0,005	0,271	0,345	0,374	0,194	0,162	0,488	0,006	0,603	0,000
N° of Females - Phase 1	Corr. Coeff.	0,186	1,000	0,305	,433*	0,097	-0,063	0,250	-0,049	0,106	0,195	,457**	-0,070	-0,215	-0,200	0,070	-0,019	,365*	0,313	,346*	0,120
	Sig.	0,293	.	0,079	0,010	0,585	0,723	0,153	0,783	0,551	0,270	0,007	0,693	0,221	0,257	0,708	0,921	0,034	0,071	0,045	0,499
N° of Family Members - Phase 1	Corr. Coeff.	0,081	0,305	1,000	0,077	0,023	0,025	-0,223	-0,226	-0,031	-0,209	0,065	0,037	-0,285	-0,293	0,162	-0,131	-0,275	-0,032	-0,127	0,101
	Sig.	0,651	0,079	.	0,665	0,897	0,888	0,206	0,199	0,862	0,235	0,713	0,834	0,102	0,092	0,384	0,489	0,115	0,859	0,474	0,571
N° of Friends - Phase 1	Corr. Coeff.	,470**	,433*	0,077	1,000	0,243	0,123	0,108	-0,202	,475**	0,251	,419*	-0,035	-0,313	-0,331	-0,009	0,063	0,160	,487**	0,331	0,232
	Sig.	0,005	0,010	0,665	.	0,167	0,489	0,543	0,253	0,004	0,152	0,014	0,845	0,071	0,056	0,962	0,742	0,365	0,003	0,056	0,186
N° of Academics - Phase 1	Corr. Coeff.	,362*	0,097	0,023	0,243	1,000	0,207	-0,163	-0,263	,429*	-0,160	0,123	0,049	,513**	,499**	-0,346	,403*	0,031	0,102	0,001	,363*
	Sig.	0,035	0,585	0,897	0,167	.	0,241	0,356	0,132	0,011	0,365	0,489	0,781	0,002	0,003	0,056	0,027	0,861	0,565	0,997	0,035
N° of Industry Partners - Phase 1	Corr. Coeff.	,382*	-0,063	0,025	0,123	0,207	1,000	0,159	-,344*	,475**	0,108	0,265	-0,180	0,117	0,104	0,022	-0,005	0,258	,401*	0,234	0,252
	Sig.	0,026	0,723	0,888	0,489	0,241	.	0,370	0,046	0,004	0,543	0,130	0,309	0,509	0,558	0,907	0,980	0,141	0,019	0,183	0,151
N° of Funding Partners - Phase 1	Corr. Coeff.	,457**	0,250	-0,223	0,108	-0,163	0,159	1,000	0,262	,359*	,572**	,473**	0,005	0,061	0,078	-0,008	0,058	,566**	,500**	0,337	,482**
	Sig.	0,007	0,153	0,206	0,543	0,356	0,370	.	0,134	0,037	0,000	0,005	0,977	0,733	0,661	0,968	0,762	0,000	0,003	0,051	0,004
N° of Others - Phase 1	Corr. Coeff.	0,206	-0,049	-0,226	-0,202	-0,263	-,344*	0,262	1,000	-0,078	,360*	0,032	0,172	0,001	0,013	-0,084	0,003	0,031	-0,004	0,008	0,153
	Sig.	0,243	0,783	0,199	0,253	0,132	0,046	0,134	.	0,662	0,037	0,855	0,330	0,995	0,941	0,653	0,987	0,864	0,981	0,964	0,389

Spearman's Rho		Net-work Size in Phase 1	N° of Females - Phase 1	N° of Family Members - Phase 1	N° of Friends - Phase 1	N° of Academics - Phase 1	N° of Industry Partners - Phase 1	N° of Funding Partners - Phase 1	N° of Others - Phase 1	Informational Content - Phase 1	Provision of Resources - Phase 1	Support - Phase 1	Other Relational Content - Phase 1	N° of Structural Holes - Phase 1	Sparsity - Phase 1	Density - Phase 1	Centrality - Phase 1	Multi-plexity - Phase 1	Strong Ties - Phase 1	Closure - Phase 1	Weak Ties - Phase 1
Informational Content - Phase 1	Corr. Coeff.	,816**	0,106	-0,031	,475**	,429*	,475**	,359*	-0,078	1,000	0,149	,368*	0,118	0,218	0,198	-0,087	0,158	0,253	,579**	0,264	,549**
	Sig.	0,000	0,551	0,862	0,004	0,011	0,004	0,037	0,662	.	0,400	0,032	0,505	0,215	0,263	0,643	0,403	0,148	0,000	0,132	0,001
Provision of Resources - Phase 1	Corr. Coeff.	0,285	0,195	-0,209	0,251	-0,160	0,108	,572**	,360*	0,149	1,000	0,175	-0,108	-0,115	-0,100	-0,048	-0,028	,379*	,441**	,341*	0,284
	Sig.	0,103	0,270	0,235	0,152	0,365	0,543	0,000	0,037	0,400	.	0,322	0,543	0,517	0,574	0,796	0,884	0,027	0,009	0,049	0,104
Support - Phase 1	Corr. Coeff.	,474**	,457**	0,065	,419*	0,123	0,265	,473**	0,032	,368*	0,175	1,000	-0,079	-0,028	-0,033	-0,083	0,145	,710**	,693**	,593**	0,245
	Sig.	0,005	0,007	0,713	0,014	0,489	0,130	0,005	0,855	0,032	0,322	.	0,657	0,874	0,855	0,657	0,443	0,000	0,000	0,000	0,163
Other Relational Content - Phase 1	Corr. Coeff.	0,194	-0,070	0,037	-0,035	0,049	-0,180	0,005	0,172	0,118	-0,108	-0,079	1,000	0,227	0,226	-0,269	0,292	-0,099	-0,182	-0,223	,344*
	Sig.	0,271	0,693	0,834	0,845	0,781	0,309	0,977	0,330	0,505	0,543	0,657	.	0,197	0,198	0,143	0,118	0,577	0,302	0,206	0,047
N° of Structural Holes - Phase 1	Corr. Coeff.	0,167	-0,215	-0,285	-0,313	,513**	0,117	0,061	0,001	0,218	-0,115	-0,028	0,227	1,000	,998**	-,514**	,554**	0,137	-0,123	-0,144	,359*
	Sig.	0,345	0,221	0,102	0,071	0,002	0,509	0,733	0,995	0,215	0,517	0,874	0,197	.	0,000	0,003	0,001	0,441	0,488	0,418	0,037
Sparsity - Phase 1	Corr. Coeff.	0,157	-0,200	-0,293	-0,331	,499**	0,104	0,078	0,013	0,198	-0,100	-0,033	0,226	,998**	1,000	-,497**	,537**	0,142	-0,138	-0,155	,362*
	Sig.	0,374	0,257	0,092	0,056	0,003	0,558	0,661	0,941	0,263	0,574	0,855	0,198	0,000	.	0,004	0,002	0,422	0,438	0,382	0,035
Density - Phase 1	Corr. Coeff.	-0,240	0,070	0,162	-0,009	-0,346	0,022	-0,008	-0,084	-0,087	-0,048	-0,083	-0,269	-,514**	-,497**	1,000	1,000**	0,072	0,106	0,193	-0,327
	Sig.	0,194	0,708	0,384	0,962	0,056	0,907	0,968	0,653	0,643	0,796	0,657	0,143	0,003	0,004	.	0,000	0,702	0,570	0,299	0,073
Centrality - Phase 1	Corr. Coeff.	0,262	-0,019	-0,131	0,063	,403*	-0,005	0,058	0,003	0,158	-0,028	0,145	0,292	,554**	,537**	1,000**	1,000	-0,024	-0,064	-0,158	,394*
	Sig.	0,162	0,921	0,489	0,742	0,027	0,980	0,762	0,987	0,403	0,884	0,443	0,118	0,001	0,002	0,000	.	0,898	0,738	0,406	0,031

Spearman's Rho		Network Size in Phase 1	N° of Females - Phase 1	N° of Family Members - Phase 1	N° of Friends - Phase 1	N° of Academics - Phase 1	N° of Industry Partners - Phase 1	N° of Funding Partners - Phase 1	N° of Others - Phase 1	Informational Content - Phase 1	Provision of Resources - Phase 1	Support - Phase 1	Other Relational Content - Phase 1	N° of Structural Holes - Phase 1	Sparsity - Phase 1	Density - Phase 1	Centrality - Phase 1	Multiplexity - Phase 1	Strong Ties - Phase 1	Closure - Phase 1	Weak Ties - Phase 1
Multiplexity - Phase 1	Corr. Coeff.	0,123	,365*	-0,275	0,160	0,031	0,258	,566**	0,031	0,253	,379*	,710**	-0,099	0,137	0,142	0,072	-0,024	1,000	,710**	,797**	0,128
	Sig.	0,488	0,034	0,115	0,365	0,861	0,141	0,000	0,864	0,148	0,027	0,000	0,577	0,441	0,422	0,702	0,898	.	0,000	0,000	0,472
Strong Ties - Phase 1	Corr. Coeff.	,458**	0,313	-0,032	,487**	0,102	,401*	,500**	-0,004	,579**	,441**	,693**	-0,182	-0,123	-0,138	0,106	-0,064	,710**	1,000	,881**	0,079
	Sig.	0,006	0,071	0,859	0,003	0,565	0,019	0,003	0,981	0,000	0,009	0,000	0,302	0,488	0,438	0,570	0,738	0,000	.	0,000	0,657
Closure - Phase 1	Corr. Coeff.	0,092	,346*	-0,127	0,331	0,001	0,234	0,337	0,008	0,264	,341*	,593**	-0,223	-0,144	-0,155	0,193	-0,158	,797**	,881**	1,000	-0,197
	Sig.	0,603	0,045	0,474	0,056	0,997	0,183	0,051	0,964	0,132	0,049	0,000	0,206	0,418	0,382	0,299	0,406	0,000	0,000	.	0,264
Weak Ties - Phase 1	Corr. Coeff.	,705**	0,120	0,101	0,232	,363*	0,252	,482**	0,153	,549**	0,284	0,245	,344*	,359*	,362*	-0,327	,394*	0,128	0,079	-0,197	1,000
	Sig.	0,000	0,499	0,571	0,186	0,035	0,151	0,004	0,389	0,001	0,104	0,163	0,047	0,037	0,035	0,073	0,031	0,472	0,657	0,264	.

**. Correlation is significant at the 0.01 level 2-tailed.
*. Correlation is significant at the 0.05 level 2-tailed.

Table 5-4: Spearman Correlations for Relational Content in Phase 1

Phase 2 – Formation

In the second phase, the correlation between *network size* and amount of friends has slightly increased to 0.569 as opposed to the correlation of 0.47 of the first phase. Moreover, number of academics and industry partners no longer correlate with the size of the network. The correlation coefficient between size of network and number of funding partners remained relatively stable at 0.441. For the first time a positive correlation of 0.35 between network size and the number of females in the personal networks of entrepreneurs can be ascertained; however, at the lower significance level of 0.05.

As to the content transferred, correlation between the amount of informational ties and network size has slightly decreased from 0.816 in the first phase to 0.703 in the second phase. Support in turn has decreased from 0.474 to 0.499. New, is the correlation between the provision of resources and the size of the network, which is rather strong with a correlation coefficient of 0.627. This development can be seen as expression of the advanced materialisation of the academics' spin-off intention entailing an increased need for financial and physical assets.

On the structural side, correlation coefficients between network size and strong or weak ties, respectively, diminish a little with strong ties even featuring significant positive correlation only on a reduced level of significance of 0.05. This can be interpreted as indication of a lower dominance of strong ties in entrepreneurs' networks of the second phase.

Looking at the number of *female contacts* in entrepreneurs' networks and their correlation with other network parameters, still they are positively related to friendship (0.36) and support (0.492) as already observed for the first phase, but now in addition featuring a positive correlation with a coefficient of 0.358 with the number of family contacts. However, only support is correlated at the higher level of significance of 0.01.

Family members, in addition, are at this point in time increasingly considered as friends displaying a positive correlation of 0.397 to number of friends in the network, still at a lower level of significance of 0.05. Nevertheless, in the first phase they exhibited to no significant correlation at all.

Coming to the number of *friends* in academic entrepreneurs" personal networks, beyond the correlations already cited, they are strongly related to informational content (0.641) and support (0.621), representing an increase to the respective correlation coefficients of 0.475 and 0.419 of the first phase. The same holds true for the correlation between number of friends and number of strong ties rising from a coefficient of 0.487 in the first phase to 0.512 in the second phase. While at a lower level of significance, for the first time the number of friends is also positively correlated to the degree of multiplexity and closure in entrepreneurs' networks.

The number of *academics* still related to the number of informational ties in entrepreneurs' networks yet it is reduced to a coefficient of 0.368 at a significance level of 0.05. While in the first phase weakly related to the number of weak ties, in the second phase the number of academics in the network is positively correlated (0.451) with the number of strong ties.

As to the amount of *industry contacts*, they still seem to crowd out other contacts (-0.411) and to some extent (0.353) tend to appear in the constellation of structural holes. Yet, they have lost their interdependence with the number of informational ties, as ascertained for phase 1.

Being strongly positively related to the amount of resource exchanges and the provision of support in connection with a correlation to multiplexity and strong and weak ties in the first phase, in the second phase the number of *funding partners* only correlates at the higher level of significance with the number of structural holes (0.469) and the number of weak ties (0.446), and at the lower level of significance with sparsity (0.436) and centrality (0.362) in the personal networks; however, as the values show only to a rather weak extent. The funding contacts seem to be rather isolated and not well connected to other members in the network. This also explains their positive relationship to an increased central position of ego in a rather sparse network. In addition, they tend to be characterised by infrequent, weak relations to ego.

As regards the relational content, starting with the number of *informational* ties, as already mentioned, the contacts to friends are most strongly correlated 0.641, while other types of alters do not correlate at all or only to a smaller extent at a lower level of significance, as the case with the number of academics (0.368). The rather small correlation to the provision of support as diagnosed for the first phase with a correlation coefficient of 0.368, has increased to 0.516 even at a higher significance level. In addition, as opposed to the first phase in the second phase informational content also relates to the provision of resources with a correlation coefficient of 0.578. This may be interpreted as the effect of increased information transported in the course of providing support or material or financial resources, with resources being for the first time also positively correlated with network size, as indicator of increased relevance and consequently argumentation for its sudden presence in this context. New is also the correlation of informational content with multiplexity (0.559) and closure (0.578). As opposed to the previous phase in the second phase information is only related to strong and no longer to weak ties, however with a much higher correlation coefficient of 0.723 compared to the previous correlation coefficient of 0.579.

The provision of *resources* is no longer related to funding partners, as in the first phase. New is the rather strong correlation to informational ties, as just described, and a weaker (0.373) correlation at a lower level of significance with the number of supportive ties. Analogue to the previous phase, the number of resource-related contacts is positively related to multiplexity, closure and the number of strong ties. Corresponding correlation coefficients, however, have slightly increased to 0.4, 0.381 and 0.511, respectively.

Support is still related to the number of women and friends, yet, to a higher extent, in particular concerning the number of friends, where not only the correlation coefficient has increased from 0.419 to 0.621, but also moved to a higher level of significance. In turn, funding partners are no longer correlated to support, similar as just observed for the provision of resources. Yet, as detailed above, the correlation to informational ties is fortified (0.516) and resources now correlate with support (0.373), even tough at to a small extent and at the lower level of significance. As already noticed for the provision of resources, analogue to the first phase, multiplexity (0.626), closure (0.496) and number of strong ties (0.517) are positively related, however with slightly smaller correlation coefficients.

While in the first phase, the number of *structural holes* was mainly related to the number of academics in a network, it now comes with industry (0.353) and funding partners, with funding partners featuring a greater (0.469) interrelation at a higher level of significance of 0.01. Again strong correlations with sparsity, density and centrality exist, whose interpretation however renders no reasonable additional information, since, as already explained in the previous section, they are the result of the underlying computational construction of the respective parameters.

Also sparsity has shifted away from correlations with academics to correlations with funding partners (0.436), still at a lower level of significance. As to the correlations with the number of structural holes and the density and centrality of the personal networks, the same argumentation as followed in the previous paragraph applies.

A likewise relationship can be diagnosed for *sparsity* and *density*. Interestingly, both of them, sparsity and density, feature a highly significant correlation with the number of weak ties. While density decreases with the number of weak ties, i.e. correlation coefficient of -0.487, centrality increases, i.e. 0.571. As for the first phase again this can be interpreted as weak ties between egos and alters resulting in alters that tend to be stranger, since they do not belong to the same community.

While having been strongly related to the number of funding partners and to some extent to the amount of female contacts in entrepreneurs' personal networks in the first phase, in the second phase *multiplexity* correlates with the number of friends (0.345). New is also the rather strong (0.559) interrelation with the number of informational ties. As in the first phase, a strong relation to supportive contacts (0.626) exists, which however has slightly decreased. In turn, the correlation with resource-providing ties has increased a little to a coefficient of 0.4. Due to the way strong ties and closure have been methodologically constructed as function of multiplex ties, strong correlations exist though without offering any meaningful additional interpretation.

Strong ties still relate to the number of friends in the network with a slightly higher correlation coefficient of 0.512, but no longer to the number of funding and industry partners. In exchange, a strong correlation of 0.451 with the number of academics has emerged. Correlations with informational, supportive and resource-related ties still exist. Regarding the provision of information the correlation coefficient has strongly increased to 0.723. The relation to resource-providing contacts has also intensified (0.511), while the correlation coefficient regarding support has decreased a little from 0.693 in the first phase to 0.517 for the present phase. Again significant correlations with ties strength and closure can be traced back to their methodological construction.

The remaining significant correlation coefficients have already been described in the course of analysing their corresponding pendants.

Spearman's Rho		Network Size in Phase 2	N° of Females - Phase 2	N° of Family Members - Phase 2	N° of Friends - Phase 2	N° of Academics - Phase 2	N° of Industry Partners - Phase 2	N° of Funding Partners - Phase 2	N° of Others - Phase 2	Informational Content - Phase 2	Provision of Resources - Phase 2	Support - Phase 2	Other Relational Content - Phase 2	N° of Structural Holes - Phase 2	Sparsity - Phase 2	Density - Phase 2	Centrality - Phase 2	Multiplexity - Phase 2	Strong Ties - Phase 2	Closure - Phase 2	Weak Ties - Phase 2
Network Size in Phase 2	Corr. Coeff.	1,000	,350*	0,321	,569**	0,304	0,058	,441**	0,270	,703**	,627**	,499**	0,056	0,225	0,143	-0,323	0,342	0,184	,390*	0,193	,603**
	Sig.	.	0,042	0,065	0,000	0,080	0,744	0,009	0,123	0,000	0,000	0,003	0,754	0,200	0,421	0,067	0,055	0,299	0,022	0,274	0,000
N° of Females - Phase 2	Corr. Coeff.	,350*	1,000	,358*	,360*	0,068	-0,088	0,054	0,038	0,317	0,189	,492**	-0,273	-0,123	-0,142	0,002	0,048	0,213	0,165	0,180	-0,030
	Sig.	0,042	.	0,038	0,037	0,704	0,623	0,761	0,831	0,067	0,285	0,003	0,119	0,487	0,423	0,991	0,794	0,227	0,350	0,308	0,866
N° of Family Members - Phase 2	Corr. Coeff.	0,321	,358*	1,000	,397*	0,058	-0,066	-0,243	-0,162	0,229	0,141	0,226	-0,149	0,034	-0,047	-0,023	0,044	-0,082	0,179	0,116	0,133
	Sig.	0,065	0,038	.	0,020	0,743	0,713	0,166	0,359	0,194	0,426	0,199	0,400	0,850	0,794	0,897	0,810	0,646	0,312	0,513	0,455
N° of Friends - Phase 2	Corr. Coeff.	,569**	,360*	,397*	1,000	0,181	0,144	-0,042	-0,164	,641**	0,338	,621**	-0,256	-0,074	-0,131	0,071	-0,030	,345*	,512**	,389*	0,283
	Sig.	0,000	0,037	0,020	.	0,306	0,417	0,812	0,354	0,000	0,051	0,000	0,144	0,676	0,460	0,695	0,870	0,046	0,002	0,023	0,105
N° of Academics - Phase 2	Corr. Coeff.	0,304	0,068	0,058	0,181	1,000	0,089	-0,083	-0,335	,368*	0,171	0,077	0,089	-0,067	-0,074	0,046	-0,019	0,142	,451**	0,287	0,156
	Sig.	0,080	0,704	0,743	0,306	.	0,618	0,642	0,053	0,032	0,332	0,667	0,619	0,706	0,678	0,797	0,918	0,422	0,007	0,100	0,380
N° of Industry Partners - Phase 2	Corr. Coeff.	0,058	-0,088	-0,066	0,144	0,089	1,000	0,044	-,411*	0,263	-0,010	0,122	-0,148	,353*	0,323	-0,203	0,227	0,189	0,176	0,120	0,098
	Sig.	0,744	0,623	0,713	0,417	0,618	.	0,805	0,016	0,133	0,955	0,490	0,403	0,041	0,063	0,258	0,211	0,285	0,320	0,498	0,579
N° of Funding Partners - Phase 2	Corr. Coeff.	,441**	0,054	-0,243	-0,042	-0,083	0,044	1,000	0,269	0,289	0,304	0,108	0,153	,469**	,436*	-0,289	,362*	0,170	0,067	-0,005	,446**
	Sig.	0,009	0,761	0,166	0,812	0,642	0,805	.	0,125	0,097	0,081	0,543	0,389	0,005	0,010	0,103	0,042	0,336	0,706	0,980	0,008
N° of Others - Phase 2	Corr. Coeff.	0,270	0,038	-0,162	-0,164	-0,335	-,411*	0,269	1,000	-0,083	0,279	-0,016	0,296	-0,067	-0,076	-0,211	0,147	-0,074	-0,154	-0,097	0,198
	Sig.	0,123	0,831	0,359	0,354	0,053	0,016	0,125	.	0,639	0,110	0,928	0,089	0,705	0,670	0,239	0,421	0,679	0,385	0,585	0,262

Spearman's Rho		Net-work Size in Phase 2	N° of Females - Phase 2	N° of Family Members - Phase 2	N° of Friends - Phase 2	N° of Academics - Phase 2	N° of Industry Partners - Phase 2	N° of Funding Partners - Phase 2	N° of Others - Phase 2	Informational Content - Phase 2	Provision of Resources - Phase 2	Support - Phase 2	Other Relational Content - Phase 2	N° of Structural Holes - Phase 2	Sparsity - Phase 2	Density - Phase 2	Centrality - Phase 2	Multiplexity - Phase 2	Strong Ties - Phase 2	Closure - Phase 2	Weak Ties - Phase 2
Informational Content - Phase 2	Corr. Coeff.	,703**	0,317	0,229	,641**	,368*	0,263	0,289	-0,083	1,000	,578**	,516**	-0,220	0,096	0,029	0,012	0,045	,559**	,723**	,578**	0,244
	Sig.	0,000	0,067	0,194	0,000	0,032	0,133	0,097	0,639	.	0,000	0,002	0,212	0,588	0,870	0,949	0,808	0,001	0,000	0,000	0,165
Provision of Resources - Phase 2	Corr. Coeff.	,627**	0,189	0,141	0,338	0,171	-0,010	0,304	0,279	,578**	1,000	,373*	-0,076	-0,007	-0,056	-0,118	0,080	,400*	,511**	,381*	0,288
	Sig.	0,000	0,285	0,426	0,051	0,332	0,955	0,081	0,110	0,000	.	0,030	0,670	0,970	0,754	0,513	0,665	0,019	0,002	0,026	0,098
Support - Phase 2	Corr. Coeff.	,499**	,492**	0,226	,621**	0,077	0,122	0,108	-0,016	,516**	,373*	1,000	-0,183	0,093	0,113	0,017	0,047	,626**	,517**	,496**	0,171
	Sig.	0,003	0,003	0,199	0,000	0,667	0,490	0,543	0,928	0,002	0,030	.	0,300	0,602	0,524	0,927	0,799	0,000	0,002	0,003	0,333
Other Relational Content - Phase 2	Corr. Coeff.	0,056	-0,273	-0,149	-0,256	0,089	-0,148	0,153	0,296	-0,220	-0,076	-0,183	1,000	0,113	0,137	0,021	-0,007	-0,099	0,013	0,019	0,178
	Sig.	0,754	0,119	0,400	0,144	0,619	0,403	0,389	0,089	0,212	0,670	0,300	.	0,526	0,439	0,908	0,969	0,577	0,940	0,915	0,313
N° of Structural Holes - Phase 2	Corr. Coeff.	0,225	-0,123	0,034	-0,074	-0,067	,353*	,469**	-0,067	0,096	-0,007	0,093	0,113	1,000	,983**	-,759**	,801**	0,030	-0,051	-0,122	,364*
	Sig.	0,200	0,487	0,850	0,676	0,706	0,041	0,005	0,705	0,588	0,970	0,602	0,526	.	0,000	0,000	0,000	0,867	0,772	0,494	0,034
Sparsity - Phase 2	Corr. Coeff.	0,143	-0,142	-0,047	-0,131	-0,074	0,323	,436*	-0,076	0,029	-0,056	0,113	0,137	,983**	1,000	-,740**	,785**	0,056	-0,082	-0,130	0,323
	Sig.	0,421	0,423	0,794	0,460	0,678	0,063	0,010	0,670	0,870	0,754	0,524	0,439	0,000	.	0,000	0,000	0,752	0,643	0,462	0,063
Density - Phase 2	Corr. Coeff.	-0,323	0,002	-0,023	0,071	0,046	-0,203	-0,289	-0,211	0,012	-0,118	0,017	0,021	-,759**	-,740**	1,000	-,740**	0,196	0,253	0,343	-,487**
	Sig.	0,067	0,991	0,897	0,695	0,797	0,258	0,103	0,239	0,949	0,513	0,927	0,908	0,000	0,000	.	0,000	0,274	0,156	0,051	0,004
Centrality - Phase 2	Corr. Coeff.	0,342	0,048	0,044	-0,030	-0,019	0,227	,362*	0,147	0,045	0,080	0,047	-0,007	,801**	,785**	-1,000**	1,000	-0,157	-0,219	-0,312	,571**

Spearman's Rho		Net-work Size in Phase 2	N° of Females - Phase 2	N° of Family Members - Phase 2	N° of Friends - Phase 2	N° of Academics - Phase 2	N° of Industry Partners - Phase 2	N° of Funding Partners - Phase 2	N° of Others - Phase 2	Informational Content - Phase 2	Provision of Resources - Phase 2	Support - Phase 2	Other Relational Content - Phase 2	N° of Structural Holes - Phase 2	Sparsity - Phase 2	Density - Phase 2	Centrality - Phase 2	Multiplexity - Phase 2	Strong Ties - Phase 2	Closure - Phase 2	Weak Ties - Phase 2
Multiplexity - Phase 2	Sig.	0,055	0,794	0,810	0,870	0,918	0,211	0,042	0,421	0,808	0,665	0,799	0,969	0,000	0,000	0,000	.	0,391	0,229	0,082	0,001
	Corr. Coeff.	0,184	0,213	-0,082	,345*	0,142	0,189	0,170	-0,074	,559**	,400*	,626**	-0,099	0,030	0,056	0,196	-0,157	1,000	,760**	,828**	-0,077
	Sig.	0,299	0,227	0,646	0,046	0,422	0,285	0,336	0,679	0,001	0,019	0,000	0,577	0,867	0,752	0,274	0,391	.	0,000	0,000	0,663
Strong Ties - Phase 2	Corr. Coeff.	,390*	0,165	0,179	,512**	,451**	0,176	0,067	-0,154	,723**	,511**	,517**	0,013	-0,051	-0,082	0,253	-0,219	,760**	1,000	,940**	-0,117
	Sig.	0,022	0,350	0,312	0,002	0,007	0,320	0,706	0,385	0,000	0,002	0,002	0,940	0,772	0,643	0,156	0,229	0,000	.	0,000	0,509
Closure - Phase 2	Corr. Coeff.	0,193	0,180	0,116	,389*	0,287	0,120	-0,005	-0,097	,578**	,381*	,496**	0,019	-0,122	-0,130	0,343	-0,312	,828**	,940**	1,000	-0,260
	Sig.	0,274	0,308	0,513	0,023	0,100	0,498	0,980	0,585	0,000	0,026	0,003	0,915	0,494	0,462	0,051	0,082	0,000	0,000	.	0,138
Weak Ties - Phase 2	Corr. Coeff.	,603**	-0,030	0,133	0,283	0,156	0,098	,446**	0,198	0,244	0,288	0,171	0,178	,364*	0,323	-,487**	,571**	-0,077	-0,117	-0,260	1,000
	Sig.	0,000	0,866	0,455	0,105	0,380	0,579	0,008	0,262	0,165	0,098	0,333	0,313	0,034	0,063	0,004	0,001	0,663	0,509	0,138	.

**. Correlation is significant at the 0.01 level 2-tailed.
*. Correlation is significant at the 0.05 level 2-tailed.

Table 5-5: Spearman Correlations for Relational Content in Phase 2

Phase 3 – Establishment

In the third phase the correlation between number of female contacts and *network size* has slightly increased to a coefficient of 0.431, while the correlation with the number of friends as dropped to 0.409, which is even lower than at the start in the first phase, and to a lower level of significance. After a period of no significant correlation in the second phase, industry partners have regained influence and now have with a positive correlation of 0.483 the highest impact ever. The same holds true for the number of funding partners, which after a short collapse in the second phase now feature the highest correlation coefficient of 0.534 across all three phase. For the first time, the number of other contacts positively correlates (0.378) with the size of the network; yet, a low level of significance. To put it together, in the last phase funding partners, closely followed by industry contacts, are the ones that have the highest impact of all relational roles on network size.

Coming to the content transferred, still informational ties feature the highest correlation impact (0.806), also compared to all other variables related to network size, which has even increased as opposed to the previous phases. Informational ties are closely followed by the correlation coefficient between network size and provision of resources (0.706), which, too, has reached its peak in the third phase. The highest growth, however, can be observed for the correlation between network size and number of supportive contacts that at this point has reached a coefficient of 0.647, which represents a large leap compared to the previous two phases. Now, all three types of content transferred feature an almost equal degree of interrelation with network size.

For the first time strong ties have outrun weak ties in their correlation with network size, both of them levelling off at a correlation coefficient of 0.584 and 0.545, respectively. New, is also a less significant correlation between multiplexity and network size with a coefficient of 0.413.

Regarding the number of *females* they could enforce their interrelation with the number of friendship ties (0.496) even bringing it to a higher level of significance, while loosing the correlation with the number of family members established in the second phase. As to the content provided, while in the first two phases correlation between women and content was restricted to the provision of support of whatever type, in the third phase significant correlations to number of informational (0.571) and supportive (0.481) ties exist instead. As already observed at the beginning, again small positive correlations with multiplexity (0.346) and closure (0.374) emerge. Moreover, for the first time a strong association (0.459) between the number of women and the number of strong ties can be determined.

The number of *family members* keeps positively related to the number of friends to some extent (0.383) and at the lower level of significance.

The number of *friends* still features a strong correlation with the number of informational ties, yet, compared to the previous phase a slight decrease from 0.641 to a current coefficient of 0.573 can be observed. Of almost equal dimension is the correlation with resource-providing contacts (0.557) that has replaced the correlation between friends and supportive contacts of the previous phases. The associations with multiplexity (0.403), tie strength (0.581) and closure (0.413), as detected for the second phase, remain and even slightly increase in strength.

The role of *academics* seems to have changed. While in the first two phases they were associated with the provision of information, in the third phase this correlation no longer exists. Instead the number of academics is related to the provision of resources (0.372), yet, at the lower level of significance. The correlation with the number of strong ties persists, but slightly decreases to a correlation coefficient of 0.348 on the lower significance level.

In the third phase, *industry contacts* are positively related to the number of supportive ties (0.348), however, on the lower level of significance. Moreover, they have lost their crowding-out effect on other contacts of the previous phases.

In the last phase *funding partners* reach the level of interaction with content-related variables of the first phase. Yet, while in the first phase the largest correlation coefficient in this context could be determined for the provision of resources, followed by support and then information, in the third phase it is the correlation with informational ties that assumes the largest value of 0.474, followed by a correlation coefficient of 0.473 for the number of resource-related interactions, and 0.356 for the correlation with supportive-ties. Compared to the first phase the association of the number of funding partners with resource- and support-related interactions has dropped in general. As in the second phase, a slight correlation between the number of funding partners and the number of structural holes (0.381) and the degree of sparsity (0.372) in the networks prevails, however, at an reduced level and with lower significance. Centrality no longer correlates with the number of funding partners. Strong ties are again positively correlated with funding partners, as already the case in the first phase, yet, with a lower correlation coefficient of 0.357 and a lower level of significance. While in the first phase this association was accompanied by a positive correlation with multiplexity, this trend cannot be confirmed for the third phase. In return, the coefficient for the correlation with the number of weak ties (0.482) is the same as in the first phase, and as such remained comparably stable over time. Relative to the development of the correlation coefficient for the number of strong ties this means that in the third phase the association between the number of funding partners and weak ties has become stronger than the association with the number of strong ties, as determined for the first phase.

The number of contacts to *other types of alters* is now positively correlated with the number of funding partners (0.414), other relational content (0.423) and the number of structural holes (0.353), however at a lower level of significance.

Looking at the content exchanged, *information* is almost equally associated with the number of female contacts and friends in the network, and to a slightly lower degree with the number of funding partners. Academics no longer correlate to informational ties. A very strong correlation (0.796) to the number of resource-related ties can be diagnosed, which is followed with some gap by the coefficient (0.521) for the correlation with the number of supportive interactions. By far the strongest, however, is the association between informational and strong ties with a correlation coefficient of 0.834, followed by a positive correlation of 0.705 to multiplexity and of 0.618 to network closure. This means the correlation structure of the previous phase remained basically unchanged, but moved to higher level of association.

As mentioned the provision of *resources* is strongly associated with the number of friends (0.557), the number of females (0.481) and the number of funding partners (0.473) in the network. Friends also play when however a minor role (0.372) in this context. In total this represents an increased level of association between the number of resource-related contacts and different types of alters that could not be detected for the previous phases. Besides the above-described strong correlation to the number of informational ties, resource contacts are also positively related (0.527) to the number of supportive ties, which represents a considerable increase as opposed to the second phase both in the strength of association and the level of significance. The same holds true for the correlations with multiplexity (0.674), strong ties (0.844), and network closure (0.632), which could already be observed for the second phase, but now in the third phase have considerably increased in strength and significance.

Supportive relations are no longer related to the number of women or friends, as the case in the previous phases, but for the first time correlate with industry (0.348) and funding (0.356) partners, however at a lower level of significance. As mentioned, they are also positively associated with the number of informational and resource-related ties. The strong relationship between the provision of support and multiplexity (0.618) and strong ties (0.567) remains relatively stable over time. The correlation to closure continues to exist, yet a lower level of significance and at a reduced extent (0.42). For the first time, the number of weak ties is also (0.358) positively associated with the amount support provided, however at a lower level of significance.

The amount of *other relational content* transferred beyond the number of other types of alters (0.423) positively correlates with the number of structural holes (0.367). For the first time in the spin-off process, significant correlation for this type of content can be determined, albeit at the lower level of significance.

Regarding the number of *structural holes* in egos' networks, as mentioned above they positively relate to the number of funding partners (0.381), other contacts (0.353), other types of content exchanged (0.367), and of course, as in the previous phases, due to their methodological construction, to sparsity (0.997) and centrality, (0.632) and with a negative sign to density (-0.572). Similar applies to sparsity, density, and centrality, which are all some function of the degree of connectedness among egos' alters, and thus correlated.

The degree of multiplexity is weakly and at a lower level of significance correlated with the number of women (0.346) and friends (0.403) in egos' networks. This represents an increase in correlation for the number of friends as opposed to the second phase. Female contacts, in turn, were already positively related to the degree of multiplexity in the first phase, but not in the second phase. A strong correlation at a higher level of significance can be diagnosed for the number of informational (0.705), resource-related (0.674) and supportive ties (0.618). Compared to the respective correlation coefficients of the second phase, i.e. 0.559, 0.4, and 0.626, this represents a strong increase for information- and resource related contacts, yet, a slight weakening in association for the provision of support, which was already very strong in the second phase. As to the correlation with the number of strong ties and closure this again can be attributed to the way the respective parameters were constructed, and as such allows for no reasonable interpretation.

As depicted for the respective antagonistic variables, the number of strong ties is for the first time positively correlated with the number of females (0.512). The correlation with the number of friends (0.451) has slightly decreased as compared to the previous phase. At a lower level of significance the number of academics (0.348) and funding partners (0.357) is also positively correlated. In addition, a strong association to all types of content exists, i.e. information (0.834), resources (0.844), and support (0.567) representing a strong intensification of association as opposed to the values of the previous phase.

In the third phase, closure again is positively related to the number of females (0.374) and friends (0.413), yet at a lower level of significance. The same as the number of strong ties it positively correlates to the number of informational (0.618), resource-related (0.632), and supportive ties (0.42), which represents a strong increase for information-related interactions, and an even stronger increase for the provision of resources when comparing the coefficients of the previous phase. As regards the number of supportive ties their correlation with the degree of multiplexity as weakened as opposed to the second phase.

Weak ties still interrelate with funding partners (0.482) and at a lower level of significance for the first time also with the number of supportive ties (0.358). All other significant associations of the previous phases are no longer given.

Spearman's Rho		Network Size in Phase 3	N° of Females - Phase 3	N° of Family Members - Phase 3	N° of Friends - Phase 3	N° of Academics Phase 3	N° of Industry Partners Phase 3	N° of Funding Partners Phase 3	N° of Others - Phase 3	Informational Content - Phase 3	Provision of Resources - Phase 3	Support - Phase 3	Other Relational Content - Phase 3	N° of Structural Holes - Phase 3	Sparsity - Phase 3	Density - Phase 3	Centrality - Phase 3	Multiplexity - Phase 3	Strong Ties - Phase 3	Closure - Phase 3	Weak Ties - Phase 3
Network Size in Phase 3	Corr. Coeff.	1,000	,431*	0,025	,409*	0,206	,483**	,534**	,378*	,806**	,706**	,647**	0,282	0,306	0,290	-0,138	0,184	,413*	,584**	0,299	,545**
	Sig.	.	0,011	0,887	0,016	0,242	0,004	0,001	0,027	0,000	0,000	0,000	0,106	0,078	0,097	0,458	0,330	0,015	0,000	0,085	0,001
N° of Females - Phase 3	Corr. Coeff.	,431*	1,000	0,290	,496**	0,106	0,120	0,262	0,214	,571**	,481**	0,277	0,086	0,142	0,148	-0,098	0,074	,346*	,459**	,374*	0,039
	Sig.	0,011	.	0,096	0,003	0,552	0,499	0,134	0,224	0,000	0,004	0,113	0,628	0,422	0,404	0,600	0,696	0,045	0,006	0,030	0,827
N° of Family Members - Phase 3	Corr. Coeff.	0,025	0,290	1,000	,383*	-0,129	-0,214	-0,254	-0,180	0,107	0,045	-0,032	0,096	0,185	0,177	-0,185	0,122	-0,010	0,144	0,155	-0,123
	Sig.	0,887	0,096	.	0,025	0,468	0,224	0,146	0,308	0,547	0,799	0,859	0,589	0,294	0,318	0,319	0,521	0,954	0,416	0,382	0,489
N° of Friends - Phase 3	Corr. Coeff.	,409*	,496**	,383*	1,000	0,253	0,309	0,014	-0,233	,573**	,557**	0,311	0,090	0,167	0,183	-0,161	0,120	,403*	,581**	,413*	0,082
	Sig.	0,016	0,003	0,025	.	0,149	0,075	0,937	0,184	0,000	0,001	0,073	0,612	0,346	0,299	0,388	0,527	0,018	0,000	0,015	0,646
N° of Academics - Phase 3	Corr. Coeff.	0,206	0,106	-0,129	0,253	1,000	0,044	0,045	-0,278	0,293	,372*	0,139	-0,148	-0,299	-0,291	0,167	-0,135	0,270	,348*	0,240	0,084
	Sig.	0,242	0,552	0,468	0,149	.	0,805	0,799	0,111	0,092	0,030	0,432	0,404	0,086	0,095	0,370	0,475	0,123	0,044	0,172	0,635
N° of Industry Partners - Phase 3	Corr. Coeff.	,483**	0,120	-0,214	0,309	0,044	1,000	0,073	-0,171	0,316	0,251	,348*	-0,002	0,115	0,135	0,033	0,005	0,108	0,206	0,021	0,255
	Sig.	0,004	0,499	0,224	0,075	0,805	.	0,683	0,333	0,069	0,153	0,044	0,993	0,518	0,448	0,860	0,981	0,544	0,242	0,906	0,146
N° of Funding Partners - Phase 3	Corr. Coeff.	,534**	0,262	-0,254	0,014	0,045	0,073	1,000	,414*	,474**	,473**	,356*	0,083	,381*	,372*	-0,253	0,359	0,336	,357*	0,260	,482**
	Sig.	0,001	0,134	0,146	0,937	0,799	0,683	.	0,015	0,005	0,005	0,039	0,643	0,026	0,030	0,170	0,051	0,052	0,038	0,137	0,004
N° of Others - Phase 3	Corr. Coeff.	,378*	0,214	-0,180	-0,233	-0,278	-0,171	,414*	1,000	0,280	0,160	0,241	,423*	,353*	0,331	-0,163	0,223	0,200	0,147	0,134	0,208
	Sig.	0,027	0,224	0,308	0,184	0,111	0,333	0,015	.	0,109	0,365	0,169	0,013	0,040	0,055	0,381	0,237	0,258	0,408	0,451	0,238
Informational Content - Phase 3	Corr. Coeff.	,806**	,571**	0,107	,573**	0,293	0,316	,474**	0,280	1,000	,796**	,521**	0,185	0,215	0,216	-0,094	0,125	,705**	,834**	,618**	0,327
	Sig.	0,000	0,000	0,547	0,000	0,092	0,069	0,005	0,109	.	0,000	0,002	0,295	0,222	0,221	0,617	0,509	0,000	0,000	0,000	0,059

153

Spearman's Rho		Net-work Size in Phase 3	N° of Females - Phase 3	N° of Family Members - Phase 3	N° of Friends - Phase 3	N° of Academics - Phase 3	N° of Industry Partners - Phase 3	N° of Funding Partners - Phase 3	N° of Others - Phase 3	Informational Content - Phase 3	Provision of Resources - Phase 3	Support - Phase 3	Other Relational Content - Phase 3	N° of Structural Holes - Phase 3	Sparsity - Phase 3	Density - Phase 3	Centrality - Phase 3	Multiplexity - Phase 3	Strong Ties - Phase 3	Closure - Phase 3	Weak Ties - Phase 3
Provision of Resources - Phase 3	Corr. Coeff.	,706**	,481**	0,045	,557**	,372*	0,251	,473**	0,160	,796**	1,000	,527**	0,079	0,157	0,167	0,074	-0,007	,674**	,844**	,632**	0,269
	Sig.	0,000	0,004	0,799	0,001	0,030	0,153	0,005	0,365	0,000	.	0,001	0,657	0,376	0,345	0,691	0,969	0,000	0,000	0,000	0,125
Support - Phase 3	Corr. Coeff.	,647**	0,277	-0,032	0,311	0,139	,348*	,356*	0,241	,521**	,527**	1,000	-0,016	0,162	0,169	-0,092	0,135	,618**	,567**	,420*	,358*
	Sig.	0,000	0,113	0,859	0,073	0,432	0,044	0,039	0,169	0,002	0,001	.	0,928	0,360	0,338	0,624	0,477	0,000	0,000	0,013	0,038
Other Relational Content - Phase 3	Corr. Coeff.	0,282	0,086	0,096	0,090	-0,148	-0,002	0,083	,423*	0,185	0,079	-0,016	1,000	,367*	0,321	-0,166	0,205	0,029	0,024	-0,066	0,124
	Sig.	0,106	0,628	0,589	0,612	0,404	0,993	0,643	0,013	0,295	0,657	0,928	.	0,033	0,064	0,371	0,277	0,871	0,895	0,710	0,486
N° of Structural Holes - Phase 3	Corr. Coeff.	0,306	0,142	0,185	0,167	-0,299	0,115	,381*	,353*	0,215	0,157	0,162	,367*	1,000	,997**	-,572**	,632**	0,139	0,128	0,089	0,263
	Sig.	0,078	0,422	0,294	0,346	0,086	0,518	0,026	0,040	0,222	0,376	0,360	0,033	.	0,000	0,001	0,000	0,434	0,471	0,617	0,133
Sparsity - Phase 3	Corr. Coeff.	0,290	0,148	0,177	0,183	-0,291	0,135	,372*	0,331	0,216	0,167	0,169	0,321	,997**	1,000	-,562**	,620**	0,153	0,142	0,107	0,255
	Sig.	0,097	0,404	0,318	0,299	0,095	0,448	0,030	0,055	0,221	0,345	0,338	0,064	0,000	.	0,001	0,000	0,387	0,424	0,548	0,146
Density - Phase 3	Corr. Coeff.	-0,138	-0,098	-0,185	-0,161	0,167	0,033	-0,253	-0,163	-0,094	0,074	-0,092	-0,166	-,572**	-,562**	1,000	-1,000**	-0,063	0,005	0,008	-0,314
	Sig.	0,458	0,600	0,319	0,388	0,370	0,860	0,170	0,381	0,617	0,691	0,624	0,371	0,001	0,001	.	0,000	0,735	0,977	0,964	0,085
Centrality - Phase 3	Corr. Coeff.	0,184	0,074	0,122	0,120	-0,135	0,005	0,359	0,223	0,125	-0,007	0,135	0,205	,632**	,620**	-1,000**	1,000	0,130	0,055	0,054	0,287
	Sig.	0,330	0,696	0,521	0,527	0,475	0,981	0,051	0,237	0,509	0,969	0,477	0,277	0,000	0,000	0,000	.	0,494	0,772	0,778	0,124
Multiplexity - Phase 3	Corr. Coeff.	,413*	,346*	-0,010	,403*	0,270	0,108	0,336	0,200	,705**	,674**	,618**	0,029	0,139	0,153	-0,063	0,130	1,000	,850**	,876**	0,049
	Sig.	0,015	0,045	0,954	0,018	0,123	0,544	0,052	0,258	0,000	0,000	0,000	0,871	0,434	0,387	0,735	0,494	.	0,000	0,000	0,783

Spearman's Rho		Net-work Size in Phase 3	N° of Females - Phase 3	N° of Family Members - Phase 3	N° of Friends - Phase 3	N° of Academics - Phase 3	N° of Industry Partners - Phase 3	N° of Funding Partners - Phase 3	N° of Others - Phase 3	Informational Content - Phase 3	Provision of Resources - Phase 3	Support - Phase 3	Other Relational Content - Phase 3	N° of Structural Holes - Phase 3	Sparsity - Phase 3	Density - Phase 3	Centrality - Phase 3	Multiplexity - Phase 3	Strong Ties - Phase 3	Closure - Phase 3	Weak Ties - Phase 3
Strong Ties - Phase 3	Corr. Coeff.	,584**	,459**	0,144	,581**	,348*	0,206	,357*	0,147	,834**	,844**	,567**	0,024	0,128	0,142	0,005	0,055	,850**	1,000	,896**	-0,040
	Sig.	0,000	0,006	0,416	0,000	0,044	0,242	0,038	0,408	0,000	0,000	0,000	0,895	0,471	0,424	0,977	0,772	0,000	.	0,000	0,824
Closure - Phase 3	Corr. Coeff.	0,299	,374*	0,155	,413*	0,240	0,021	0,260	0,134	,618**	,632**	,420*	-0,066	0,089	0,107	0,008	0,054	,876**	,896**	1,000	-0,263
	Sig.	0,085	0,030	0,382	0,015	0,172	0,906	0,137	0,451	0,000	0,000	0,013	0,710	0,617	0,548	0,964	0,778	0,000	0,000	.	0,133
Weak Ties - Phase 3	Corr. Coeff.	,545**	0,039	-0,123	0,082	0,084	0,255	,482**	0,208	0,327	0,269	,358*	0,124	0,263	0,255	-0,314	0,287	0,049	-0,040	-0,263	1,000
	Sig.	0,001	0,827	0,489	0,646	0,635	0,146	0,004	0,238	0,059	0,125	0,038	0,486	0,133	0,146	0,085	0,124	0,783	0,824	0,133	.

**. Correlation is significant at the 0.01 level 2-tailed.
*. Correlation is significant at the 0.05 level 2-tailed.

Table 5-6: Spearman Correlations for Relational Content in Phase 3

Summary

Putting together the first indications rendered by Spearman's Rho coefficients, one could say that regarding types of alters friends play a continuous role for network size over all three phases of the spin-off process. Their influence is the largest in the second phase and decreases a bit, also in significance, in the last phase. On contrary, funding partners, who seem to have an almost equally influence on network size than the number of friends in the first phase, lose part of their influence in the second phase, but regain influence, beyond that of friends, in the last phase. Industry partners have a smaller and less significant effect on network size, which disappears for the second phase, but returns for the last phase. Academics correlation with network size is restricted to the first phase.

Concerning the content transferred, the strongest influence on network size is exerted by the number of informational ties. It is the highest in the first phase, drops a bit in the second phase, and recovers in the third phase, without achieving the all-time high of the first phase. Resources come first into play in the second phase and increase in correlation with network size in the last phase. Support on the other hand is already significantly correlated to network size in the first phase, yet, to a small extent compared to the number of informational ties. Correlation increases a bit in the second phase. In the last phase, support has reached a rather high significant correlation with network size.

At the beginning weak ties influence on network size exceeds the one of strong ties. Both correlation coefficients drop a bit in the second phase. In the last phase, strong ties have caught up and now even slightly exceed the influence of weak ties.

Regarding the provision of information, in the first phase it is mainly associated with the number of friends and industry partners, and to a weaker extent with the number of academics in the networks. Resources and support are closely correlated with the number funding partners. Friends and academics keep their influence on information in the second phase, while industry partners loose their influence. Friends, in addition, are now also closely correlated with the provision of support. Funding partners feature no correlation at all in the second phase. In the last phase, friends have lost a bit of their correlation with informational ties in favour of an additional correlation with the provision of resources. Still their influence on information provision is the highest among all types of alters. Funding partners have regained their strong influence on informational, resource-related, and supportive ties, yet with a lower emphasis on support. Academics are now slightly correlated with the provision of resources, industry partners with the provision of support.

In the first phase, strong ties are strongly related to all types of content, while weak ties only correlate with information provided. In the second phase weak ties loose their correlation with information-related ties, to gain some, but less significant correlation with supportive interactions in the last phase. In general, the influence of strong ties on the amount of different contents exchanged strongly increases over time.

Closures and multiplexity feature a rather strong and significant correlation with support and to a lesser extent also with resources in the first phase. In the second phase, a strong association with informational ties adds to this structure. In the last phase, both closure and multiplexity gain considerable influence on the provision of information and resources. Yet, in particular as regards closure, influence on support drops sharply for the last phase.

5.2.2.2 Informational Content

In the following chapter a closer look is made at the detailed composition of the information transferred at the different phases of the spin-off process. To this end the Spearman correlation coefficients for the individual sub-categories of information content identified are computed for all other network variables. To facilitate representation associated correlation matrices are reduced to the correlation coefficients of interest with rows containing redundant information having been deleted. Relevant correlations discussed are highlighted in bold letters.

Phase 1 – Opportunity Recognition

Starting with the correlation with and thus contribution of the different types of information exchanged to the total amount of *informational content* of entrepreneurs' networks, the most potent association (0.708) exists between informational content and the residual category of other information. It is followed with some distance by the correlation coefficients for the number of ties providing economic information (0.579) and legal information (0.483). At a lower level of significance are the Spearman correlations for ties providing information about potential customers (0.378) and the ones informing about appropriate employees for the planned venture (0.371).

Looking at the individual information categories in more depth, *legal information* is significantly correlated with ties providing economic information (0.534). Moreover, at a lower level of significance legal information also relates to multiplex (0.409) and strong ties (0.384) and the number of funding partners (0.345).

Economic information is positively associated with the number of industry partners (0.503) and to lesser extent and lower level of significance to the number of funding partners (0.392). Regarding the interrelation with other types of information, as mentioned legal information (0.534) and information about potential customers (0.561) seem to come together with economic information. The number of ties providing economic information is furthermore correlated with the number of strong ties (0.638) and increasing network closure (0.554) and multiplexity (0.535).

Information about customers correlates with the number of friends (0.479) and the number of industry partners (0.51). On the level of other information provided client information is also positively associated with economic (0.561) and personnel-related (0.46) information. Furthermore, the number of ties providing market information increases with the number strong ties (0.573) and the degree of network closure (0.53).

In the first phase, ties furnishing *information about potential employees* relate to the number of industry partners (0.564). Other types of information seem to come from academics (0.45) and are basically related to the number of weak ties (0.5).

Spearman's Rho		N° of Females - Phase 1	N° of Family Members - Phase 1	N° of Friends - Phase 1	N° of Academics - Phase 1	N° of Industry Partners - Phase 1	N° of Funding Partner s - Phase 1	N° of Others - Phase 1	Informational Content - Phase 1	Legal Information - Phase 1	Economic Information - Phase 1	Information about Potential Customers - Phase 1	Information about Potential Employees - Phase 1	Other Information - Phase 1	N° of Structural Holes - Phase 1	Sparsity - Phase 1	Density - Phase 1	Centrality - Phase 1	Multiplexity - Phase 1	Strong Ties - Phase 1	Closure - Phase 1	Weak Ties - Phase 1
Informational Content - Phase 1	Corr. Coeff.	,106	-,031	,475**	,429*	,475**	,359*	-,078	1,000	,483**	,579**	,378*	,371*	,708**	,218	,198	-,087	,158	,253	,579**	,264	,549**
	Sig.	,551	,862	,004	,011	,004	,037	,662	.	,004	,000	,027	,031	,000	,215	,263	,643	,403	,148	,000	,132	,001
Legal Information - Phase 1	Corr. Coeff.	-,176	-,167	,059	,148	,321	,345*	,048	,483**	1,000	,534**	,290	,207	,190	,298	,277	,035	,010	,409*	,384*	,311	,247
	Sig.	,321	,346	,742	,402	,064	,046	,789	,004	.	,001	,096	,240	,282	,087	,113	,852	,957	,016	,025	,074	,160
Economic Information - Phase 1	Corr. Coeff.	-,016	-,050	,215	-,116	,503**	,392*	-,093	,579**	,534**	1,000	,561**	,327	-,003	,014	,002	,151	-,112	,535**	,638**	,554**	,133
	Sig.	,930	,778	,221	,513	,002	,022	,601	,000	,001	.	,001	,059	,987	,937	,993	,418	,555	,001	,000	,001	,453
Information about Potential Customers - Phase 1	Corr. Coeff.	,054	,076	,479**	,006	,510**	-,054	-,235	,378*	,290	,561**	1,000	,460**	,001	-,171	-,190	,240	-,220	,299	,573**	,530**	-,086
	Sig.	,760	,669	,004	,974	,002	,760	,181	,027	,096	,001	.	,006	,994	,333	,282	,193	,243	,086	,000	,001	,628
Information about Potential Employees - Phase 1	Corr. Coeff.	-,026	,096	,200	-,036	,564**	,174	-,179	,371*	,207	,327	,460**	1,000	,060	-,233	-,233	,219	-,216	,116	,323	,147	,205
	Sig.	,883	,588	,256	,840	,001	,325	,310	,031	,240	,059	,006	.	,736	,184	,185	,236	,251	,515	,063	,405	,246
Other Information - Phase 1	Corr. Coeff.	,108	-,168	,320	,450**	,207	,155	,034	,708**	,190	-,003	,001	,060	1,000	,307	,294	-,234	,289	-,062	,157	-,125	,500**
	Sig.	,542	,342	,065	,008	,240	,380	,846	,000	,282	,987	,994	,736	.	,077	,092	,204	,121	,728	,374	,480	,003

**. Correlation is significant at the 0.01 level (2-tailed).
*. Correlation is significant at the 0.05 level (2-tailed).

Table 5-7: Spearman Correlations for Informational Content in Phase 1

Phase 2 – Formation

In the second phase economic information has gained in strength as regards its correlation with the total amount of *informational content* (0.767). The correlation coefficient for other information has dropped from 0.708 in the first phase to only 0.586 in the second phase. Legal information has also lost in association strength, but only minorily by minus 0.015 down to a value of 0.468. While information about potential customers has increased in correlation strength up to a coefficient of 0.506 as opposed to 0.378 in the first phase, information about potential employees in turn is no longer significantly correlated to informational content.

Legal information is no longer positively correlated with the number of funding partners. The correlation with the number of ties providing economic information however increased to 0.647, while contacts delivering information about future customers now for the first time relate to the number of ties exchanging legal information with a correlation coefficient of 0.403, yet on a lower level of significance. The number of strong ties and for the first time the degree of network closure feature strong associations with the number of ties containing legal knowledge, i.e. 0.596 and 0.571, respectively. The correlation with multiplexity has lost in strength, its coefficient decreasing to 0.373.

In the second phase *economic information* corresponds to the number of friends with a correlation coefficient of 0.513 and not to the number of industry partners as at the previous stage. Funding partners have gained a little in correlation strength arriving at a value of 0.404; however could not raise the level of significance. Correlation with the number of ties providing legal information has increased to 0.647, while the association with the number of customer-related ties decreased to 0.497. Multiplexity (0.587), number of strong ties (0.616), and network closure (0.571) gained in strength as to their correlation with economic information compared to the previous phase.

Information about *customers*, now is no longer only related to the number of friends (0.69), but also strongly associated with the number of family members (0.516), and the number of women (0.457) in egos' networks. Moreover, as already mentioned a positive correlation to economic (0.497) and for the first time, however at a lower level of significance, also to legal (0.403) and personnel-related (0.374) information exists. Beyond positive correlation coefficients to the number of strong ties and degree of network closure, which have remained comparatively stable at 0.555 and 0.553, respectively, multiplexity now also features a strong correlation (0.486) with the number of ties delivering information about possible clients.

Beyond its just-mentioned correlation with customer-related information that has suffered in significance, ties transferring information about adequate *employees* no longer relate to the number of industry partners. In turn, in the second phase they are now weakly correlated with the number of strong ties (0.364) and network closure (0.434) on the lower level of significance.

Other information has lost its positive correlation with the number of academics and the number of weak ties and at the current stage features no significant association at all.

Spearman's Rho		N° of Females - Phase 2	N° of Family Members - Phase 2	N° of Friends - Phase 2	N° of Academics - Phase 2	N° of Industry Partners - Phase 2	N° of Funding Partners - Phase 2	N° of Others - Phase 2	Informational Content - Phase 2	Legal Information - Phase 2	Economic Information - Phase 2	Information about Potential Customers - Phase 2	Information about Potential Employees - Phase 2	Other Information - Phase 2	N° of Structural Holes - Phase 2	Sparsity - Phase 2	Density - Phase 2	Centrality - Phase 2	Multiplexity - Phase 2	Strong Ties - Phase 2	Closure - Phase 2	Weak Ties - Phase 2
Informational Content - Phase 2	Corr. Coeff.	,317	,229	,641**	,368*	,263	,289	-,083	1,000	,468**	,767**	,506**	,236	,586**	,096	,029	,012	,045	,559**	,723**	,578**	,244
	Sig.	,067	,194	,000	,032	,133	,097	,639	.	,005	,000	,002	,179	,000	,588	,870	,949	,808	,001	,000	,000	,165
Legal Information - Phase 2	Corr. Coeff.	-,095	,211	,337	,099	-,047	,185	-,014	,468**	1,000	,647**	,403*	,288	,019	,122	,076	,077	-,031	,373*	,573**	,596**	,032
	Sig.	,592	,231	,051	,579	,793	,295	,935	,005	.	,000	,018	,099	,917	,490	,671	,669	,864	,030	,000	,000	,857
Economic Information - Phase 2	Corr. Coeff.	,046	,081	,513**	,112	,301	,404*	,061	,767**	,647**	1,000	,497**	,211	,201	,203	,144	-,070	,124	,587**	,616**	,571**	,338
	Sig.	,797	,649	,002	,528	,084	,018	,731	,000	,000	.	,003	,231	,255	,249	,417	,700	,500	,000	,000	,000	,051
Information about Potential Customers - Phase 2	Corr. Coeff.	,457**	,516**	,690**	,191	,261	-,150	-,245	,506**	,403*	,497**	1,000	,374*	,068	-,003	-,050	,021	,019	,486**	,555**	,553**	,120
	Sig.	,007	,002	,000	,279	,136	,396	,163	,002	,018	,003	.	,029	,702	,987	,780	,908	,917	,004	,001	,001	,500
Information about Potential Employees - Phase 2	Corr. Coeff.	,258	,229	,148	-,076	,108	-,008	-,133	,236	,288	,211	,374*	1,000	,061	,094	,094	,088	-,075	,189	,364*	,434*	-,314
	Sig.	,141	,192	,404	,670	,544	,965	,454	,179	,099	,231	,029	.	,733	,596	,596	,626	,685	,284	,034	,010	,071
Other Information - Phase 2	Corr. Coeff.	,313	,034	,278	,337	,020	,151	,122	,586**	,019	,201	,068	,061	1,000	-,059	-,092	,149	-,133	,252	,338	,223	,177
	Sig.	,072	,847	,111	,051	,909	,393	,490	,000	,917	,255	,702	,733	.	,739	,605	,407	,469	,150	,050	,206	,318

**. Correlation is significant at the 0.01 level (2-tailed).
*. Correlation is significant at the 0.05 level (2-tailed).

Table 5-8: Spearman Correlations for Informational Content in Phase 2

Phase 3 – Establishment

In the third phase legal information has reached its all time high as to its contribution to overall information provided. It positively correlates with *informational content* with a value of 0.611 and as such has almost caught up with other information that has slightly regained in strength as opposed to the first phase (0.682). Information about potential customers, too, has at this point of time reached its maximal association with the amount of information provided (0.516). Economic information in turn has suffered a slight loss in association of 0.015 down to a coefficient value of 0.752 after its correlative peak of 0.767 in the second phase still featuring the highest correlation out of all sub-categories of information with the overall amount of information transported in the last phase.

As to *legal information*, it is now associated with the number of friends (0.448) and at a lower level of significance with the number of family members (0.35) in entrepreneurs' networks. Moreover, it positively associates with economic (0.658), client- (0.527) and for the first time with personnel-related (0.563) information. In addition, the positive correlation with the number of strong ties has further increased to 0.634, and even more with the degree of multiplexity, which jumped to a value of 0.509 and a higher level of significance. On contrary, the positive relationship between the number of legal information ties and closure as slightly decreased from a correlation coefficient of 0.596 in the second, to 0.571 in the last phase.

Ties providing *economic information* for the first time positively relate to the number of ego's female contacts with a correlation coefficient of 0.525. The association with the number of friends has further increased to a value of 0.593. Correlation with the number of funding partners in turn has dropped to 0.356 compared to the previous phase. Looking at the correlations with other types of information, a considerable increase in strength can be diagnosed for the association with client-related information up to a value of 0.618, and to a lower extent also for legal information, which increased from 0.647 in the second phase to 0.658 in the third phase. For the first time, the number of ties providing economic information is also positively correlated with the number of ties transferring personnel-related information (0.597).

Regarding *information about potential customers* women's influence of the second phase has slightly decreased to 0.448. The correlation with the number of family members has disappeared at all. The correlation with the number of friends has also slightly reduced to 0.675. Yet, the associations with the other types of information have strongly increased to 0.527 for the correlation with legal information, 0.618 for the correlation with economic information, and 0.708 for the correlation with information about possible customers. The correlation coefficients for multiplexity (0.512), number of strong ties (0.637), and network closure (0.532) have almost equally increased compared to the values of the previous phase.

Information about adequate staff for the first time positively correlates with the number of female contacts (0.383), however, at the lower level of significance. Moreover, it is also for the first time positively related to the amount of legal (0.563) and economic (0.597) information. A significant increase in correlation strength can be observed for the relationship between personnel-related information and information about potential customers, which rises to a value of 0.708 as opposed to 0.374 in the previous phase. Correlations with the number of strong ties and network closure also increase also in significance to 0.492 and 0.508, respectively. For the first time also a positive correlation of 0.438 with the degree of multiplexity can be determined in the last phase.

In the last phase, ties that provide *other types of information* than the pre-defined ones gain in influence. They are now for the first time positively associated with the number of funding partners (0.454) and the number of others (0.483); on a lower level of significance also with the number of females (0.387). This development is accompanied by a positive association with the number of strong ties of 0.567, and also on a lower level of significance with the degree of multiplexity (0.409) and the degree of network closure (0.378).

Spearman's Rho		N° of Females - Phase 3	N° of Family Members - Phase 3	N° of Friends - Phase 3	N° of Academics - Phase 3	N° of Industry Partners - Phase 3	N° of Funding Partners - Phase 3	N° of Others - Phase 3	Informational Content - Phase 3	Legal Information - Phase 3	Economic information - Phase 3	Information about Potential Customers - Phase 3	Information about Potential Employees - Phase 3	Other Information - Phase 3	N° of Structural Holes - Phase 3	Sparsity - Phase 3	Density - Phase 3	Centrality - Phase 3	Multiplexity - Phase 3	Strong Ties - Phase 3	Closure - Phase 3	Weak Ties - Phase 3
Informational Content - Phase 3	Corr. Coeff.	,571**	,107	,573**	,293	,316	,474**	,280	1,000	,611**	,752**	,516**	,290	,682**	,215	,216	-,094	,125	,705**	,834**	,618**	,327
	Sig.	,000	,547	,000	,092	,069	,005	,109	.	,000	,000	,002	,096	,000	,222	,221	,617	,509	,000	,000	,000	,059
Legal Information - Phase 3	Corr. Coeff.	,173	,350*	,448**	,162	,134	,257	,263	,611**	1,000	,658**	,527**	,563**	,210	,331	,335	-,157	,237	,509**	,634**	,571**	,157
	Sig.	,327	,043	,008	,360	,449	,142	,132	,000	.	,000	,001	,001	,234	,056	,053	,399	,207	,002	,000	,000	,375
Economic information - Phase 3	Corr. Coeff.	,525**	,192	,593**	,316	,156	,356*	,151	,752**	,658**	1,000	,618**	,597**	,280	,251	,259	-,110	,193	,742**	,774**	,696**	,074
	Sig.	,001	,277	,000	,069	,377	,039	,394	,000	,000	.	,000	,000	,109	,153	,139	,557	,307	,000	,000	,000	,676
Information about Potential Customers - Phase 3	Corr. Coeff.	,448**	,282	,675**	,186	,196	,068	-,204	,516**	,527**	,618**	1,000	,708**	,058	,075	,111	-,065	,003	,512**	,637**	,532**	,015
	Sig.	,008	,107	,000	,292	,266	,703	,248	,002	,001	,000	.	,000	,744	,675	,531	,730	,986	,002	,000	,001	,932
Information about Potential Employees - Phase 3	Corr. Coeff.	,383*	,293	,298	,231	-,053	,149	,004	,290	,563**	,597**	,708**	1,000	-,066	,138	,157	-,117	,155	,438**	,492**	,508**	-,093
	Sig.	,025	,092	,087	,188	,766	,399	,982	,096	,001	,000	,000	.	,709	,437	,376	,531	,414	,010	,003	,002	,603
Other Information - Phase 3	Corr. Coeff.	,387*	-,088	,128	,197	,024	,454**	,483**	,682**	,210	,280	,058	-,066	1,000	,082	,077	,225	-,183	,409*	,567**	,378*	,182
	Sig.	,024	,620	,471	,264	,891	,007	,004	,000	,234	,109	,744	,709	.	,644	,666	,224	,334	,016	,000	,028	,302

**. Correlation is significant at the 0.01 level (2-tailed).
*. Correlation is significant at the 0.05 level (2-tailed).

Table 5-9: Spearman Correlations for Informational Content in Phase 3

Summary

Taking the above described correlations as first indications of possible dependencies, the following assumptions can be drawn up. In the *first phase*, the amount of information exchanged is predominantly determined by ties providing information other than legal, economic, client- and personnel-related. Having in mind that in the first phase the idea to spin-off has just emerged and only vague conceptions of the future business exist, it seems to be natural that information exchanged does not directly deal with aspects relating to the foundation of a company such as legal aspects or future customers, but centre more on general and technical terms. It has to be noted that respondents, when specifying what they meant with "other information", frequently referred to technical and scientific know-how. This other information in turn features as strong positive correlation with the number of academics and the number of weak ties in the network.

Still first considerations concerning the future business are already made at this point, which materialises in smaller, yet still highly significant correlations between total amount of information and economic as well as legal information. Information about potential customers and future employees play only a minor role. Economic information is primarily associated with industry partners, and is accompanied by legal and customer-related information. Moreover, it comes with strong ties and closed networks with a high degree of multiplexity. Information about possible employees, too, seems to come from industry partners and also positively relates to information about future customers.

In the *second phase* other information has lost a part of its impact on information exchanged in favour of economic information, which considerably gained in correlation strength and now features the highest association with informational content in general. Information concerning future staff is no longer significantly correlated to the amount of information transferred. While legal information has slightly lost in associative power, information about potential customers as further gained in relevance. Opposed to the previous phase other information features no significant correlation to any other variable. Economic information, on the other hand, no longer correlates to the number of industry partners, but is now strongly and significantly associated with the number of friends. Its correlation to funding partners has increased in strength, but could not improve in significance. The connection to legal information is reinforced; the connection to personnel-related information has slightly weakened. The association between economic information and the degree of multiplexity and closure of the networks has gained a bit in strength; the association with the number of strong ties is slightly reduced.

In the *last phase*, economic information is still the informational category the strongest correlated with the overall amount of information in the network. However, at this stage legal information has caught up and almost equals other information in its influence on network size. The correlation between information about possible customers and informational content has further increased reaching its maximum in the last phase. At this stage, other information is positively associated with the number of funding partners and other types of contact. Moreover, but to a smaller extent and on a lower level of significance is also positively relates to the number of women in the network. It is further strongly associated with the number of strong ties, and to some extent and at the lower level of significance with the degree of multiplexity and closure of the personal networks. The amount of legal information is positively related to the number of friends in the network, and at a lower level of significance and to a lesser extent also to the number of family members. Legal information comes together with all other types of information. The strongest association exists with economic information, followed by information concerning personnel, and by information about potential customers. Furthermore, it is strongly correlated with the number of strong ties, and the degree of multiplexity and closure of the network. The influence of friends on the amount of economic information has increased compared to the previous phase. The impact of funding partners in return decreased still at the lower level of significance. For the first time a highly significant positive association between the number of women in the network and the amount of economic information can be observed that considerably exceeds the impact of funding partners not only as regards the level of significance. Compared to the previous phase, economic information lost in association strength with legal information in favour of information about potential customers and for the first time a positive correlation with information about adequate employees. Regarding its association with the number of strong ties, multiplexity and closure, as opposed to the second phase a considerable further increase can be determined.

5.2.2.3 Resource Content

The next type of content to be analysed with respect to significant correlations of its sub-categories to other parameters is the provision of resources. According to the specification of the research model possible content realisation of interest in this context were specified with financial resources, tangible assets, infrastructure, human resources, and other, non-specified resource-related content.

Again respective correlation matrices were reduced to their essentials in terms of the indented analysis, and correlation coefficients to be described in more detail were highlighted bold.

Phase 1 – Opportunity Recognition

As Table 5-10 reveals, in the first phase the type of resource featuring the strongest correlation with the *general amount of resources* provided are financial resources with a correlation coefficient of 0.754, closely followed by the provision of personnel with a Spearman value of 0.727. Infrastructure also exhibits a rather strong association (0.620) with resources obtained at this stage. At some distance follow material resources with a correlation coefficient of 0.443. Other resources display no significant relation to overall resources at all.

As to be expected, *financial resources* are strongly associated with the number of funding partners (0.705), followed by other types of contacts (0.376) on a lower level of significance. Moreover, they are positively related to infrastructure (0.531) and material resources (0.46) provided. In addition, the provision of capital seems to require strong ties (0.436), and at a lower level of significance to some extent also weak ties (0.389) and multiplex networks (0.399).

Material resources are with a coefficient of 0.36 associated on a lower level of significance with the number of friends in a network. They seem to come together with financial resources and infrastructure featuring a correlation coefficient of 0.46 and 0.505, respectively. But primarily, they are positively related with the number of strong ties (0.531), followed by network closure with 0.469. On a lower level of significance they also weakly correspond with 0.345 to the degree of multiplexity in the personal networks.

Infrastructure is positively correlated with the number of friends (0.45) and funding partners (0.438), to some extent (0.349) but with lesser significance also with the number of women in the network. As mentioned it positively corresponds to material (0.505) and financial resources (0.531). Moreover, it is rather strongly correlated with the number of strong ties (0.523), network closure (0.491) and multiplexity (0.531).

Spearman's Rho		N° of Females - Phase 1	N° of Family Members - Phase 1	N° of Friends - Phase 1	N° of Academics - Phase 1	N° of Industry Partners - Phase 1	N° of Funding Partners - Phase 1	N° of Others - Phase 1	Provision of Resources - Phase 1	Material Resources - Phase 1	Financial Resources - Phase 1	Provision of Infrastructure - Phase 1	Provision of Personnel - Phase 1	Other Resources - Phase 1	N° of Structural Holes - Phase 1	Sparsity - Phase 1	Density - Phase 1	Centrality - Phase 1	Multiplexity - Phase 1	Strong Ties - Phase 1	Closure - Phase 1	Weak Ties - Phase 1
Provision of Resources - Phase 1	Corr. Coeff.	,195	-,209	,251	-,160	,108	,572**	,360*	1,000	,443**	,754**	,620**	,727**	,281	-,115	-,100	-,048	-,028	,379*	,441**	,341*	,284
	Sig.	,270	,235	,152	,365	,543	,000	,037	.	,009	,000	,000	,000	,108	,517	,574	,796	,884	,027	,009	,049	,104
Material Resources - Phase 1	Corr. Coeff.	,233	,069	,360*	,154	,263	,092	-,028	,443**	1,000	,460**	,505**	,175	-,080	-,061	-,072	,201	-,187	,345*	,531**	,469**	-,031
	Sig.	,184	,698	,037	,384	,133	,603	,876	,009	.	,006	,002	,322	,651	,732	,688	,279	,323	,046	,001	,005	,860
Financial Resources - Phase 1	Corr. Coeff.	,181	-,082	,191	-,100	,008	,705**	,376*	,754**	,460**	1,000	,531**	,316	,261	,007	,025	,025	,007	,399*	,436**	,292	,389*
	Sig.	,306	,643	,279	,575	,962	,000	,028	,000	,006	.	,001	,068	,136	,968	,890	,894	,971	,019	,010	,094	,023
Provision of Infrastructure - Phase 1	Corr. Coeff.	,349*	-,196	,450**	,023	,139	,438**	-,111	,620**	,505**	,531**	1,000	,335	-,118	-,045	-,044	-,090	,118	,531**	,523**	,491**	,131
	Sig.	,043	,267	,008	,899	,434	,010	,534	,000	,002	,001	.	,053	,505	,798	,803	,629	,533	,001	,002	,003	,459
Provision of Personnel - Phase 1	Corr. Coeff.	,154	-,263	,113	-,096	,137	,302	,291	,727**	,175	,316	,335	1,000	,333	-,188	-,175	-,120	,035	,122	,242	,178	,085
	Sig.	,385	,133	,523	,591	,439	,083	,095	,000	,322	,068	,053	.	,055	,287	,322	,521	,855	,491	,168	,314	,632
Other Resources - Phase 1	Corr. Coeff.	,109	,208	,028	-,163	-,125	,291	,216	,281	-,080	,261	-,118	,333	1,000	-,088	-,088	-,064	,078	-,225	-,028	-,090	,211
	Sig.	,541	,238	,875	,356	,481	,095	,221	,108	,651	,136	,505	,055	.	,620	,621	,733	,680	,201	,877	,612	,231

*. Correlation is significant at the 0.05 level (2-tailed).

**. Correlation is significant at the 0.01 level (2-tailed).

Table 5-10: Spearman Correlations for Resource Content in Phase 1

Phase 2 – Formation

In the second phase emphasis has shifted between the mobilisation of capital and the provision of relevant staff and their correlation with the *overall amount of resources* exchanged in favour of the human resources aspect. Yet as Table 5-11 illustrates, both of them have lost in associative power with slightly decreased correlation values – 0.628 for financial resources and 0.636 for the provision of personnel – as opposed to the previous phase. Material resources are the only type of asset that has increased in association (0.478) with the overall amount of resources obtained at this stage. Infrastructure, in contrast as suffered a significant reduction in correlation strength, its coefficient dropping from 0.62 in the first to only 0.38 at the present phase at an even lower level of significance.

The influence of funding partners on the amount of *financial resources* has weakened to 0.563 as opposed to the coefficient of 0.705 of the previous phase. The same holds true for the provision of infrastructure which decreased to 0.461, while the correlation with material resources slightly increased to a value of 0.471. The association between the provision of capital and multiplexity and strong ties no longer exists. Yet, the influence of weak ties has increased in strength to 0.523 and significance.

Beyond the positive correlation of 0.475 with the number of friends that has further increased also in significance, in the second phase the amount of *material resources* provided also positively relates (0.401), yet on a lower level of significance, to the number of family member in egos' networks. The association with financial resources has further increased to 0.471, the one with the provision of infrastructure slightly decreased also in significance to 0.407 from the previous to the current stage. While the correlation between material resources and the degree of multiplexity further deepened (0.412), strong ties and network closure lost part of their influence of the first phase, now amounting to 0.527 and 0.444, respectively.

Apart from the association with the number of female contacts in the network, which gained in strength (0.374), all other correlations relating to the *provision of infrastructure* weakened compared to the previous phase. The correlation with the number of friends now amounts to 0.443, the one with material resources to 0.407, and the interrelation with the provision of capital is reduced to 0.461. Strong ties and network closure are no longer significantly correlated at all. The correlation coefficient for multiplexity decreased from 0.531 to 0.513.

In turn, the *provision of personnel* for the first time shows, yet small and less significant correlations with the number of strong ties (0.42) and network closure (0.341).

Spearman's Rho		N° of Females - Phase 2	N° of Family Members - Phase 2	N° of Friends - Phase 2	N° of Academics - Phase 2	N° of Industry Partners - Phase 2	N° of Funding Partners - Phase 2	N° of Others - Phase 2	Provision of Resources - Phase 2	Material Resources - Phase 2	Financial Resources - Phase 2	Provision of Infrastructure - Phase 2	Provision of Personnel - Phase 2	Other Resources - Phase 2	N° of Structural Holes - Phase 2	Sparsity - Phase 2	Density - Phase 2	Centrality - Phase 2	Multiplexity - Phase 2	Strong Ties - Phase 2	Closure - Phase 2	Weak Ties - Phase 2
Provision of Resources - Phase 2	Corr. Coeff.	,189	,141	,338	,171	-,010	,304	,279	1,000	,478**	,628**	,380*	,636**	,164	-,007	-,056	-,118	,080	,400*	,511**	,381*	,288
	Sig.	,285	,426	,051	,332	,955	,081	,110		,004	,000	,027	,000	,354	,970	,754	,513	,665	,019	,002	,026	,098
Material Resources - Phase 2	Corr. Coeff.	,299	,401*	,475**	,259	,068	-,003	-,054	,478**	1,000	,471**	,407*	,092	-,103	-,191	-,221	,162	-,143	,412*	,527**	,444**	,165
	Sig.	,086	,019	,005	,139	,702	,989	,763	,004		,005	,017	,605	,562	,278	,209	,367	,434	,016	,001	,008	,352
Financial Resources - Phase 2	Corr. Coeff.	,193	,053	,182	,169	-,139	,563**	,224	,628**	,471**	1,000	,461**	,048	,028	,130	,096	-,109	,159	,223	,230	,152	,523**
	Sig.	,274	,766	,304	,340	,434	,001	,204	,000	,005		,006	,787	,875	,465	,590	,546	,384	,206	,190	,389	,002
Provision of Infrastructure - Phase 2	Corr. Coeff.	,374*	,070	,443**	-,041	-,036	,106	,121	,380*	,407*	,461**	1,000	-,092	-,142	,048	,066	-,102	,145	,513**	,218	,286	,323
	Sig.	,029	,694	,009	,818	,841	,552	,494	,027	,017	,006		,606	,422	,787	,710	,573	,428	,002	,216	,101	,062
Provision of Personnel - Phase 2	Corr. Coeff.	,151	,005	,147	,059	,037	,086	,156	,636**	,092	,048	-,092	1,000	,300	-,092	-,141	-,026	-,048	,265	,420*	,341*	-,231
	Sig.	,394	,979	,407	,740	,836	,630	,377	,000	,605	,787	,606		,085	,604	,428	,884	,795	,130	,013	,048	,189
Other Resources - Phase 2	Corr. Coeff.	,238	-,104	,206	,153	-,103	,207	,139	,164	-,103	,028	-,142	,300	1,000	-,096	-,095	,160	-,158	,036	,196	,074	-,219
	Sig.	,176	,558	,242	,387	,562	,241	,433	,354	,562	,875	,422	,085		,591	,591	,374	,387	,838	,267	,676	,214

*. Correlation is significant at the 0.05 level (2-tailed).

**. Correlation is significant at the 0.01 level (2-tailed).

Table 5-11: Spearman Correlations for Resource Content in Phase 2

Phase 3 – Establishment

In the last phase the strongest influence with a correlation coefficient of 0.747 on the *amount of resources transferred* can be observed for the number of ties providing material resources. It is followed by the provision of infrastructure that at this stage has considerably regained in associative power as opposed to the previous phases reaching an all time high of 0.721 and coming back to the higher level of significance. After their slump in the second phase financial resources and the provision of personnel have picked up in correlation strength with coefficient values of 0.691 and 0.65, respectively.

The influence of friends on the amount of *material resources* has increased over time and in the third phase reached its all time high of a correlation coefficient of 0.67. It is followed by the number of female contacts in the network that for the first time also features a positive effect of 0.586 on material resources provided. A smaller and less significant influence emanates from the number of industry contacts (0.405). Moreover, material resources positively interact with the provision of infrastructure (0.585) and financial resources (0.468), which represents an increase compared to the previous stage. For the first time positively related with the amount of material resources is the provision of personnel (0.365), yet, on a lower level of significance. A significant increase in correlation strength can be diagnosed for the relationship between material resources and the number of strong ties (0.772), closure (0.607), and multiplexity (0.602).

As to *financial resources*, influence of funding partners has basically remained the same with a value of 0.569. The same holds true for the association between the provision of capital and material resources (0.468) and infrastructure (0.474). For the first time, however, on a lower level of significance a positive association of 0.394 between the provision of financial resources and personnel can be observed. The relationship to weak ties of the second phase has disappeared, and the association with the number of strong ties of the first phase turned up again featuring a positive coefficient of 0.581. The association with multiplexity, as observed in the first phase, has re-emerged, too, and now amounts to a strong and highly significant correlation coefficient of 0.532. Closure for the first time positively relates to financial resources with a value of 0.482.

Regarding the *provision of infrastructure* the influence of women in entrepreneurs" networks has suffered in favour of the impact of the number of friends, which has considerably increased from a correlation coefficient of 0.443 in the previous to 0.61 in the present stage. For the first time, also the number of academics and the number of industry partners have a positive, even tough lesser significant influence on the provision of infrastructure. As in the second phase infrastructure is positively correlated to financial resources. Its relationship to material resources increased both in correlation strength and significance as opposed to the previous phase. Again for the first time a positive, but less significant association with the number of ties providing human resources exists. The influence of multiplexity has further increased to 0.598. As in the first phase, also in the last phase the provision of infrastructure seem to be enabled by strong ties and closed networks.

Spearman's Rho		N° of Females - Phase 3	N° of Family Members - Phase 3	N° of Friends - Phase 3	N° of Academics - Phase 3	N° of Industry Partners - Phase 3	N° of Funding Partners - Phase 3	N° of Others - Phase 3	Provision of Resources - Phase 3	Material Resources - Phase 3	Financial Resources - Phase 3	Provision of Infrastructure - Phase 3	Provision of Personnel - Phase 3	Other Resources - Phase 3	N° of Structural Holes - Phase 3	Sparsity - Phase 3	Density - Phase 3	Centrality - Phase 3	Multiplexity - Phase 3	Strong Ties - Phase 3	Closure - Phase 3	Weak Ties - Phase 3
Provision of Resources - Phase 3	Corr. Coeff.	,481**	,045	,557**	,372*	,251	,473**	,160	1,000	,747**	,691**	,721**	,650**	,162	,157	,167	,074	-,007	,674**	,844**	,632**	,269
	Sig.	,004	,799	,001	,030	,153	,005	,365		,000	,000	,000	,000	,359	,376	,345	,691	,969	,000	,000	,000	,125
Material Resources - Phase 3	Corr. Coeff.	,586**	,202	,670**	,276	,405*	,095	,087	,747**	1,000	,468**	,585**	,365*	,220	-,051	-,030	,063	-,025	,602**	,772**	,607**	-,007
	Sig.	,000	,252	,000	,114	,018	,594	,626	,000		,005	,000	,034	,211	,776	,868	,737	,896	,000	,000	,000	,970
Financial Resources - Phase 3	Corr. Coeff.	,270	,007	,194	,201	,040	,569**	,192	,691**	,468**	1,000	,474**	,394*	,124	,052	,040	,000	,064	,532**	,581**	,482**	,203
	Sig.	,123	,967	,272	,254	,824	,000	,277	,000	,005		,005	,021	,486	,771	,824	,998	,738	,001	,000	,004	,251
Provision of Infrastructure - Phase 3	Corr. Coeff.	,364*	,000	,610**	,366*	,351*	,198	-,091	,721**	,585**	,474**	1,000	,419*	,064	,149	,168	,007	,032	,598**	,699**	,517**	,199
	Sig.	,034	,999	,000	,033	,042	,261	,610	,000	,000	,005		,014	,721	,400	,344	,970	,865	,000	,000	,002	,260
Provision of Personnel - Phase 3	Corr. Coeff.	,353*	,250	,361*	,207	,071	,258	-,059	,650**	,365*	,394*	,419*	1,000	-,234	,163	,182	-,064	,120	,295	,474*	,410*	,071
	Sig.	,041	,154	,036	,240	,691	,141	,741	,000	,034	,021	,014		,183	,357	,304	,730	,529	,090	,005	,016	,690
Other Resources - Phase 3	Corr. Coeff.	,097	-,149	,020	-,214	,296	,316	,115	,162	,220	,124	,064	-,234	1,000	-,127	-,126	,047	-,034	,053	,157	,065	,218
	Sig.	,584	,399	,911	,225	,090	,069	,519	,359	,211	,486	,721	,183		,476	,477	,800	,860	,768	,374	,714	,215

**. Correlation is significant at the 0.01 level (2-tailed).
*. Correlation is significant at the 0.05 level (2-tailed).

Table 5-12: Spearman Correlations for Resource Content in Phase 3

Finally, the *provision of personnel* has gained in importance as to the number of significant correlations with other variables. First, as to the possible sources of personnel, they seem to almost equally relate to the number of women (0.353) and friends (0.361) in the network, however, at the lower level of significance. Moreover, ties providing access to personnel are positively associated with contacts supplying material (0.365) and financial (0.394) resources, as well as infrastructure (0.419). The association with the number of strong ties already noticed in the previous phase has further increased, also in significance, up to a correlation coefficient of 0.474. The same applies to the correlation between the provision of personnel and network closure, which climbed up to 0.41.

Summary

Comparing the influence of different types of resources on the total amount of resources, one can observe that the ties featuring the strongest association in the first phase, i.e. those supplying financial resources and human resources, both loose in correlation strength during the formation phase in order to regain in strength when it comes to establishing on the market in the course of the last phase. An even more drastic break-in of associative power with the total amount of resources is suffered by the provision of infrastructure, whose correlation coefficient is not only reduced by half, but also moves to a lower level of significance in the second phase. Yet, in the third stage ties furnishing infrastructure considerably recapture correlative power reaching their maximum and even almost surpassing ties providing material resources that have the strongest association with the total among of resources exchanged at this stage.

In the first phase, financial support is primarily derived from funding partners, and accompanied by the provision of material and infrastructure. Moreover, ties supplying capital seem to be predominantly strong and multiplex, yet to some extent also weak.

In the second phase, the influence of funding partners on the provision of capital decreases. Yet, it is still accompanied by material resources and infrastructure. Most interestingly, at this stage ties delivering financial support are only related to the number of weak ties. The association with the number of strong ties and multiplexity has disappeared. Material resources are provided by friends and also to some part by family members. As mentioned they come with financial support and the provision of infrastructure. Moreover, they feature a strong association with the number of strong ties, closure and multiplexity.

In the last phase, the positive effect of funding partners on the amount of financial resources has recovered a bit, however, without achieving the strong correlation of the first phase. In addition to material resources and infrastructure at this stage capital also seems to be positively related with the provision of personnel. Likewise as in the first stage ties supplying capital are positively associated with the number of strong ties, multiplexity, and, for the first time, also with network closure. The influence of friendship ties on material resources has increased. The positive correlation between the provision of tangible assets and the number of family members has shifted in favour of the number of female and, to a lesser extent, of industry contacts in the network. The positive associations with the provision of capital and infrastructure have been enlarged by the provision of personnel. The relationship with tie strength, closure and multiplexity has reinforced. The provision of infrastructure is primarily related to the number of friends, followed with some distance by the number of women, academics, and industry partners in the network. As already mentioned, infrastructure-related ties are positively associated with the provision of material, financial, and human resources. Furthermore, they correspond to the number of strong ties, multiplexity and network closure.

5.2.2.4 Supportive Content

The last type of content to be analysed in more detail with respect to its correlations with other network parameters are the two forms of support provided. While emotional support refers to different forms of psychological encouragement in the course of the spin-off process, institutional support refers to the transfer of trust and positive reputation, and/or increased visibility and recognition on the market due to a relation with a certain contact.

Phase 1 – Opportunity Recognition

In the first phase the strongest correlation of the two different types of support with the overall amount of *support* can be observed for emotional encouragement with highly significant correlation coefficient of 0.901. Institutional support is also correlated with a coefficient of 0.889.

Regarding the association of *emotional support* to different types of contact, the strongest correlation of 0.494 can be found with the number of females in researchers' networks at this stage, closely followed by the number of friends with a coefficient of 0.491. Moreover, emotional support is also rather strongly associated (0.669) with the number of ties providing institutional support. As to be expected, another strong relation exists to the number of strong ties (0.685), and with some distance to the degree of closure (0.572) and multiplexity (0.536) in the network.

Looking at *institutional support*, it also features a correlation to the number of females and to the number of friends, yet a lower level of significance and to a lesser extent, i.e. 0.415 and 0.363, respectively. However, in contrast to emotional backing, funding partners are significantly associated with institutional support with a correlation coefficient of 0.446. Compared to emotional support the relationship to strong ties and closure is lower, however still rather strong with 0.577 and 0.534. In turn, multiplexity features a very strong correlation of 0.712 with institutional support as opposed to an emotional support coefficient of 0.536.

Spearman's Rho		N° of Females - Phase 1	N° of Family Members - Phase 1	N° of Friends - Phase 1	N° of Academics - Phase 1	N° of Industry Partners - Phase 1	N° of Funding Partners - Phase 1	N° of Others - Phase 1	Support - Phase 1	Emotional Support - Phase 1	Institutional Support - Phase 1	N° of Structural Holes - Phase 1	Sparsity - Phase 1	Density - Phase 1	Centrality - Phase 1	Multiplexity - Phase 1	Strong Ties - Phase 1	Closure - Phase 1	Weak Ties - Phase 1
Support - Phase 1	Corr. Coeff.	,457**	,065	,419*	,123	,265	,473**	,032	1,000	,901**	,889**	-,028	-,033	-,083	,145	,710**	,693**	,593**	,245
	Sig.	,007	,713	,014	,489	,130	,005	,855	.	,000	,000	,874	,855	,657	,443	,000	,000	,000	,163
Emotional Support - Phase 1	Corr. Coeff.	,494**	,262	,491**	,137	,269	,285	,032	,901**	1,000	,669**	-,159	-,172	-,052	,105	,536**	,685**	,572**	,167
	Sig.	,003	,135	,003	,440	,124	,102	,858	,000	.	,000	,370	,331	,780	,581	,001	,000	,000	,345
Institutional Support - Phase 1	Corr. Coeff.	,415*	-,115	,363*	,067	,184	,446**	,029	,889**	,669**	1,000	,045	,045	-,032	,080	,712**	,577**	,534**	,145
	Sig.	,015	,519	,035	,708	,298	,008	,869	,000	,000	.	,799	,802	,864	,675	,000	,000	,001	,413

*. Correlation is significant at the 0.05 level (2-tailed).

**. Correlation is significant at the 0.01 level (2-tailed).

Table 5-13: Spearman Correlations for Supportive Content in Phase 1

Phase 2 – Formation

In the second phase institutional support gains increased importance as regards is influence on the overall amount of *support* entrepreneurs' obtained with a correlation coefficient of 0.873. Emotional support's correlation (0.702) with support as such has dropped more considerably than emotional support and ranks now behind the latter as regards its contribution to the overall amount of support as compared to the previous phase.

This increased importance is also reflected by the correlations of *emotional support* to different types of partner roles. The most considerable increase in association of the amount of emotional support can be observed with the number of friends from 0.491 to 0.633, and for the number of females from 0.494 to 0.577, as opposed to the previous phase. Family members for the first time also feature a weak correlation of 0.424 at the lower level of significance. The association between emotional and institutional support has weakened both, in strength (0.387) and in significance, which can be interpreted as a result of the reduced significance of intuitional support expressed in terms of influence on network size in the second phase. Moreover, emotional ties are not so strongly associated with strong ties and network closure anymore, the respective correlation coefficients having dropped to 0.514 and 0.533 as opposed to 0.685 and 0.572 of the previous phase. The correlation with multiplexity, in turn, has increased to a value of 0.558.

As mentioned, *institutional support* seems to have lost in its relevance in the second phase; this is reflected by the lower number of significant correlation coefficients at this stage. As such, the significant associations between intuitional support and number of females and funding partners as observed in the previous phase have disappeared. Correlation with the number of friends is slightly strengthened to 0.424, however still at the lower level of significance. The interrelations with the number of strong ties, network closure and multiplexity are also weakened, and at this stage amount to 0.457, 0.461 and 0.557, respectively.

Spearman's Rho		N° of Females - Phase 2	N° of Family Members - Phase 2	N° of Friends - Phase 2	N° of Academics - Phase 2	N° of Industry Partners - Phase 2	N° of Funding Partners - Phase 2	N° of Others - Phase 2	Support - Phase 2	Emotional Support - Phase 2	Institutional Support - Phase 2	N° of Structural Holes - Phase 2	Sparsity - Phase 2	Density - Phase 2	Centrality - Phase 2	Multiplexity - Phase 2	Strong Ties - Phase 2	Closure - Phase 2	Weak Ties - Phase 2
Support - Phase 2	Corr. Coeff.	,492**	,226	,621**	,077	,122	,108	-,016	1,000	,702**	,873**	,093	,113	,017	,047	,626**	,517**	,496**	,171
	Sig.	,003	,199	,000	,667	,490	,543	,928	.	,000	,000	,602	,524	,927	,799	,000	,002	,003	,333
Emotional Support - Phase 2	Corr. Coeff.	,577**	,424*	,633**	,032	-,056	,045	,219	,702**	1,000	,387*	-,091	-,129	,112	-,067	,558**	,514**	,533**	,230
	Sig.	,000	,012	,000	,858	,753	,803	,214	,000	.	,024	,610	,467	,536	,717	,001	,002	,001	,191
Institutional Support - Phase 2	Corr. Coeff.	,297	,039	,424*	,125	,148	,015	-,131	,873**	,387*	1,000	,061	,114	,030	,030	,557**	,457**	,461**	,021
	Sig.	,088	,829	,012	,482	,404	,935	,461	,000	,024	.	,732	,519	,867	,869	,001	,007	,006	,904

*. Correlation is significant at the 0.05 level (2-tailed).

**. Correlation is significant at the 0.01 level (2-tailed).

Table 5-14: Spearman Correlations for Supportive Content in Phase 2

Phase 3 – Establishment

In the last phase the association between the different types of support and the overall amount of *support* exchanged in the personal networks has taken different directions. While emotional support has encountered an increase in correlation from 0.702 in the previous phase to a current coefficient of 0.829, institutional support has lost part of the significance gained in the second phase and now again ranks behind emotional support with a correlation value of 0.871 similar to the first stage.

As to the sources of *emotional support*, the friends have lost their influence on the amount of emotional backing, now featuring a correlation coefficient of only 0.356 at the lower level of significance as opposed to a highly significant coefficient of 0.633 in the previous phase. For the first time the number of funding partners positively correlates (0.415) with the amount of emotional support, however only at the lower level of significance. Furthermore, both types of support are still interrelated with a correlation coefficient of 0.596, which could not reach the high association of the first phase, but has regained in strength and in addition also in significance compared to the previous phase. After a slight drop in the second phase, both, the correlation with strong ties and with network closure, have picked up, now featuring coefficients of 0.675 and 0.563, respectively. The association between emotional support and multiplexity of the entrepreneurial networks has continuously increased in strength to a current value of 0.679. In the last stage for the first time, the number of weak ties and ties providing emotional support are positively correlated (0.393), however, at the lower level of significance.

Coming to *institutional support*, it is no longer correlated with the number of friends, but again, as in the first phase, with the number of funding partners, yet, at a reduced strength of 0.364 and the lower level of significance. For the first time, also the number of industry partners positively influences the amount of institutional support obtained. Network closure no longer correlates with institutional support as in the previous two phases. The association with the number of strong ties and network closure has further decreased to values of 0.452 and 0.474, respectively.

Spearman's Rho		N° of Females - Phase 3	N° of Family Members - Phase 3	N° of Friends - Phase 3	N° of Academics - Phase 3	N° of Industry Partners - Phase 3	N° of Funding Partners - Phase 3	N° of Others - Phase 3	Support - Phase 3	Emotional Support - Phase 3	Institutional Support - Phase 3	N° of Structural Holes - Phase 3	Sparsity - Phase 3	Density - Phase 3	Centrality - Phase 3	Multiplexity - Phase 3	Strong Ties - Phase 3	Closure - Phase 3	Weak Ties - Phase 3
Support - Phase 3	Corr. Coeff.	,277	-,032	,311	,139	,348*	,356*	,241	1,000	,829**	,871**	,162	,169	-,092	,135	,618**	,567**	,420*	,358*
	Sig.	,113	,859	,073	,432	,044	,039	,169	.	,000	,000	,360	,338	,624	,477	,000	,000	,013	,038
Emotional Support - Phase 3	Corr. Coeff.	,319	,169	,356*	,235	,111	,415*	,208	,829**	1,000	,596**	,123	,124	-,208	,226	,679**	,675**	,563**	,393*
	Sig.	,066	,339	,039	,180	,530	,015	,238	,000	.	,000	,487	,485	,262	,230	,000	,000	,001	,022
Institutional Support - Phase 3	Corr. Coeff.	,253	-,237	,195	,132	,358*	,364*	,330	,871**	,596**	1,000	,208	,220	,025	,043	,474**	,452**	,289	,332
	Sig.	,149	,178	,270	,458	,037	,035	,056	,000	,000	.	,238	,210	,896	,823	,005	,007	,097	,055

**. Correlation is significant at the 0.01 level (2-tailed).
*. Correlation is significant at the 0.05 level (2-tailed).

Table 5-15: Spearman Correlations for Supportive Content in Phase 3

Summary

In general, the effect of both types of support on *support* in general is the largest in the first phase of the spin-off process. Following further trends can be observed. While the impact of the number of ties providing emotional support on the overall amount of support looses a bit in strength in the second phase, however regaining it in the last stage of the process, institutional support features a continuous slight reduction in influence over time. Yet, in principle both types of support exhibit the same magnitude of correlation without any large differences.

Emotional support is primarily provided by female contacts and friends. Both types increase in influence from the first to the second phase. In the last phase, however, women loose their role provider of emotional support, which to some extent is taken over by funding partners, who feature a first significant correlation. Friends influence remains, but drops considerably as opposed to the previous phases. While the association between emotional support and multiplexity continuously increases, the relationship of emotional backing and tie strength and network closure suffers a bit in the second phase without regaining its original strength in the last phase.

Institutional support is provided by changing types of contact over time. In the first phase, it is derived from women, friends and funding partners; in the second phase only by friends; and in the last phase by industry and funding partners. As to the correlation with multiplexity, number of strong ties and network closure, it is the highest in the first phase, and continuously declines over time with network closure loosing its positive relation to instructional support at all in the last phase.

In general, a positive association between emotional and institutional support can be observed for all three phases, which collapses a bit in the second phase without regaining its original strength in the last phase.

5.2.3 Testing for Interrelations

In the following chapter the indicative coherences derived from the correlation matrices of Chapter 5.2.2 are subjected to causal-analytical examinations by means of regression analysis. However, it has to be noted that the relations analysed in the course of the subsequent multiple regression models cannot be qualified as causal relations in the narrower sense as defined by Blalock (1985). As no definite time relations exist, they rather have to be regarded as indications of common occurrence in terms of associetation.

As mentioned, regression analysis with not normally distributed variables, as the case in this thesis, has to be considered with caution, since infringements of the assumption of normal distribution bear the risk of reduced standard errors and thus of allegedly significant coefficients. The prepended distribution-free correlation analysis after Spearman allows validly detecting significant relations between non-normally distributed variables. In the following, only correlations regarded as significant after Spearman are subjected to regression analysis. Following the underlying research model again results of analysis are structured along the three assumed phases of the spin-off process: opportunity recognition, formation, and establishment.

Multiple regression analysis with stepwise inclusion of independent variables is applied. In the course of this approach only those explanatory variables are kept in the regression equation that have proven to be of statistical relevance. To this end variables with the largest respective partial correlation coefficient with the dependent variable are consecutively included in the regression equation. Criterion for including or eliminating an independent variable is the significance level of the partial correlation coefficient. In the subsequent regression analyses variables are kept in the equation, if the significance level of their partial correlation coefficient is lower than 0.05 and eliminated if it is larger than 0.1. Respective iterations are terminated when no more explanatory variables can be considered for the analysis. As such the order of inclusion provides an indication of the statistical importance of the variables.

Following the hypotheses established in Chapter 5.1.2, analysis first concentrates on the dominance of certain contact and content types as well as structural characteristics on the size of the overall personal network in the individual phase. This analysis is followed by a detailed look on the dependencies of certain relational contents on different network parameters. Only those relational contents are further taken into account that turned out to be of relevance for network size in the first instance. This approach accounts for the hypotheses relating to the different network characteristics of the content-related network types: information, resource, and support network.

5.2.3.1 Opportunity Recognition

As mentioned, analysis starts with network size as dependent variable. Only those variables are considered as predictors that have proven to be significant in the course of the previous correlation analysis according to Spearman. In the following, those relational contents that turned out to be of significant influence on network size and thus of dominance in the first phase are subjected to further regressions as dependent variables.

5.2.3.1.1 Network Size

In accordance with the correlation analysis of Chapter 5.2.2 only those variables are considered as predictors of network size in the course of the following regression analyses that featured significant correlations with the dependent variable. To model the influence of the different types of network parameters independently, individual regression equations are set up accounting for the totality of significant alter-specific, content-specific, and structure-specific variables of the previous correlation analysis.

Table 5-16 summarises the results of the first regression model that captures the influence of the amount of different contact types on the size of entrepreneurs' personal networks as dependent variable in the first phase. In accordance with the correlation analysis after Spearman the number of friends, academics, industry partners, and funding partners are subjected to regression analysis as explanatory variables for network size.

To come to the final regression equation, in the first step the SPSS chooses the variable with the highest correlation coefficient with the dependent variable; in this case the *Number of Funding Partners*. At each step the beta value (*Beta In*) variables would obtain in case of a possible inclusion in the next step is given for all variables not yet considered. In the neighbouring column the partial correlation coefficients (*Partial*) applied to the selection are provided. At the end of all iterations all contact types featuring significant correlations after Spearman have been included in the regression model. This means that the number of funding partners, the number of friends, the number of industry partners, and the number of academics in the network have a significant influence on network size.

The *Model Summary* provides an overview of the changes in the coefficient of determination resulting from the inclusion of an additional predictor in the equation. In general the coefficient of determination or *R Square* expresses the proportion of explained deviations of all deviations of the observed values from the sample mean, and thus serves as indicator of the explanatory power of the regression model as regards the prediction of the dependent variable. The *Adjusted R Square* accounts for the fact that each additional independent variable automatically leads to a larger R square independent from its actual explanatory relevance. As a result, it reduces the simple coefficient of determination by a correction parameter that increases with increasing number of regressors and with decreasing number of degrees of freedom. The *Adjusted R Square* of 0.622 of the final regression model with all four contact types implies that 62% of the variability in network size is explained by the four predictors.

ANOVA or the analysis of variance indicates the estimation quality of the entire regression function for each model with a new additional predictor. *df* denotes the degrees of freedom, *Sum of Squares* reports the variance explained by the assumed regression model, while *Mean Square* is the quotient of variance and degrees of freedom. The F-test renders the statistical significance of the model. A significance value (Sig.) smaller than 0.05 denotes that the probability of error, i.e. the probability that a relation between the regressors and the dependent variable is detected that is only random, is neglectable small, and that the assumed relationship can be deemed as significant.

Model Summary

Model	R	R Square	Adjusted R Square	Std. Error of the Estimate
1	,505[a]	,255	,232	1,810
2	,679[b]	,461	,426	1,564
3	,762[c]	,581	,539	1,401
4	,817[d]	,668	,622	1,270

a. Predictors: (Constant), N° of Funding Partners - Phase 1
b. Predictors: (Constant), N° of Funding Partners - Phase 1, N° of Friends - Phase 1
c. Predictors: (Constant), N° of Funding Partners - Phase 1, N° of Friends - Phase 1, N° of Industry Partners - Phase 1
d. Predictors: (Constant), N° of Funding Partners - Phase 1, N° of Friends - Phase 1, N° of Industry Partners - Phase 1, N° of Academics - Phase 1

ANOVA[e]

Model		Sum of Squares	df	Mean Square	F	Sig.
1	Regression	35,883	1	35,883	10,951	,002[a]
	Residual	104,852	32	3,277		
	Total	140,735	33			
2	Regression	64,887	2	32,443	13,260	,000[b]
	Residual	75,849	31	2,447		
	Total	140,735	33			
3	Regression	81,818	3	27,273	13,887	,000[c]
	Residual	58,917	30	1,964		
	Total	140,735	33			
4	Regression	93,961	4	23,490	14,564	,000[d]
	Residual	46,774	29	1,613		
	Total	140,735	33			

a. Predictors: (Constant), N° of Funding Partners - Phase 1
b. Predictors: (Constant), N° of Funding Partners - Phase 1, N° of Friends - Phase 1
c. Predictors: (Constant), N° of Funding Partners - Phase 1, N° of Friends - Phase 1, N° of Industry Partners - Phase 1
d. Predictors: (Constant), N° of Funding Partners - Phase 1, N° of Friends - Phase 1, N° of Industry Partners - Phase 1, N° of Academics - Phase 1
e. Dependent Variable: Network Size in Phase 1

Coefficients[a]

Model		Unstandardized Coefficients		Standardized Coefficients	t	Sig.
		B	Std. Error	Beta		
1	(Constant)	2,913	,433		6,730	,000
	N° of Funding Partners - Phase 1	1,061	,321	,505	3,309	,002
2	(Constant)	2,072	,447		4,636	,000
	N° of Funding Partners - Phase 1	1,029	,277	,490	3,712	,001
	N° of Friends - Phase 1	,723	,210	,454	3,443	,002
3	(Constant)	1,779	,413		4,312	,000
	N° of Funding Partners - Phase 1	,937	,250	,446	3,742	,001
	N° of Friends - Phase 1	,707	,188	,444	3,758	,001
	N° of Industry Partners - Phase 1	,542	,185	,350	2,936	,006

4	(Constant)	1,348	,406		3,323	,002
	N° of Funding Partners - Phase 1	,991	,228	,472	4,354	,000
	N° of Friends - Phase 1	,609	,174	,383	3,495	,002
	N° of Industry Partners - Phase 1	,568	,168	,367	3,392	,002
	N° of Academics - Phase 1	,542	,198	,302	2,744	,010

a. Dependent Variable: Network Size in Phase 1

Excluded Variables[d]

Model		Beta In	t	Sig.	Partial Correlation	Collinearity Statistics Tolerance
1	N° of Friends - Phase 1	,454[a]	3,443	,002	,526	,999
	N° of Academics - Phase 1	,361[a]	2,556	,016	,417	,992
	N° of Industry Partners - Phase 1	,363[a]	2,552	,016	,417	,984
2	N° of Academics - Phase 1	,280[b]	2,198	,036	,372	,951
	N° of Industry Partners - Phase 1	,350[b]	2,936	,006	,472	,983
3	N° of Academics - Phase 1	,302[c]	2,744	,010	,454	,948

a. Predictors in the Model: (Constant), N° of Funding Partners - Phase 1
b. Predictors in the Model: (Constant), N° of Funding Partners - Phase 1, N° of Friends - Phase 1
c. Predictors in the Model: (Constant), N° of Funding Partners - Phase 1, N° of Friends - Phase 1, N° of Industry Partners - Phase 1
d. Dependent Variable: Network Size in Phase 1

Table 5-16: Multiple Regression Analysis - Phase 1: Network Size with Functional Roles

Coefficients contains the results of the regression analysis in the form of the regression coefficients (*B*) of the regression equation for all variables finally included in the model. As the values indicate the *Number of Funding Partners* features the largest contribution to the prediction of the dependent variable *Network Size*. An additional contact in this domain increases the size of the personal network by almost one tie (0.991). *Number of Funding Partners* is followed by the *Number of Friends* with a regression coefficient of 0.609, and the *Number of Industry Partners* with a coefficient of 0.568 and the *Number of Academics* with a value of 0.542. In the last case the value of 0.542 implies that an additional contact to an academic colleague increase the size of entrepreneurs' personal networks in average by a half tie. While *B* depends on the dimension of the underlying variables and is thus not really comparable when it comes to the relative importance of the respective predictors, the standardised regression coefficient *Beta* reflects the dimensionless regression coefficients. The standardised regression constant always assumes the value 0. The standardised and consequently dimensionless partial regression coefficients render an important indication as to the relevance of the individual independent variables on estimating the dependent variable. The larger the value *Beta* the more important is the respective predictor. In the current model of Table 5-16 due to the same dimension, i.e. number of ties, the relative importance of the types of contacts for the size of ego's personal network does not really change. Still *Number of Funding Partners* features the highest influence with a *Beta* value of 0.472 followed by the *Number of Friends* with a *Beta* of 0.383, *Number of Industry Partners* (*Beta* 0.367) and *Number of Academics* (*Beta* 0.302).

The next model as depicted in Table 5-17 captures the influence of the different types of content exchanged between ego and their alters on the overall size of ego's personal network in the first phase of the spin-off process. Again only those content types were subjected to analysis that already proved to feature significant interrelations in the course of the previous correlation analysis of Chapter 5.2.2. This applies to *Informational Content* and *Support*.

The results of the correlation analysis are supported by regression analysis. Both variables, *Informational Content* and *Support* exhibit significant regression coefficients that explain more than the half of the variance in the dependent variable *Network Size (Adjusted R Square* of 0.698). The most important influence on network size, however, is realised by *Informational Content* with a standardised regression coefficient *Beta* of 0.707. Opposed to that alters offering support only display a limited effect on the size of the personal network with a rather small *Beta* of 0.259.

Model Summary

Model	R	R Square	Adjusted R Square	Std. Error of the Estimate
1	,813[a]	,660	,650	1,222
2	,846[b]	,717	,698	1,134

a. Predictors: (Constant), Informational Content - Phase 1
b. Predictors: (Constant), Informational Content - Phase 1, Support - Phase 1

ANOVA[c]

Model		Sum of Squares	df	Mean Square	F	Sig.
1	Regression	92,950	1	92,950	62,245	,000[a]
	Residual	47,785	32	1,493		
	Total	140,735	33			
2	Regression	100,838	2	50,419	39,175	,000[b]
	Residual	39,898	31	1,287		
	Total	140,735	33			

a. Predictors: (Constant), Informational Content - Phase 1
b. Predictors: (Constant), Informational Content - Phase 1, Support - Phase 1
c. Dependent Variable: Network Size in Phase 1

Coefficients[a]

Model		Unstandardized Coefficients		Standardized Coefficients	t	Sig.
		B	Std. Error	Beta		
1	(Constant)	1,329	,389		3,419	,002
	Informational Content - Phase 1	,869	,110	,813	7,890	,000
2	(Constant)	1,096	,373		2,937	,006
	Informational Content - Phase 1	,756	,112	,707	6,741	,000
	Support - Phase 1	,273	,110	,259	2,476	,019

a. Dependent Variable: Network Size in Phase 1

Excluded Variables[b]

Model		Beta In	t	Sig.	Partial Correlation	Collinearity Statistics Tolerance
1	Support - Phase 1	,259[a]	2,476	,019	,406	,833

a. Predictors in the Model: (Constant), Informational Content - Phase 1
b. Dependent Variable: Network Size in Phase 1

Table 5-17: Multiple Regression Analysis - Phase 1: Network Size with Relational Content

5.2.3.1.2 Informational Content

Following the overall assumption of the research model structuring around three main types of networks, i.e. information, resource and support network, those content types that were found to be significantly related to network size in the first phase are now looked at in more detail as to the influence of the other network parameters on their composition. Respective results of these regression models are summarised in the subsequent tables.

As revealed in the previous section information is one of the content types of significant influence on network size in the first phase. Consequently it is indicated to have a closer look as to the dominant sources of information and the quality of the associated relations. To this end, those alter types that according to the results of the correlation analysis of Chapter 5.2.2 featured significant interrelations with the provision of information are included as potential predictors in a regression analysis with Informational Content as dependent variable (see Table 5-18).

Model Summary

Model	R	R Square	Adjusted R Square	Std. Error of the Estimate
1	,433[a]	,187	,162	1,76731
2	,624[b]	,389	,350	1,55692
3	,708[c]	,501	,451	1,42991
4	,769[d]	,591	,535	1,31635

a. Predictors: (Constant), N° of Industry Partners - Phase 1
b. Predictors: (Constant), N° of Industry Partners - Phase 1, N° of Academics - Phase 1
c. Predictors: (Constant), N° of Industry Partners - Phase 1, N° of Academics - Phase 1, N° of Funding Partners - Phase 1
d. Predictors: (Constant), N° of Industry Partners - Phase 1, N° of Academics - Phase 1, N° of Funding Partners - Phase 1, N° of Friends - Phase 1

ANOVA[e]

Model		Sum of Squares	df	Mean Square	F	Sig.
1	Regression	23,022	1	23,022	7,371	,011[a]
	Residual	99,949	32	3,123		
	Total	122,971	33			
2	Regression	47,827	2	23,913	9,865	,000[b]
	Residual	75,144	31	2,424		
	Total	122,971	33			
3	Regression	61,631	3	20,544	10,048	,000[c]
	Residual	61,339	30	2,045		
	Total	122,971	33			
4	Regression	72,720	4	18,180	10,492	,000[d]
	Residual	50,250	29	1,733		
	Total	122,971	33			

a. Predictors: (Constant), N° of Industry Partners - Phase 1
b. Predictors: (Constant), N° of Industry Partners - Phase 1, N° of Academics - Phase 1
c. Predictors: (Constant), N° of Industry Partners - Phase 1, N° of Academics - Phase 1, N° of Funding Partners - Phase 1
d. Predictors: (Constant), N° of Industry Partners - Phase 1, N° of Academics - Phase 1, N° of Funding Partners - Phase 1, N° of Friends - Phase 1
e. Dependent Variable: Informational Content - Phase 1

Coefficients[a]

Model		Unstandardized Coefficients		Standardized Coefficients	t	Sig.
		B	Std. Error	Beta		
1	(Constant)	2,510	,347		7,225	,000
	N° of Industry Partners - Phase 1	,627	,231	,433	2,715	,011
2	(Constant)	1,813	,376		4,829	,000
	N° of Industry Partners - Phase 1	,666	,204	,460	3,270	,003
	N° of Academics - Phase 1	,756	,236	,450	3,199	,003
3	(Constant)	1,191	,420		2,837	,008
	N° of Industry Partners - Phase 1	,606	,189	,419	3,217	,003
	N° of Academics - Phase 1	,802	,218	,477	3,680	,001
	N° of Funding Partners - Phase 1	,665	,256	,339	2,598	,014
4	(Constant)	,773	,420		1,839	,076
	N° of Industry Partners - Phase 1	,589	,174	,407	3,391	,002
	N° of Academics - Phase 1	,695	,205	,414	3,393	,002
	N° of Funding Partners - Phase 1	,637	,236	,325	2,701	,011
	N° of Friends - Phase 1	,457	,181	,307	2,530	,017

a. Dependent Variable: Informational Content - Phase 1

Excluded Variables[d]

Model		Beta In	t	Sig.	Partial Correlation	Collinearity Statistics Tolerance
1	N° of Friends - Phase 1	,400[a]	2,756	,010	,444	,999
	N° of Academics - Phase 1	,450[a]	3,199	,003	,498	,996
	N° of Funding Partners - Phase 1	,300[a]	1,950	,060	,331	,984
2	N° of Friends - Phase 1	,323[b]	2,418	,022	,404	,958
	N° of Funding Partners - Phase 1	,339[b]	2,598	,014	,429	,978
3	N° of Friends - Phase 1	,307[c]	2,530	,017	,425	,956

a. Predictors in the Model: (Constant), N° of Industry Partners - Phase 1
b. Predictors in the Model: (Constant), N° of Industry Partners - Phase 1, N° of Academics - Phase 1
c. Predictors in the Model: (Constant), N° of Industry Partners - Phase 1, N° of Academics - Phase 1, N° of Funding Partners - Phase 1
d. Dependent Variable: Informational Content - Phase 1

Table 5-18: Multiple Regression Analysis Phase 1: Informational Content with Functional Roles

As can be learned from Table 5-18 all variables on the functional role of alters included in the analysis are kept in the equation model. This means that in the first phase industry and funding partners, academics, and friends can be deemed as important sources of information. Comparing their relevance by looking at the standardised regression coefficients, it can be stated that the most important source of information are academic colleagues (*Beta* 0.414) closely followed by industry partners (*Beta* 0.407) and with some distance by funding partners (*Beta* 0.325) and friends (*Beta* 0.307).

Regarding the different contents of the information exchanged, the regression model depicted in Table 5-19 reveals that the major part can be attributed to *Other Information*, which, as referenced above, primarily relates to information of technical and scientific nature. It is followed by business-related topics with a regression coefficient of 0.627 or *Beta* of 0.425 for *Economic Information*. Far behind, rank *Legal Information* (Beta 0.247) and *Information about Potential Employees*, i.e. *Beta* of 0.189.

Model Summary

Model	R	R Square	Adjusted R Square	Std. Error of the Estimate
1	,625[a]	,390	,371	1,53079
2	,882[b]	,778	,764	,93753
3	,912[c]	,831	,814	,83240
4	,928[d]	,862	,843	,76560

a. Predictors: (Constant), Other Information - Phase 1
b. Predictors: (Constant), Other Information - Phase 1, Economic Information - Phase 1
c. Predictors: (Constant), Other Information - Phase 1, Economic Information - Phase 1, Legal Information - Phase 1
d. Predictors: (Constant), Other Information - Phase 1, Economic Information - Phase 1, Legal Information - Phase 1, Information about Potential Employees - Phase 1

ANOVA[e]

Model		Sum of Squares	df	Mean Square	F	Sig.
1	Regression	47,984	1	47,984	20,477	,000[a]
	Residual	74,986	32	2,343		
	Total	122,971	33			
2	Regression	95,723	2	47,861	54,452	,000[b]
	Residual	27,248	31	,879		
	Total	122,971	33			
3	Regression	102,184	3	34,061	49,158	,000[c]
	Residual	20,787	30	,693		
	Total	122,971	33			
4	Regression	105,972	4	26,493	45,198	,000[d]
	Residual	16,998	29	,586		
	Total	122,971	33			

a. Predictors: (Constant), Other Information - Phase 1
b. Predictors: (Constant), Other Information - Phase 1, Economic Information - Phase 1
c. Predictors: (Constant), Other Information - Phase 1, Economic Information - Phase 1, Legal Information - Phase 1
d. Predictors: (Constant), Other Information - Phase 1, Economic Information - Phase 1, Legal Information - Phase 1, Information about Potential Employees - Phase 1
e. Dependent Variable: Informational Content - Phase 1

Coefficients[a]

Model		Unstandardized Coefficients		Standardized Coefficients	t	Sig.
		B	Std. Error	Beta		
1	(Constant)	1,987	,341		5,830	,000
	Other Information - Phase 1	,904	,200	,625	4,525	,000
2	(Constant)	,535	,287		1,865	,072
	Other Information - Phase 1	,944	,122	,653	7,712	,000
	Economic Information - Phase 1	,920	,125	,624	7,370	,000
3	(Constant)	,390	,259		1,506	,143
	Other Information - Phase 1	,905	,109	,625	8,263	,000
	Economic Information - Phase 1	,702	,132	,476	5,316	,000
	Legal Information - Phase 1	,555	,182	,274	3,054	,005
4	(Constant)	,392	,238		1,643	,111
	Other Information - Phase 1	,912	,101	,631	9,056	,000
	Economic Information - Phase 1	,627	,125	,425	5,021	,000
	Legal Information - Phase 1	,500	,169	,247	2,968	,006
	Information about Potential Employees - Phase 1	,589	,232	,189	2,542	,017

a. Dependent Variable: Informational Content - Phase 1

Excluded Variables[e]

Model		Beta In	t	Sig.	Partial Correlation	Collinearity Statistics Tolerance
1	Legal Information - Phase 1	,532[a]	5,156	,000	,679	,994
	Economic Information - Phase 1	,624[a]	7,370	,000	,798	,998
	Information about Potential Customers - Phase 1	,390[a]	3,199	,003	,498	,997
	Information about Potential Employees - Phase 1	,406[a]	3,388	,002	,520	,999
2	Legal Information - Phase 1	,274[b]	3,054	,005	,487	,702
	Information about Potential Customers - Phase 1	,040[b]	,378	,708	,069	,659
	Information about Potential Employees - Phase 1	,217[b]	2,624	,014	,432	,880
3	Information about Potential Customers - Phase 1	,061[c]	,653	,519	,120	,655
	Information about Potential Employees - Phase 1	,189[c]	2,542	,017	,427	,866
4	Information about Potential Customers - Phase 1	-,026[d]	-,276	,785	-,052	,559

a. Predictors in the Model: (Constant), Other Information - Phase 1
b. Predictors in the Model: (Constant), Other Information - Phase 1, Economic Information - Phase 1
c. Predictors in the Model: (Constant), Other Information - Phase 1, Economic Information - Phase 1, Legal Information - Phase 1
d. Predictors in the Model: (Constant), Other Information - Phase 1, Economic Information - Phase 1, Legal Information - Phase 1, Information about Potential Employees - Phase 1
e. Dependent Variable: Informational Content - Phase 1

Table 5-19: Multiple Regression Analysis Phase 1: Informational Content with Information Types

Looking at the quality of informational ties, weak ties and strong ties, both significantly correlated with informational content as to be inferred from the correlation analysis of Chapter 5.2.2, keep their influence also according to the subsequent regression analysis (see Table 5-20). While both are positively related to *Informational Content*, *Weak Ties* feature a larger influence on the provision of information according to its larger Beta value of 0.619 as opposed to 0.469 for *Strong Ties*.

Model Summary

Model	R	R Square	Adjusted R Square	Std. Error of the Estimate
1	,647[a]	,419	,400	1,49465
2	,799[b]	,638	,615	1,19829

a. Predictors: (Constant), Weak Ties - Phase 1
b. Predictors: (Constant), Weak Ties - Phase 1, Strong Ties - Phase 1

ANOVA[c]

Model		Sum of Squares	df	Mean Square	F	Sig.
1	Regression	51,483	1	51,483	23,045	,000[a]
	Residual	71,488	32	2,234		
	Total	122,971	33			
2	Regression	78,457	2	39,229	27,320	,000[b]
	Residual	44,513	31	1,436		
	Total	122,971	33			

a. Predictors: (Constant), Weak Ties - Phase 1
b. Predictors: (Constant), Weak Ties - Phase 1, Strong Ties - Phase 1
c. Dependent Variable: Informational Content - Phase 1

Coefficients[a]

Model		Unstandardized Coefficients		Standardized Coefficients	t	Sig.
		B	Std. Error	Beta		
1	(Constant)	1,827	,350		5,219	,000
	Weak Ties - Phase 1	,763	,159	,647	4,801	,000
2	(Constant)	,897	,353		2,541	,016
	Weak Ties - Phase 1	,730	,128	,619	5,721	,000
	Strong Ties - Phase 1	,723	,167	,469	4,334	,000

a. Dependent Variable: Informational Content - Phase 1

Excluded Variables[b]

Model		Beta In	t	Sig.	Partial Correlation	Collinearity Statistics
						Tolerance
1	Strong Ties - Phase 1	,469[a]	4,334	,000	,614	,997

a. Predictors in the Model: (Constant), Weak Ties - Phase 1
b. Dependent Variable: Informational Content - Phase 1

Table 5-20: Multiple Regression Analysis Phase 1: Informational Content with Structural Parameters

In the following the sources of different types of information are analysed in more detail. To this end those informational contents identified as relevant for the first phase of the spin-off process according to the results of the regression analysis with different types of information as predictors of the total amount of information (see Table 5-19), are now as dependent variables subjected to regression analysis with different functional roles of alters. In line with the usual procedure, only those relational role variables are included in the analysis that have already been proven as of possible significant relevance in the course of the correlation analysis of Chapter 5.2.2.2.

As Table 5-19 reveals the first type of information significantly contributing to the overall amount of information transferred during the opportunity recognition phase of the start-up process is economic information. In line with the results of the correlation analysis after Spearman industry contacts and funding partners are potential relevant sources of economic know-how at this stage. Yet, they are subjected as possible predictors to multivariate regression analysis with economic information as depicted in the subsequent summary of the regression model of Table 5-21. Yet, in the end only the number of industry partners is kept in the regression equation as significant explanatory variable and thus source of economic know-how.

Model Summary

Model	R	R Square	Adjusted R Square	Std. Error of the Estimate
1	,569^a	,324	,303	1,09197

a. Predictors: (Constant), N° of Industry Partners - Phase 1

ANOVA^b

Model		Sum of Squares	df	Mean Square	F	Sig.
1	Regression	18,314	1	18,314	15,359	,000^a
	Residual	38,157	32	1,192		
	Total	56,471	33			

a. Predictors: (Constant), N° of Industry Partners - Phase 1
b. Dependent Variable: Economic Information - Phase 1

Coefficients^a

Model		Unstandardized Coefficients		Standardized Coefficients	t	Sig.
		B	Std. Error	Beta		
1	(Constant)	1,118	,215		5,211	,000
	N° of Industry Partners - Phase 1	,559	,143	,569	3,919	,000

a. Dependent Variable: Economic Information - Phase 1

Excluded Variables^b

Model		Beta In	t	Sig.	Partial Correlation	Collinearity Statistics
						Tolerance
1	N° of Funding Partners - Phase 1	,168^a	1,150	,259	,202	,984

a. Predictors in the Model: (Constant), N° of Industry Partners - Phase 1
b. Dependent Variable: Economic Information - Phase 1

Table 5-21: Multiple Regression Analysis Phase 1: Sources of Economic Information

The next form of informational content identified as relevant contributor to information at large, refers to information about potential employees. In this context, only network partners coming from business interactions featured significant associations according to the correlation analysis of Chapter 5.2.2.2. When subjected to regression analysis their influence on *Information about Potential Customers* is confirmed as presented in Table 5-22.

Model Summary

Model	R	R Square	Adjusted R Square	Std. Error of the Estimate
1	,492[a]	,242	,218	,54665

a. Predictors: (Constant), N° of Industry Partners - Phase 1

ANOVA[b]

Model		Sum of Squares	df	Mean Square	F	Sig.
1	Regression	3,055	1	3,055	10,224	,003[a]
	Residual	9,562	32	,299		
	Total	12,618	33			

a. Predictors: (Constant), N° of Industry Partners - Phase 1
b. Dependent Variable: Information about Potential Employees - Phase 1

Coefficients[a]

Model		Unstandardized Coefficients		Standardized Coefficients	t	Sig.
		B	Std. Error	Beta		
1	(Constant)	,097	,107		,901	,374
	N° of Industry Partners - Phase 1	,228	,071	,492	3,197	,003

a. Dependent Variable: Information about Potential Employees - Phase 1

Table 5-22: Regression Analysis Phase 1: Sources of Information about Potential Employees

The last type of information primarily transferred in the first phase relates to *Other Information*. According to the correlation analysis after Spearman this is significantly associated with the number of academic contacts in entrepreneur's personal network. This discovery is confirmed by the results of regression analysis of Table 5-23.

Model Summary

Model	R	R Square	Adjusted R Square	Std. Error of the Estimate
1	,640[a]	,409	,391	1,04117

a. Predictors: (Constant), N° of Academics - Phase 1

ANOVA[b]

Model		Sum of Squares	df	Mean Square	F	Sig.
1	Regression	24,046	1	24,046	22,182	,000[a]
	Residual	34,689	32	1,084		
	Total	58,735	33			

a. Predictors: (Constant), N° of Academics - Phase 1
b. Dependent Variable: Other Information - Phase 1

Coefficients[a]

Model		Unstandardized Coefficients		Standardized Coefficients	t	Sig.
		B	Std. Error	Beta		
1	(Constant)	,432	,226		1,910	,065
	N° of Academics - Phase 1	,743	,158	,640	4,710	,000

a. Dependent Variable: Other Information - Phase 1

Table 5-23: Regression Analysis Phase 1: Sources of Other Information

5.2.3.1.3 Resources

According to the correlation analysis of Chapter 5.2.2, when looking at the possible suppliers of resources, the types of alters with significant correlation coefficients to be subjected to regression analysis relate to the *Number of Funding Partners* and *Others*. Yet, as Table 5-24 illustrates both variables are analysed with respect to their explanatory power on the amount of resources provided at the first stage of the spin-off process. Only *Number of Funding Partners* is kept in the regression equation with a regression coefficient of 0.889. In this case, the coefficient of determination is rather low, i.e. *Adjusted R Square* of 0.316.

Model Summary

Model	R	R Square	Adjusted R Square	Std. Error of the Estimate
1	,580[a]	,336	,316	1,24658

a. Predictors: (Constant), N° of Funding Partners - Phase 1

ANOVA[b]

Model		Sum of Squares	df	Mean Square	F	Sig.
1	Regression	25,214	1	25,214	16,226	,000[a]
	Residual	49,727	32	1,554		
	Total	74,941	33			

a. Predictors: (Constant), N° of Funding Partners - Phase 1
b. Dependent Variable: Provision of Resources - Phase 1

Coefficients[a]

Model		Unstandardized Coefficients		Standardized Coefficients	t	Sig.
		B	Std. Error	Beta		
1	(Constant)	,339	,298		1,139	,263
	N° of Funding Partners - Phase 1	,889	,221	,580	4,028	,000

a. Dependent Variable: Provision of Resources - Phase 1

Excluded Variables[b]

Model		Beta In	t	Sig.	Partial Correlation	Collinearity Statistics
						Tolerance
1	N° of Others - Phase 1	,207[a]	1,459	,155	,253	,995

a. Predictors in the Model: (Constant), N° of Funding Partners - Phase 1
b. Dependent Variable: Provision of Resources - Phase 1

Table 5-24: Multiple Regression Analysis Phase 1: Resources with Functional Roles

Table 5-25 informs about the major types of resources exchanged at this stage of the new venture creation process. According to the regression analysis the *Provision of Personnel* and *Financial Resources* have the largest impact on the overall amount of resources exchanged at this time with regression coefficients of 0.528 and 0.48. Also of significant influence, yet with a lower share (0.267), is the *Provision of Infrastructure*. *Material Resources* that featured a significant Spearman correlation beforehand could not be kept in the regression equation.

Model Summary

Model	R	R Square	Adjusted R Square	Std. Error of the Estimate
1	,765[a]	,585	,572	,98545
2	,931[b]	,866	,857	,56901
3	,958[c]	,917	,909	,45562

a. Predictors: (Constant), Financial Resources - Phase 1
b. Predictors: (Constant), Financial Resources - Phase 1, Provision of Personnel - Phase 1
c. Predictors: (Constant), Financial Resources - Phase 1, Provision of Personnel - Phase 1, Provision of Infrastructure - Phase 1

ANOVA[d]

Model		Sum of Squares	df	Mean Square	F	Sig.
1	Regression	43,866	1	43,866	45,171	,000[a]
	Residual	31,075	32	,971		
	Total	74,941	33			
2	Regression	64,904	2	32,452	100,230	,000[b]
	Residual	10,037	31	,324		
	Total	74,941	33			
3	Regression	68,713	3	22,904	110,333	,000[c]
	Residual	6,228	30	,208		
	Total	74,941	33			

a. Predictors: (Constant), Financial Resources - Phase 1
b. Predictors: (Constant), Financial Resources - Phase 1, Provision of Personnel - Phase 1
c. Predictors: (Constant), Financial Resources - Phase 1, Provision of Personnel - Phase 1, Provision of Infrastructure - Phase 1
d. Dependent Variable: Provision of Resources - Phase 1

Coefficients[a]

Model		Unstandardized Coefficients		Standardized Coefficients	t	Sig.
		B	Std. Error	Beta		
1	(Constant)	,430	,202		2,126	,041
	Financial Resources - Phase 1	1,209	,180	,765	6,721	,000
2	(Constant)	,116	,123		,945	,352
	Financial Resources - Phase 1	,973	,108	,616	9,014	,000
	Provision of Personnel - Phase 1	1,116	,138	,551	8,061	,000
3	(Constant)	,015	,101		,147	,884
	Financial Resources - Phase 1	,759	,100	,480	7,610	,000
	Provision of Personnel - Phase 1	1,071	,111	,528	9,615	,000
	Provision of Infrastructure - Phase 1	,571	,133	,267	4,284	,000

a. Dependent Variable: Provision of Resources - Phase 1

Excluded Variables[a]

Model		Beta In	t	Sig.	Partial Correlation	Collinearity Statistics
						Tolerance
1	Material Resources - Phase 1	-,005[a]	-,038	,970	-,007	,858
	Provision of Infrastructure - Phase 1	,324[a]	2,627	,013	,427	,720
	Provision of Personnel - Phase 1	,551[a]	8,061	,000	,823	,926
2	Material Resources - Phase 1	,000[b]	-,002	,998	,000	,858
	Provision of Infrastructure - Phase 1	,267[b]	4,284	,000	,616	,714
3	Material Resources - Phase 1	-,084[c]	-1,424	,165	-,256	,775

a. Predictors in the Model: (Constant), Financial Resources - Phase 1
b. Predictors in the Model: (Constant), Financial Resources - Phase 1, Provision of Personnel - Phase 1
c. Predictors in the Model: (Constant), Financial Resources - Phase 1, Provision of Personnel - Phase 1, Provision of Infrastructure - Phase 1
d. Dependent Variable: Provision of Resources - Phase 1

Table 5-25: Multiple Regression Analysis Phase 1: Resources with Resource Types

The nature of resource providing ties is analysed in Table 5-26. Out of the three parameters with significant Spearman correlations (see Chapter 5.2.2), i.e. *Closure*, *Multiplexity* and *Strong Ties*, only *Strong Ties* exhibit a significant relation, i.e. regression coefficient of 0.597, with the amount of resources furnished at this stage.

Model Summary

Model	R	R Square	Adjusted R Square	Std. Error of the Estimate
1	,496[a]	,246	,222	1,32897

a. Predictors: (Constant), Strong Ties - Phase 1

ANOVA[b]

Model		Sum of Squares	df	Mean Square	F	Sig.
1	Regression	18,424	1	18,424	10,432	,003[a]
	Residual	56,517	32	1,766		
	Total	74,941	33			

a. Predictors: (Constant), Strong Ties - Phase 1
b. Dependent Variable: Provision of Resources - Phase 1

Coefficients[a]

Model		Unstandardized Coefficients		Standardized Coefficients	t	Sig.
		B	Std. Error	Beta		
1	(Constant)	,369	,338		1,092	,283
	Strong Ties - Phase 1	,597	,185	,496	3,230	,003

a. Dependent Variable: Provision of Resources - Phase 1

Excluded Variables[b]

Model		Beta In	t	Sig.	Partial Correlation	Collinearity Statistics Tolerance
1	Multiplexity - Phase 1	-,020[a]	-,094	,926	-,017	,547
	Closure - Phase 1	-,538[a]	-1,913	,065	-,325	,275

a. Predictors in the Model: (Constant), Strong Ties - Phase 1
b. Dependent Variable: Provision of Resources - Phase 1

Table 5-26: Multiple Regression Analysis Phase 1: Resources with Structural Parameters

To determine the predominant suppliers of resources exchanged during the pre-start-up phase, analogue to the procedure presented in the previous section about informational content those types of resources determined as significant contributors to the overall amount of resources transferred at this stage as presented in Table 5-25 are subjected to regression analysis with significantly correlated alter types according to the exploratory analysis of Chapter 5.2.2.3.

According to the results of the previous regression analysis of Table 5-25, financial resources are one of the assets primarily requested and exchanged during the first phase. Possible financial backers identified at this stage of the process are funding partners and the category of other alters as indicated by the Spearman correlations of Chapter 5.2.2.3. Subjected to regression analysis (see Table 5-27), this indicative association is only confirmed for *Funding Partners* therewith being the predominant source of money in the opportunity recognition phase.

Model Summary

Model	R	R Square	Adjusted R Square	Std. Error of the Estimate
1	,654[a]	,428	,410	,73285

a. Predictors: (Constant), N° of Funding Partners - Phase 1

ANOVA[b]

Model		Sum of Squares	df	Mean Square	F	Sig.
1	Regression	12,843	1	12,843	23,913	,000[a]
	Residual	17,186	32	,537		
	Total	30,029	33			

a. Predictors: (Constant), N° of Funding Partners - Phase 1
b. Dependent Variable: Financial Resources - Phase 1

Coefficients[a]

Model		Unstandardized Coefficients		Standardized Coefficients	t	Sig.
		B	Std. Error	Beta		
1	(Constant)	,020	,175		,116	,909
	N° of Funding Partners - Phase 1	,635	,130	,654	4,890	,000

a. Dependent Variable: Financial Resources - Phase 1

Excluded Variables[b]

Model		Beta In	t	Sig.	Partial Correlation	Collinearity Statistics
						Tolerance
1	N° of Others - Phase 1	,112[a]	,829	,414	,147	,995

a. Predictors in the Model: (Constant), N° of Funding Partners - Phase 1
b. Dependent Variable: Financial Resources - Phase 1

Table 5-27: Multiple Regression Analysis Phase 1: Sources of Financial Resources

The second important asset at this time is infrastructure that according to the correlation analysis after Spearman may be obtained from friends, women or funding partners. Yet, as revealed by regression analysis as presented in Table 5-28, the influence of female contacts in this context is not strong enough to persist respective significance tests. Consequently, only *Friends* and *Funding Partners* are kept in the regression equation as relevant sources of infrastructure with almost equal influence.

Model Summary

Model	R	R Square	Adjusted R Square	Std. Error of the Estimate
1	,494[a]	,244	,221	,62200
2	,593[b]	,352	,310	,58509

a. Predictors: (Constant), N° of Friends - Phase 1
b. Predictors: (Constant), N° of Friends - Phase 1, N° of Funding Partners - Phase 1

ANOVA[c]

Model		Sum of Squares	df	Mean Square	F	Sig.
1	Regression	4,002	1	4,002	10,345	,003[a]
	Residual	12,380	32	,387		
	Total	16,382	33			
2	Regression	5,770	2	2,885	8,428	,001[b]
	Residual	10,612	31	,342		
	Total	16,382	33			

a. Predictors: (Constant), N° of Friends - Phase 1
b. Predictors: (Constant), N° of Friends - Phase 1, N° of Funding Partners - Phase 1
c. Dependent Variable: Provision of Infrastructure - Phase 1

Coefficients[a]

Model		Unstandardized Coefficients		Standardized Coefficients	t	Sig.
		B	Std. Error	Beta		
1	(Constant)	,118	,147		,801	,429
	N° of Friends - Phase 1	,268	,083	,494	3,216	,003
2	(Constant)	-,097	,167		-,580	,566
	N° of Friends - Phase 1	,262	,079	,483	3,341	,002
	N° of Funding Partners - Phase 1	,236	,104	,329	2,273	,030

a. Dependent Variable: Provision of Infrastructure - Phase 1

Excluded Variables[c]

Model		Beta In	t	Sig.	Partial Correlation	Collinearity Statistics
						Tolerance
1	N° of Funding Partners - Phase 1	,329[a]	2,273	,030	,378	,999
	N° of Females - Phase 1	-,097[a]	-,570	,573	-,102	,831
2	N° of Females - Phase 1	-,192[b]	-1,183	,246	-,211	,786

a. Predictors in the Model: (Constant), N° of Friends - Phase 1
b. Predictors in the Model: (Constant), N° of Friends - Phase 1, N° of Funding Partners - Phase 1
c. Dependent Variable: Provision of Infrastructure - Phase 1

Table 5-28: Multiple Regression Analysis Phase 1: Sources of Infrastructure

As to the third type of resource requested in the first phase of the spin-off process, namely personnel, no particular type of source could be identified in the correlation analysis after Spearman of Chapter 5.2.2.3. Consequently, no further examination is carried out in this regard.

5.2.3.1.4 Support

Besides information another important content provided in the first phase of the spin-off process is support of different kinds as confirmed by the regression analysis of Chapter 5.2.3.1.1. In the following it will be analysed, which types of alters are mainly responsible for the provision of support and how the nature of supporting relations looks like.

Model Summary

Model	R	R Square	Adjusted R Square	Std. Error of the Estimate
1	,603[a]	,364	,344	1,58723
2	,704[b]	,496	,463	1,43551

a. Predictors: (Constant), N° of Females - Phase 1
b. Predictors: (Constant), N° of Females - Phase 1, N° of Funding Partners - Phase 1

ANOVAc

Model		Sum of Squares	df	Mean Square	F	Sig.
1	Regression	46,118	1	46,118	18,306	,000[a]
	Residual	80,617	32	2,519		
	Total	126,735	33			
2	Regression	62,854	2	31,427	15,251	,000[b]
	Residual	63,882	31	2,061		
	Total	126,735	33			

a. Predictors: (Constant), N° of Females - Phase 1
b. Predictors: (Constant), N° of Females - Phase 1, N° of Funding Partners - Phase 1
c. Dependent Variable: Support - Phase 1

Coefficientsa

Model		Unstandardized Coefficients		Standardized Coefficients	t	Sig.
		B	Std. Error	Beta		
1	(Constant)	1,072	,361		2,968	,006
	N° of Females - Phase 1	1,329	,311	,603	4,279	,000
2	(Constant)	,514	,381		1,350	,187
	N° of Females - Phase 1	1,143	,288	,519	3,963	,000
	N° of Funding Partners - Phase 1	,744	,261	,373	2,850	,008

a. Dependent Variable: Support - Phase 1

Excluded Variables[c]

Model		Beta In	t	Sig.	Partial Correlation	Collinearity Statistics Tolerance
1	N° of Friends - Phase 1	,152[a]	,981	,334	,174	,831
	N° of Funding Partners - Phase 1	,373[a]	2,850	,008	,456	,949
2	N° of Friends - Phase 1	,179[b]	1,293	,206	,230	,828

a. Predictors in the Model: (Constant), N° of Females - Phase 1
b. Predictors in the Model: (Constant), N° of Females - Phase 1, N° of Funding Partners - Phase 1
c. Dependent Variable: Support - Phase 1

Table 5-29: Multiple Regression Analysis Phase 1: Support with Functional Roles

As to the first case, all functional roles significantly correlated to the provision of support according to the Spearman coefficients of Chapter 5.2.2 are subjected to the multiple regression analysis. As Table 5-31 reveals, in particular females and funding partners delivered support in the course of the first phase. Comparing both predictors *Females* feature the larger influence with a standardised correlation coefficient *Beta* of 0.519 opposed to a *Beta* value of 0.373 for *Funding Partners*. *Friends*, originally positively correlated with *Support* according to Chapter 5.2.2, do not feature a significant regression coefficient.

Having a closer look at the distribution between emotional and institutional support according to their share in the overall amount of support provided at this stage of the spin-off process (see Table 5-30), one will find that *Institutional Support* features a slightly larger impact on *Support* (*Beta* of 0.588) than *Emotional Support* with a *Beta* of 0.45.

Model Summary

Model	R	R Square	Adjusted R Square	Std. Error of the Estimate
1	,876[a]	,767	,760	,96106
2	,941[b]	,886	,879	,68283

a. Predictors: (Constant), Institutional Support - Phase 1
b. Predictors: (Constant), Institutional Support - Phase 1, Emotional Support - Phase 1

ANOVA[c]

Model		Sum of Squares	df	Mean Square	F	Sig.
1	Regression	97,179	1	97,179	105,215	,000[a]
	Residual	29,556	32	,924		
	Total	126,735	33			
2	Regression	112,281	2	56,141	120,408	,000[b]
	Residual	14,454	31	,466		
	Total	126,735	33			

a. Predictors: (Constant), Institutional Support - Phase 1
b. Predictors: (Constant), Institutional Support - Phase 1, Emotional Support - Phase 1
c. Dependent Variable: Support - Phase 1

Coefficients[a]

Model		Unstandardized Coefficients		Standardized Coefficients	t	Sig.
		B	Std. Error	Beta		
1	(Constant)	,577	,221		2,610	,014
	Institutional Support - Phase 1	1,049	,102	,876	10,257	,000
2	(Constant)	,281	,165		1,697	,100
	Institutional Support - Phase 1	,704	,095	,588	7,439	,000
	Emotional Support - Phase 1	,642	,113	,450	5,691	,000

a. Dependent Variable: Support - Phase 1

Excluded Variables[b]

Model		Beta In	t	Sig.	Partial Correlation	Collinearity Statistics Tolerance
1	Emotional Support - Phase 1	,450[a]	5,691	,000	,715	,590

a. Predictors in the Model: (Constant), Institutional Support - Phase 1
b. Dependent Variable: Support - Phase 1

Table 5-30: Multiple Regression Analysis Phase 1: Support with Support Types

Looking at the quality of the ties providing support as depicted in Table 5-31, they are mainly characterised by strong and multiplex relationships. According to the standardised regression coefficients *Strong Ties* display the largest influence with a *Beta* value of 0.839 followed by *Multiplexity* with 0.615. Interestingly, network closure that also significantly predicts the provision of support is negatively related with a *Beta* value of -0.719. This means that the higher the degree of network closure the smaller the amount of support provided to ego.

Model Summary

Model	R	R Square	Adjusted R Square	Std. Error of the Estimate
1	,641[a]	,411	,393	1,52702
2	,698[b]	,487	,454	1,44760
3	,773[c]	,598	,558	1,30359

a. Predictors: (Constant), Strong Ties - Phase 1
b. Predictors: (Constant), Strong Ties - Phase 1, Multiplexity - Phase 1
c. Predictors: (Constant), Strong Ties - Phase 1, Multiplexity - Phase 1, Closure - Phase 1

ANOVA[d]

Model		Sum of Squares	df	Mean Square	F	Sig.
1	Regression	52,118	1	52,118	22,351	,000[a]
	Residual	74,617	32	2,332		
	Total	126,735	33			
2	Regression	61,773	2	30,887	14,739	,000[b]
	Residual	64,962	31	2,096		
	Total	126,735	33			
3	Regression	75,755	3	25,252	14,860	,000[c]
	Residual	50,980	30	1,699		
	Total	126,735	33			

a. Predictors: (Constant), Strong Ties - Phase 1
b. Predictors: (Constant), Strong Ties - Phase 1, Multiplexity - Phase 1
c. Predictors: (Constant), Strong Ties - Phase 1, Multiplexity - Phase 1, Closure - Phase 1
d. Dependent Variable: Support - Phase 1

Coefficients[a]

Model		Unstandardized Coefficients		Standardized Coefficients	t	Sig.
		B	Std. Error	Beta		
1	(Constant)	,731	,389		1,880	,069
	Strong Ties - Phase 1	1,003	,212	,641	4,728	,000
2	(Constant)	,362	,406		,892	,379
	Strong Ties - Phase 1	,610	,272	,390	2,241	,032
	Multiplexity - Phase 1	1,955	,911	,373	2,146	,040
3	(Constant)	,358	,366		,979	,335
	Strong Ties - Phase 1	1,313	,346	,839	3,788	,001
	Multiplexity - Phase 1	3,221	,931	,615	3,458	,002
	Closure - Phase 1	-4,501	1,569	-,719	-2,868	,007

a. Dependent Variable: Support - Phase 1

Excluded Variables[c]

Model		Beta In	t	Sig.	Partial Correlation	Collinearity Statistics Tolerance
1	Multiplexity - Phase 1	,373[a]	2,146	,040	,360	,547
	Closure - Phase 1	-,308[a]	-1,201	,239	-,211	,275
2	Closure - Phase 1	-,719[b]	-2,868	,007	-,464	,214

a. Predictors in the Model: (Constant), Strong Ties - Phase 1
b. Predictors in the Model: (Constant), Strong Ties - Phase 1, Multiplexity - Phase 1
c. Dependent Variable: Support - Phase 1

Table 5-31: Multiple Regression Analysis Phase 1: Support with Structural Parameters

According to the Spearman correlation table of Chapter 5.2.2.4, possible providers of emotional support are friends and women. Both sources of support are confirmed by the regression analysis of Table 5-32 with *Number of Females* featuring the larger influence with a *Beta* value of 0.509 over 0.324 for *Number of Friends*.

Model Summary

Model	R	R Square	Adjusted R Square	Std. Error of the Estimate
1	,643[a]	,413	,395	1,06754
2	,707[b]	,500	,468	1,00050

a. Predictors: (Constant), N° of Females - Phase 1
b. Predictors: (Constant), N° of Females - Phase 1, N° of Friends - Phase 1

ANOVA[c]

Model		Sum of Squares	df	Mean Square	F	Sig.
1	Regression	25,649	1	25,649	22,506	,000[a]
	Residual	36,468	32	1,140		
	Total	62,118	33			
2	Regression	31,087	2	15,543	15,528	,000[b]
	Residual	31,031	31	1,001		
	Total	62,118	33			

a. Predictors: (Constant), N° of Females - Phase 1
b. Predictors: (Constant), N° of Females - Phase 1, N° of Friends - Phase 1
c. Dependent Variable: Emotional Support - Phase 1

Coefficients[a]

Model		Unstandardized Coefficients B	Std. Error	Standardized Coefficients Beta	t	Sig.
1	(Constant)	,477	,243		1,965	,058
	N° of Females - Phase 1	,991	,209	,643	4,744	,000
2	(Constant)	,221	,253		,873	,389
	N° of Females - Phase 1	,785	,215	,509	3,658	,001
	N° of Friends - Phase 1	,343	,147	,324	2,331	,026

a. Dependent Variable: Emotional Support - Phase 1

Excluded Variables[b]

Model		Beta In	t	Sig.	Partial Correlation	Collinearity Statistics
						Tolerance
1	N° of Friends - Phase 1	,324[a]	2,331	,026	,386	,831

a. Predictors in the Model: (Constant), N° of Females - Phase 1
b. Dependent Variable: Emotional Support - Phase 1

Table 5-32: Multiple Regression Analysis Phase 1: Sources of Emotional Support

In line with Spearman, institutional support may be provided by women, friends or funding partners. Interestingly after regression analysis (see Table 5-33), only the number of female contacts is maintained in the regression model as predictor of institutional support.

Model Summary

Model	R	R Square	Adjusted R Square	Std. Error of the Estimate
1	,573[a]	,328	,307	1,36204

a. Predictors: (Constant), N° of Females - Phase 1

ANOVA[b]

Model		Sum of Squares	df	Mean Square	F	Sig.
1	Regression	29,017	1	29,017	15,642	,000[a]
	Residual	59,365	32	1,855		
	Total	88,382	33			

a. Predictors: (Constant), N° of Females - Phase 1
b. Dependent Variable: Institutional Support - Phase 1

Coefficients[a]

Model		Unstandardized Coefficients		Standardized Coefficients	t	Sig.
		B	Std. Error	Beta		
1	(Constant)	,635	,310		2,049	,049
	N° of Females - Phase 1	1,054	,267	,573	3,955	,000

a. Dependent Variable: Institutional Support - Phase 1

Excluded Variables[b]

Model		Beta In	t	Sig.	Partial Correlation	Collinearity Statistics
						Tolerance
1	N° of Friends - Phase 1	,110[a]	,688	,497	,123	,831
	N° of Funding Partners - Phase 1	,199[a]	1,352	,186	,236	,949

a. Predictors in the Model: (Constant), N° of Females - Phase 1
b. Dependent Variable: Institutional Support - Phase 1

Table 5-33: Multiple Regression Analysis Phase 1: Sources of Institutional Support

5.2.3.2 Formation

As in the previous Chapter 5.2.3.1 again those relational contents that turned out to feature significant correlations according to the correlation analysis after Spearman are subjected to further regressions as dependent variables of the second phase.

5.2.3.2.1 Network Size

Starting with the influence of different types of alters on the size of egos' personal networks, out of the significantly correlated functional role of alters, i.e. friends, funding partners and women, as for the first phase *Number of Females* do not exhibit a significant regression coefficient and are thus not kept in the regression equation of the regression model as depicted in the following Table 5-34. The largest influence on network size can be attributed to *Number of Friends* with a *Beta* value of 0.7070 over *Number of Funding Partners* with a *Beta* value of 0.477.

Model Summary

Model	R	R Square	Adjusted R Square	Std. Error of the Estimate
1	,653[a]	,426	,408	1,476
2	,806[b]	,650	,628	1,171

a. Predictors: (Constant), N° of Friends - Phase 2
b. Predictors: (Constant), N° of Friends - Phase 2, N° of Funding Partners - Phase 2

ANOVA[c]

Model		Sum of Squares	df	Mean Square	F	Sig.
1	Regression	51,773	1	51,773	23,750	,000[a]
	Residual	69,757	32	2,180		
	Total	121,529	33			
2	Regression	79,042	2	39,521	28,835	,000[b]
	Residual	42,488	31	1,371		
	Total	121,529	33			

a. Predictors: (Constant), N° of Friends - Phase 2
b. Predictors: (Constant), N° of Friends - Phase 2, N° of Funding Partners - Phase 2
c. Dependent Variable: Network Size in Phase 2

Coefficients[a]

Model		Unstandardized Coefficients		Standardized Coefficients	t	Sig.
		B	Std. Error	Beta		
1	(Constant)	2,908	,323		9,011	,000
	N° of Friends - Phase 2	,737	,151	,653	4,873	,000
2	(Constant)	1,733	,367		4,719	,000
	N° of Friends - Phase 2	,798	,121	,707	6,613	,000
	N° of Funding Partners - Phase 2	1,033	,232	,477	4,460	,000

a. Dependent Variable: Network Size in Phase 2

Excluded Variables[c]

Model		Beta In	t	Sig.	Partial Correlation	Collinearity Statistics Tolerance
1	N° of Females - Phase 2	,084[a]	,556	,582	,099	,808
	N° of Funding Partners - Phase 2	,477[a]	4,460	,000	,625	,987
2	N° of Females - Phase 2	,070[b]	,587	,562	,107	,807

a. Predictors in the Model: (Constant), N° of Friends - Phase 2
b. Predictors in the Model: (Constant), N° of Friends - Phase 2, N° of Funding Partners - Phase 2
c. Dependent Variable: Network Size in Phase 2

Table 5-34: Multiple Regression Analysis Phase 2: Network Size with Functional Roles

Coming to the type of content exchanged and its effect on network size, the Spearman correlation analysis of Chapter 5.2.2 identified all three types of content, i.e. information, resources and support, as significantly related to the size of entrepreneurial researchers' personal networks. Subjecting these three related variables to multiple regression analysis, *Provision of Resources* in the following is excluded from the regression model due to its lack of significance. Remaining variables are *Informational Content* and *Support*, with *Informational Content*, similar to the first phase, featuring the larger effect on network size with a *Beta* value of 0.568 over 0.301 for *Support*. However, the gap between the two parameters has considerably been reduced indicating a decreasing dominance of informational contents in the course of the spin-off process.

Model Summary

Model	R	R Square	Adjusted R Square	Std. Error of the Estimate
1	,693[a]	,481	,464	1,405
2	,745[b]	,556	,527	1,320

a. Predictors: (Constant), Informational Content - Phase 2
b. Predictors: (Constant), Informational Content - Phase 2, Support - Phase 2

ANOVA[c]

Model		Sum of Squares	df	Mean Square	F	Sig.
1	Regression	58,398	1	58,398	29,601	,000[a]
	Residual	63,131	32	1,973		
	Total	121,529	33			
2	Regression	67,529	2	33,764	19,383	,000[b]
	Residual	54,001	31	1,742		
	Total	121,529	33			

a. Predictors: (Constant), Informational Content - Phase 2
b. Predictors: (Constant), Informational Content - Phase 2, Support - Phase 2
c. Dependent Variable: Network Size in Phase 2

Coefficients[a]

Model		Unstandardized Coefficients		Standardized Coefficients	t	Sig.
		B	Std. Error	Beta		
1	(Constant)	2,154	,399		5,405	,000
	Informational Content - Phase 2	,660	,121	,693	5,441	,000
2	(Constant)	1,767	,411		4,301	,000
	Informational Content - Phase 2	,541	,125	,568	4,315	,000
	Support - Phase 2	,345	,150	,301	2,289	,029

a. Dependent Variable: Network Size in Phase 2

Excluded Variables[c]

Model		Beta In	t	Sig.	Partial Correlation	Collinearity Statistics
						Tolerance
1	Provision of Resources - Phase 2	,263[a]	1,790	,083	,306	,706
	Support - Phase 2	,301[a]	2,289	,029	,380	,827
2	Provision of Resources - Phase 2	,236[b]	1,698	,100	,296	,701

a. Predictors in the Model: (Constant), Informational Content - Phase 2
b. Predictors in the Model: (Constant), Informational Content - Phase 2, Support - Phase 2
c. Dependent Variable: Network Size in Phase 2

Table 5-35: Multiple Regression Analysis Phase 2: Network Size with Relational Content

5.2.3.2.2 Informational Content

As potential sources of information in the second phase the correlation analysis after Spearman carried out in Chapter 5.2.2 qualified only two functional roles as significant: friends and academic contacts. Submitting these two variables to a multiple regression analysis with *Informational Content* as dependent variable, only the *Number of Friends* remains of significant influence with an unstandardised regression coefficient of 0.694 and a standardised coefficient of 0.585 (see Table 5-36). It has to be noted that the coefficient of determination is rather low for this regression model with an *Adjusted R Square* of only 0.322.

Model Summary

Model	R	R Square	Adjusted R Square	Std. Error of the Estimate
1	,585[a]	,343	,322	1,65929

a. Predictors: (Constant), N° of Friends - Phase 2

ANOVA[b]

Model		Sum of Squares	df	Mean Square	F	Sig.
1	Regression	45,926	1	45,926	16,681	,000[a]
	Residual	88,104	32	2,753		
	Total	134,029	33			

a. Predictors: (Constant), N° of Friends - Phase 2
b. Dependent Variable: Informational Content - Phase 2

Coefficients[a]

Model		Unstandardized Coefficients		Standardized Coefficients	t	Sig.
		B	Std. Error	Beta		
1	(Constant)	1,700	,363		4,687	,000
	N° of Friends - Phase 2	,694	,170	,585	4,084	,000

a. Dependent Variable: Informational Content - Phase 2

Excluded Variables[b]

Model		Beta In	t	Sig.	Partial Correlation	Collinearity Statistics
						Tolerance
1	N° of Academics - Phase 2	,237[a]	1,614	,117	,278	,905

a. Predictors in the Model: (Constant), N° of Friends - Phase 2
b. Dependent Variable: Informational Content - Phase 2

Table 5-36: Multiple Regression Analysis Phase 2: Informational Content with Functional Roles

Out of the content types deemed as relevant in the second stage according to the correlation analysis after Spearman, i.e. *Economic, Legal* and *Other Information* as well as *Information about Potential Customers*, only *Economic* and *Other Information* featured a significant contribution in explaining changes in the overall amount of information exchanged at this stage, and are thus included in the subsequent regression model of Table 5-37. The largest explanatory power can be attributed to ties providing economic knowledge with a standardised regression coefficient of 0.726 opposed to a value of 0.465 for the provision of scientific and technical know-how.

Model Summary

Model	R	R Square	Adjusted R Square	Std. Error of the Estimate
1	,810[a]	,656	,645	1,20083
2	,930[b]	,865	,856	,76485

a. Predictors: (Constant), Economic Information - Phase 2
b. Predictors: (Constant), Economic Information - Phase 2, Other Information - Phase 2

ANOVA[c]

Model		Sum of Squares	df	Mean Square	F	Sig.
1	Regression	87,886	1	87,886	60,948	,000[a]
	Residual	46,143	32	1,442		
	Total	134,029	33			
2	Regression	115,895	2	57,947	99,056	,000[b]
	Residual	18,135	31	,585		
	Total	134,029	33			

a. Predictors: (Constant), Economic Information - Phase 2
b. Predictors: (Constant), Economic Information - Phase 2, Other Information - Phase 2
c. Dependent Variable: Informational Content - Phase 2

Coefficients[a]

Model		Unstandardized Coefficients		Standardized Coefficients	t	Sig.
		B	Std. Error	Beta		
1	(Constant)	1,116	,282		3,961	,000
	Economic Information - Phase 2	,912	,117	,810	7,807	,000
2	(Constant)	,623	,193		3,226	,003
	Economic Information - Phase 2	,817	,076	,726	10,806	,000
	Other Information - Phase 2	,735	,106	,465	6,919	,000

a. Dependent Variable: Informational Content - Phase 2

Excluded Variables[c]

Model		Beta In	t	Sig.	Partial Correlation	Collinearity Statistics
						Tolerance
1	Legal Information - Phase 2	-,044[a]	-,287	,776	-,051	,463
	Information about Potential Customers - Phase 2	,152[a]	1,374	,179	,240	,851
	Other Information - Phase 2	,465[a]	6,919	,000	,779	,967
2	Legal Information - Phase 2	,048[b]	,484	,632	,088	,454
	Information about Potential Customers - Phase 2	,128[b]	1,847	,075	,320	,849

a. Predictors in the Model: (Constant), Economic Information - Phase 2
b. Predictors in the Model: (Constant), Economic Information - Phase 2, Other Information - Phase 2
c. Dependent Variable: Informational Content - Phase 2

Table 5-37: Multiple Regression Analysis Phase 2: Informational Content with Information Types

Regarding the nature of informational ties in the second phase, as opposed to the first phase weak ties no longer feature significant Spearman correlations and are thus not included in the multiple regression analysis. Remaining significantly correlated variables are *Strong Ties*, *Multiplexity* and *Closure*, out of which *Multiplexity* does not feature the required significance to be kept in the regression equation (see Table 5-38). While the number of strong ties has a positive impact on the amount of informational ties in the personal network, network closure on contrary is negatively associated with the provision of information and as such reminds us of the same negative effect of network closure on the provision of support observed for the previous phase. While the standardised regression coefficient B of 1.629 for *Strong Ties* indicates that one additional strong tie entails an increase of informational ties by more than one and a half, an increase in network closure by one unit decreases the amount of information-providing ties by more than three and a half (B of -3.523).

Model Summary

Model	R	R Square	Adjusted R Square	Std. Error of the Estimate
1	,709[a]	,503	,487	1,44289
2	,771[b]	,595	,569	1,32345

a. Predictors: (Constant), Strong Ties - Phase 2
b. Predictors: (Constant), Strong Ties - Phase 2, Closure - Phase 2

ANOVA[c]

Model		Sum of Squares	df	Mean Square	F	Sig.
1	Regression	67,407	1	67,407	32,377	,000[a]
	Residual	66,622	32	2,082		
	Total	134,029	33			
2	Regression	79,732	2	39,866	22,761	,000[b]
	Residual	54,297	31	1,752		
	Total	134,029	33			

a. Predictors: (Constant), Strong Ties - Phase 2
b. Predictors: (Constant), Strong Ties - Phase 2, Closure - Phase 2
c. Dependent Variable: Informational Content - Phase 2

Coefficients[a]

Model		Unstandardized Coefficients		Standardized Coefficients	t	Sig.
		B	Std. Error	Beta		
1	(Constant)	1,355	,332		4,076	,000
	Strong Ties - Phase 2	,895	,157	,709	5,690	,000
2	(Constant)	1,546	,313		4,935	,000
	Strong Ties - Phase 2	1,629	,312	1,291	5,219	,000
	Closure - Phase 2	-3,523	1,328	-,656	-2,653	,012

a. Dependent Variable: Informational Content - Phase 2

Excluded Variables[c]

Model		Beta In	t	Sig.	Partial Correlation	Collinearity Statistics
						Tolerance
1	Multiplexity - Phase 2	-,069[a]	-,377	,709	-,068	,474
	Closure - Phase 2	-,656[a]	-2,653	,012	-,430	,213
2	Multiplexity - Phase 2	,355[b]	1,699	,100	,296	,282

a. Predictors in the Model: (Constant), Strong Ties - Phase 2
b. Predictors in the Model: (Constant), Strong Ties - Phase 2, Closure - Phase 2
c. Dependent Variable: Informational Content - Phase 2

Table 5-38: Multiple Regression Analysis Phase 2: Informational Content with Structural Parameters

As depicted in Table 5-37, economic know-how and information of other nature, such as technical and scientific knowledge, are predominantly requested during the formation phase of the academic start-up. However, only for economic information particular sources could be identified in the course of the Spearman correlation analysis, viz. friends and funding partners. Consequently, those two types of contacts are subjected to regression analysis with economic information. As Table 5-39 discloses, both categories of alters can be deemed as significant sources of economic know-how, whereas *Number of Friends* features the stronger influence with a *Beta* value of 0.522 compared to 0.398 for *Number of Funding Partners*.

Model Summary

Model	R	R Square	Adjusted R Square	Std. Error of the Estimate
1	,477[a]	,227	,203	1,59826
2	,619[b]	,383	,343	1,45059

a. Predictors: (Constant), N° of Friends - Phase 2
b. Predictors: (Constant), N° of Friends – Phase 2, N° of Funding Partners - Phase 2

ANOVA[c]

Model		Sum of Squares	Df	Mean Square	F	Sig.
1	Regression	24,022	1	24,022	9,404	,004[a]
	Residual	81,742	32	2,554		
	Total	105,765	33			
2	Regression	40,534	2	20,267	9,632	,001[b]
	Residual	65,230	31	2,104		
	Total	105,765	33			

a. Predictors: (Constant), N° of Friends - Phase 2
b. Predictors: (Constant), N° of Friends - Phase 2, N° of Funding Partners - Phase 2
c. Dependent Variable: Economic Information - Phase 2

Coefficients[a]

Model		Unstandardized Coefficients		Standardized Coefficients	t	Sig.
		B	Std. Error	Beta		
1	(Constant)	,983	,349		2,814	,008
	N° of Friends - Phase 2	,502	,164	,477	3,067	,004
2	(Constant)	,069	,455		,151	,881
	N° of Friends - Phase 2	,549	,149	,522	3,675	,001
	N° of Funding Partners - Phase 2	,804	,287	,398	2,801	,009

a. Dependent Variable: Economic Information - Phase 2

Excluded Variables[b]

Model		Beta In	t	Sig.	Partial Correlation	Collinearity Statistics
						Tolerance
1	N° of Funding Partners - Phase 2	,398[a]	2,801	,009	,449	,987

a. Predictors in the Model: (Constant), N° of Friends - Phase 2
b. Dependent Variable: Economic Information - Phase 2

Table 5-39: Regression Analysis Phase 2: Sources of Economic Information

5.2.3.2.3 Resources

Following the correlation analysis of Chapter 5.2.2, no significant correlation between any type of functional role and the number of resource-providing ties could be detected for the second stage. As a result no regression analysis is carried out for this specific network parameter.

As regards the composition of resources obtained at this stage, similar to the previous stage *Financial* and *Material Resources* as well as the *Provision of Personnel* and *Infrastructure* are subjected to regression analysis. And as for the previous phase again all types of resources apart from *Material Resources* are found to exhibit a significant influence on the overall amount of resources provided. *Personnel* and *Financial Resources* feature the largest impact with *Beta* values of 0.662 and 0.504, respectively. At some distance follows the *Provision of Infrastructure* with a *Beta* value of 0.266.

Referring to the nature of resource-delivering ties, at this stage only *Strong Ties* are significantly related to resource provision with a regression coefficient of 0.452. *Multiplexity* and *Closure* again feature no significant regression coefficients. Again the coefficient of determination is rather low with an *Adjusted R Square* of 0.284.

Model Summary

Model	R	R Square	Adjusted R Square	Std. Error of the Estimate
1	,650^a	,422	,404	1,00822
2	,896^b	,803	,790	,59835
3	,919^c	,845	,830	,53864

a. Predictors: (Constant), Financial Resources - Phase 2
b. Predictors: (Constant), Financial Resources - Phase 2, Provision of Personnel - Phase 2
c. Predictors: (Constant), Financial Resources - Phase 2, Provision of Personnel - Phase 2, Provision of Infrastructure - Phase 2

ANOVA^d

Model		Sum of Squares	df	Mean Square	F	Sig.
1	Regression	23,736	1	23,736	23,351	,000^a
	Residual	32,528	32	1,017		
	Total	56,265	33			
2	Regression	45,166	2	22,583	63,077	,000^b
	Residual	11,099	31	,358		
	Total	56,265	33			
3	Regression	47,561	3	15,854	54,642	,000^c
	Residual	8,704	30	,290		
	Total	56,265	33			

a. Predictors: (Constant), Financial Resources - Phase 2
b. Predictors: (Constant), Financial Resources - Phase 2, Provision of Personnel - Phase 2
c. Predictors: (Constant), Financial Resources - Phase 2, Provision of Personnel - Phase 2, Provision of Infrastructure - Phase 2
d. Dependent Variable: Provision of Resources - Phase 2

Coefficients[a]

Model		Unstandardized Coefficients		Standardized Coefficients	t	Sig.
		B	Std. Error	Beta		
1	(Constant)	1,063	,238		4,468	,000
	Financial Resources - Phase 2	,707	,146	,650	4,832	,000
2	(Constant)	,471	,161		2,932	,006
	Financial Resources - Phase 2	,723	,087	,665	8,332	,000
	Provision of Personnel - Phase 2	,750	,097	,617	7,737	,000
3	(Constant)	,400	,147		2,726	,011
	Financial Resources - Phase 2	,548	,099	,504	5,523	,000
	Provision of Personnel - Phase 2	,804	,089	,662	9,006	,000
	Provision of Infrastructure - Phase 2	,404	,141	,266	2,873	,007

a. Dependent Variable: Provision of Resources - Phase 2

Excluded Variables[d]

Model		Beta In	t	Sig.	Partial Correlation	Collinearity Statistics Tolerance
1	Material Resources - Phase 2	,189[a]	1,228	,229	,215	,748
	Provision of Infrastructure - Phase 2	,091[a]	,531	,599	,095	,628
	Provision of Personnel - Phase 2	,617[a]	7,737	,000	,812	,999
2	Material Resources - Phase 2	,185[b]	2,121	,042	,361	,748
	Provision of Infrastructure - Phase 2	,266[b]	2,873	,007	,465	,600
3	Material Resources - Phase 2	,107[c]	1,182	,247	,214	,626

a. Predictors in the Model: (Constant), Financial Resources - Phase 2
b. Predictors in the Model: (Constant), Financial Resources - Phase 2, Provision of Personnel - Phase 2
c. Predictors in the Model: (Constant), Financial Resources - Phase 2, Provision of Personnel - Phase 2, Provision of Infrastructure - Phase 2
d. Dependent Variable: Provision of Resources - Phase 2

Table 5-40: Multiple Regression Analysis Phase 2: Resources with Resources Types

According to Table 5-40, in the formation phase capital, personnel and infrastructure are the resources that are primarily exchanged in the personal networks of the academic entrepreneurs. While for the provision of personnel no specific source could be identified in the course of the correlation analysis of Chapter 5.2.2.3, money and infrastructure was significantly correlated to some of the alter types.

Model Summary

Model	R	R Square	Adjusted R Square	Std. Error of the Estimate
1	,553[a]	,306	,284	1,10495

a. Predictors: (Constant), Strong Ties - Phase 2

ANOVA[b]

Model		Sum of Squares	df	Mean Square	F	Sig.
1	Regression	17,196	1	17,196	14,084	,001[a]
	Residual	39,069	32	1,221		
	Total	56,265	33			

a. Predictors: (Constant), Strong Ties - Phase 2
b. Dependent Variable: Provision of Resources - Phase 2

Coefficients[a]

Model		Unstandardized Coefficients		Standardized Coefficients	t	Sig.
		B	Std. Error	Beta		
1	(Constant)	1,215	,255		4,773	,000
	Strong Ties - Phase 2	,452	,120	,553	3,753	,001

a. Dependent Variable: Provision of Resources - Phase 2

Excluded Variables[b]

Model		Beta In	t	Sig.	Partial Correlation	Collinearity Statistics
						Tolerance
1	Multiplexity - Phase 2	,019[a]	,090	,929	,016	,474
	Closure - Phase 2	-,531[a]	-1,715	,096	-,294	,213

a. Predictors in the Model: (Constant), Strong Ties - Phase 2
b. Dependent Variable: Provision of Resources - Phase 2

Table 5-41: Multiple Regression Analysis Phase 2: Resources with Structural Parameters

As to first, the only type of alters significantly associated with financial resources was the *Number of Funding Partners*. This association is also confirmed by the subsequent regression analysis of Table 5-42. The *Number of Funding Partners* features a significant positive influence of 0.572 on the amount of money-providing contacts.

Model Summary

Model	R	R Square	Adjusted R Square	Std. Error of the Estimate
1	,421[a]	,177	,151	1,10548

a. Predictors: (Constant), N° of Funding Partners - Phase 2

ANOVA[b]

Model		Sum of Squares	df	Mean Square	F	Sig.
1	Regression	8,423	1	8,423	6,892	,013[a]
	Residual	39,107	32	1,222		
	Total	47,529	33			

a. Predictors: (Constant), N° of Funding Partners - Phase 2
b. Dependent Variable: Financial Resources - Phase 2

Coefficients[a]

Model		Unstandardized Coefficients		Standardized Coefficients	t	Sig.
		B	Std. Error	Beta		
1	(Constant)	,514	,298		1,723	,095
	N° of Funding Partners - Phase 2	,570	,217	,421	2,625	,013

a. Dependent Variable: Financial Resources - Phase 2

Table 5-42: Regression Analysis Phase 2: Sources of Financial Resources

According to Chapter 5.2.2.3, the provision of infrastructure is associated with number of females and friends in the network. This original assumption can only be corroborated for *Number of Friends*. As the regression analysis of Table 5-43 depicts, this variable exhibits a positive influence of 0.292 on the number of infrastructure-providing ties. The *Number of Females*, however, has to be excluded from the regression equation due to lack of significance.

Model Summary

Model	R	R Square	Adjusted R Square	Std. Error of the Estimate
1	,577[a]	,333	,313	,71270

a. Predictors: (Constant), N° of Friends - Phase 2

ANOVA[b]

Model		Sum of Squares	df	Mean Square	F	Sig.
1	Regression	8,128	1	8,128	16,003	,000[a]
	Residual	16,254	32	,508		
	Total	24,382	33			

a. Predictors: (Constant), N° of Friends - Phase 2
b. Dependent Variable: Provision of Infrastructure - Phase 2

Coefficients[a]

Model		Unstandardized Coefficients		Standardized Coefficients	t	Sig.
		B	Std. Error	Beta		
1	(Constant)	,173	,156		1,108	,276
	N° of Friends - Phase 2	,292	,073	,577	4,000	,000

a. Dependent Variable: Provision of Infrastructure - Phase 2

Excluded Variables[b]

Model		Beta In	t	Sig.	Partial Correlation	Collinearity Statistics
						Tolerance
1	N° of Females - Phase 2	-,079[a]	-,484	,632	-,087	,808

a. Predictors in the Model: (Constant), N° of Friends - Phase 2
b. Dependent Variable: Provision of Infrastructure - Phase 2

Table 5-43: Regression Analysis Phase 2: Sources of Infrastructure

5.2.3.2.4 Support

Sources of support to be taken into account in the second phase according the Spearman correlations are friends and women, both of which are included in the subsequent regression model depicted in Table 5-44. Both variables are kept in the regression equation, the strongest effect on the dependent variable being exerted by *Number of Friends* with a *Beta* value of 0.635 compared to the *Beta* value of 0.288 for *Number of Females*.

Model Summary

Model	R	R Square	Adjusted R Square	Std. Error of the Estimate
1	,761[a]	,579	,566	1,10597
2	,804[b]	,646	,623	1,03053

a. Predictors: (Constant), N° of Friends - Phase 2
b. Predictors: (Constant), N° of Friends - Phase 2, N° of Females - Phase 2

ANOVA[c]

Model		Sum of Squares	df	Mean Square	F	Sig.
1	Regression	53,829	1	53,829	44,008	,000[a]
	Residual	39,141	32	1,223		
	Total	92,971	33			
2	Regression	60,049	2	30,024	28,272	,000[b]
	Residual	32,922	31	1,062		
	Total	92,971	33			

a. Predictors: (Constant), N° of Friends - Phase 2
b. Predictors: (Constant), N° of Friends - Phase 2, N° of Females - Phase 2
c. Dependent Variable: Support - Phase 2

Coefficients[a]

Model		Unstandardized Coefficients		Standardized Coefficients	t	Sig.
		B	Std. Error	Beta		
1	(Constant)	1,035	,242		4,284	,000
	N° of Friends - Phase 2	,751	,113	,761	6,634	,000
2	(Constant)	,783	,248		3,154	,004
	N° of Friends - Phase 2	,626	,117	,635	5,337	,000
	N° of Females - Phase 2	,458	,189	,288	2,420	,022

a. Dependent Variable: Support - Phase 2

Excluded Variables[b]

Model		Beta In	t	Sig.	Partial Correlation	Collinearity Statistics
						Tolerance
1	N° of Females - Phase 2	,288[a]	2,420	,022	,399	,808

a. Predictors in the Model: (Constant), N° of Friends - Phase 2
b. Dependent Variable: Support - Phase 2

Table 5-44: Multiple Regression Analysis Phase 2: Support with Functional Roles

Similar to the first phase, both types of support, *Emotional* and *Institutional Support* are significantly related to the overall number of support-providing ties at this stage (see Table 5-45). *Institutional Support* further dominates with a slightly larger standardised regression coefficient of 0.636 over a *Beta* value of for *Emotional Support* 0.489.

Model Summary

Model	R	R Square	Adjusted R Square	Std. Error of the Estimate
1	,864[a]	,747	,739	,85800
2	,966[b]	,934	,929	,44618

a. Predictors: (Constant), Institutional Support - Phase 2
b. Predictors: (Constant), Institutional Support - Phase 2, Emotional Support - Phase 2

ANOVA[c]

Model		Sum of Squares	df	Mean Square	F	Sig.
1	Regression	69,413	1	69,413	94,291	,000[a]
	Residual	23,557	32	,736		
	Total	92,971	33			
2	Regression	86,799	2	43,400	218,009	,000[b]
	Residual	6,171	31	,199		
	Total	92,971	33			

a. Predictors: (Constant), Institutional Support - Phase 2
b. Predictors: (Constant), Institutional Support - Phase 2, Emotional Support - Phase 2
c. Dependent Variable: Support - Phase 2

Coefficients[a]

Model		Unstandardized Coefficients		Standardized Coefficients	t	Sig.
		B	Std. Error	Beta		
1	(Constant)	,567	,211		2,693	,011
	Institutional Support - Phase 2	1,058	,109	,864	9,710	,000
2	(Constant)	,245	,115		2,136	,041
	Institutional Support - Phase 2	,778	,064	,636	12,143	,000
	Emotional Support - Phase 2	,618	,066	,489	9,345	,000

a. Dependent Variable: Support - Phase 2

Excluded Variables[b]

Model		Beta In	t	Sig.	Partial Correlation	Collinearity Statistics
						Tolerance
1	Emotional Support - Phase 2	,489[a]	9,345	,000	,859	,782

a. Predictors in the Model: (Constant), Institutional Support - Phase 2
b. Dependent Variable: Support - Phase 2

Table 5-45: Multiple Regression Analysis Phase 2: Support with Support Types

Coming to the nature of support providing relations, the Spearman correlation analysis of Chapter 5.2.2 identified *Strong Ties*, *Multiplexity* and *Closure* as influential tie characteristics in this regard. Being subjected to multiple regression analysis only *Multiplexity* turns out to serve as significant predictor for the provision of support. As the subsequent Table 5-46 reveals, the higher the multiplexity of the network the larger the amount of support provided (B value of 2.199).

Model Summary

Model	R	R Square	Adjusted R Square	Std. Error of the Estimate
1	,525[a]	,276	,253	1,45076

a. Predictors: (Constant), Multiplexity - Phase 2

ANOVA[b]

Model		Sum of Squares	df	Mean Square	F	Sig.
1	Regression	25,620	1	25,620	12,173	,001[a]
	Residual	67,350	32	2,105		
	Total	92,971	33			

a. Predictors: (Constant), Multiplexity - Phase 2
b. Dependent Variable: Support - Phase 2

Coefficients[a]

Model		Unstandardized Coefficients		Standardized Coefficients	t	Sig.
		B	Std. Error	Beta		
1	(Constant)	1,022	,381		2,683	,011
	Multiplexity - Phase 2	2,199	,630	,525	3,489	,001

a. Dependent Variable: Support - Phase 2

Excluded Variables[b]

Model		Beta In	t	Sig.	Partial Correlation	Collinearity Statistics Tolerance
1	Strong Ties - Phase 2	,188[a]	,857	,398	,152	,474
	Closure - Phase 2	-,128[a]	-,449	,656	-,080	,285

a. Predictors in the Model: (Constant), Multiplexity - Phase 2
b. Dependent Variable: Support - Phase 2

Table 5-46: Multiple Regression Analysis Phase 2: Support with Structural Parameters

Model Summary

Model	R	R Square	Adjusted R Square	Std. Error of the Estimate
1	,662[a]	,439	,421	1,01109
2	,754[b]	,569	,541	,90029

a. Predictors: (Constant), N° of Friends - Phase 2
b. Predictors: (Constant), N° of Friends - Phase 2, N° of Females - Phase 2

ANOVA[c]

Model		Sum of Squares	df	Mean Square	F	Sig.
1	Regression	25,551	1	25,551	24,994	,000[a]
	Residual	32,714	32	1,022		
	Total	58,265	33			
2	Regression	33,139	2	16,569	20,443	,000[b]
	Residual	25,126	31	,811		
	Total	58,265	33			

a. Predictors: (Constant), N° of Friends - Phase 2
b. Predictors: (Constant), N° of Friends - Phase 2, N° of Females - Phase 2
c. Dependent Variable: Emotional Support - Phase 2

Coefficients[a]

Model		Unstandardized Coefficients		Standardized Coefficients	t	Sig.
		B	Std. Error	Beta		
1	(Constant)	,462	,221		2,092	,044
	N° of Friends - Phase 2	,517	,103	,662	4,999	,000
2	(Constant)	,183	,217		,845	,404
	N° of Friends - Phase 2	,380	,103	,486	3,704	,001
	N° of Females - Phase 2	,506	,165	,402	3,060	,005

a. Dependent Variable: Emotional Support - Phase 2

Excluded Variables[c]

Model		Beta In	t	Sig.	Partial Correlation	Collinearity Statistics Tolerance
1	N° of Females - Phase 2	,402[a]	3,060	,005	,482	,808
	N° of Family Members - Phase 2	,190[a]	1,414	,167	,246	,944
2	N° of Family Members - Phase 2	,070[b]	,538	,594	,098	,833

a. Predictors in the Model: (Constant), N° of Friends - Phase 2
b. Predictors in the Model: (Constant), N° of Friends - Phase 2, N° of Females - Phase 2
c. Dependent Variable: Emotional Support - Phase 2

Table 5-47: Regression Analysis Phase 2: Sources of Emotional Support

As the previous correlation analysis of Chapter 5.2.2.4 highlights, possible providers of emotional support are friends, women and family member. Yet, the association of the latter is not confirmed by regression analysis as Table 5-47 reveals. *Number of*

Friends and *Number of Females*, in turn, exhibit almost equally strong positive influences, i.e. *Beta* values of 0.486 and 0.402 respectively, on the amount of emotional support offered.

Regarding the provision of institutional support in the second phase of the spin-off process, friends were identified as possible source in this respect in the course of the correlation analysis of Chapter 5.2.2.4. This original assumption is confirmed by the subsequent regression analysis summarised in Table 5-48.

Model Summary

Model	R	R Square	Adjusted R Square	Std. Error of the Estimate
1	,621[a]	,386	,367	1,09109

a. Predictors: (Constant), N° of Friends - Phase 2

ANOVA[b]

Model		Sum of Squares	df	Mean Square	F	Sig.
1	Regression	23,934	1	23,934	20,104	,000[a]
	Residual	38,096	32	1,190		
	Total	62,029	33			

a. Predictors: (Constant), N° of Friends - Phase 2
b. Dependent Variable: Institutional Support - Phase 2

Coefficients[a]

Model		Unstandardized Coefficients		Standardized Coefficients	t	Sig.
		B	Std. Error	Beta		
1	(Constant)	,720	,238		3,018	,005
	N° of Friends - Phase 2	,501	,112	,621	4,484	,000

a. Dependent Variable: Institutional Support - Phase 2

Table 5-48: Regression Analysis Phase 2: Sources of Institutional Support

5.2.3.3 Establishment

Similar to the two previous phases, also in the last phase of firm establishment and its first steps on the market, those variables featuring significant correlations after Spearman are subjected to regression analysis, in order to reveal the actual nature of their relations.

5.2.3.3.1 Network Size

Starting with Network Size as dependent variable, at first a deeper look is made at the different functional roles of alters and their influence on the size of ego's personal network. Against the background of the results of the correlation analysis of Chapter 5.2.2, *Number of Others*, *Industry* and *Funding Partners*, *Friends* and *Females* are subjected to regression analysis. As a result, the number of female contacts is excluded from the regression equation due to a lack of significance (see Table 5-49).

Model Summary

Model	R	R Square	Adjusted R Square	Std. Error of the Estimate
1	,574[a]	,329	,308	1,858
2	,769[b]	,592	,565	1,472
3	,852[c]	,726	,698	1,227
4	,894[d]	,799	,771	1,069

a. Predictors: (Constant), N° of Others - Phase 3
b. Predictors: (Constant), N° of Others - Phase 3, N° of Industry Partners - Phase 3
c. Predictors: (Constant), N° of Others - Phase 3, N° of Industry Partners - Phase 3, N° of Friends - Phase 3
d. Predictors: (Constant), N° of Others - Phase 3, N° of Industry Partners - Phase 3, N° of Friends - Phase 3, N° of Funding Partners - Phase 3

ANOVA[e]

Model		Sum of Squares	df	Mean Square	F	Sig.
1	Regression	54,167	1	54,167	15,693	,000[a]
	Residual	110,451	32	3,452		
	Total	164,618	33			
2	Regression	97,419	2	48,710	22,471	,000[b]
	Residual	67,198	31	2,168		
	Total	164,618	33			
3	Regression	119,443	3	39,814	26,440	,000[c]
	Residual	45,174	30	1,506		
	Total	164,618	33			
4	Regression	131,479	4	32,870	28,764	,000[d]
	Residual	33,139	29	1,143		
	Total	164,618	33			

a. Predictors: (Constant), N° of Others - Phase 3
b. Predictors: (Constant), N° of Others - Phase 3, N° of Industry Partners - Phase 3
c. Predictors: (Constant), N° of Others - Phase 3, N° of Industry Partners - Phase 3, N° of Friends - Phase 3
d. Predictors: (Constant), N° of Others - Phase 3, N° of Industry Partners - Phase 3, N° of Friends - Phase 3, N° of Funding Partners - Phase 3
e. Dependent Variable: Network Size in Phase 3

Coefficients[a]

Model		Unstandardized Coefficients		Standardized Coefficients	t	Sig.
		B	Std. Error	Beta		
1	(Constant)	3,431	,382		8,987	,000
	N° of Others - Phase 3	,833	,210	,574	3,961	,000
2	(Constant)	2,476	,371		6,683	,000
	N° of Others - Phase 3	,932	,168	,642	5,543	,000
	N° of Industry Partners - Phase 3	,856	,192	,517	4,467	,000
3	(Constant)	1,650	,377		4,380	,000
	N° of Others - Phase 3	1,039	,143	,716	7,272	,000
	N° of Industry Partners - Phase 3	,654	,168	,395	3,886	,001
	N° of Friends - Phase 3	,728	,190	,396	3,824	,001
4	(Constant)	1,250	,351		3,564	,001
	N° of Others - Phase 3	,801	,145	,551	5,535	,000
	N° of Industry Partners - Phase 3	,597	,148	,361	4,044	,000
	N° of Friends - Phase 3	,752	,166	,409	4,529	,000
	N° of Funding Partners - Phase 3	,839	,258	,317	3,245	,003

a. Dependent Variable: Network Size in Phase 3

Excluded Variables[e]

Model		Beta In	t	Sig.	Partial Correlation	Collinearity Statistics Tolerance
1	N° of Females - Phase 3	,250[a]	1,704	,098	,293	,919
	N° of Friends - Phase 3	,522[a]	4,406	,000	,621	,948
	N° of Industry Partners - Phase 3	,517[a]	4,467	,000	,626	,983
	N° of Funding Partners - Phase 3	,360[a]	2,269	,030	,377	,737
2	N° of Females - Phase 3	,189[b]	1,603	,119	,281	,905
	N° of Friends - Phase 3	,396[b]	3,824	,001	,572	,855
	N° of Funding Partners - Phase 3	,298[b]	2,372	,024	,397	,728
3	N° of Females - Phase 3	,080[c]	,756	,456	,139	,823
	N° of Funding Partners - Phase 3	,317[c]	3,245	,003	,516	,726
4	N° of Females - Phase 3	-,021[d]	-,208	,837	-,039	,733

a. Predictors in the Model: (Constant), N° of Others - Phase 3
b. Predictors in the Model: (Constant), N° of Others - Phase 3, N° of Industry Partners - Phase 3
c. Predictors in the Model: (Constant), N° of Others - Phase 3, N° of Industry Partners - Phase 3, N° of Friends - Phase 3
d. Predictors in the Model: (Constant), N° of Others - Phase 3, N° of Industry Partners - Phase 3, N° of Friends - Phase 3, N° of Funding Partners - Phase 3
e. Dependent Variable: Network Size in Phase 3

Table 5-49: Multiple Regression Analysis Phase 3: Network Size with Functional Roles

Having a look at the relative influence of the remaining regressors on *Network Size* as expressed by the standardised regression coefficients in Table 5-49, one may state the it is the *Number of Others* that features the largest impact on the size of entrepreneurs' personal networks with a *Beta* value of 0.551, followed by the *Number of Friends* with a *Beta* value of 0.409, and finally the *Number of Industry Partners* with 0.361 and *Funding Partners* with 0.317. In this context, *B* and *Beta* values, i.e. unstandardised and standardised regression coefficients, do not render the same result. Yet, when intending to determine the relative impact on network size among the predictors, one has to take the standardised values into account that allow for a dimensionless comparison. When requiring information on the actual estimated impact of one additional unit of one of the independent variables on the dependent variable unstandardised B values are indicated. Since we are only interested in the relative importance of the individual explanatory network parameters, referring to Beta values is sufficient.

As *Number of Others* is the predominant type of functional role in the last phase, at this point it seems to be necessary to commemorate what types of alters are subsumed under this heading. As presented in Chapter 4.3.8.3, the variable primarily referred to consultants, lawyers, notaries, accountants, funding relations, i.e. banks, and particular in the last phase to strategic co-operation partners and associates.

Model Summary

Model	R	R Square	Adjusted R Square	Std. Error of the Estimate
1	,793[a]	,629	,618	1,381
2	,830[b]	,689	,669	1,285

a. Predictors: (Constant), Informational Content - Phase 3
b. Predictors: (Constant), Informational Content - Phase 3, Support - Phase 3

ANOVA[c]

Model		Sum of Squares	df	Mean Square	F	Sig.
1	Regression	103,623	1	103,623	54,365	,000[a]
	Residual	60,994	32	1,906		
	Total	164,618	33			
2	Regression	113,461	2	56,731	34,378	,000[b]
	Residual	51,157	31	1,650		
	Total	164,618	33			

a. Predictors: (Constant), Informational Content - Phase 3
b. Predictors: (Constant), Informational Content - Phase 3, Support - Phase 3
c. Dependent Variable: Network Size in Phase 3

Coefficients[a]

Model		Unstandardized Coefficients		Standardized Coefficients	t	Sig.
		B	Std. Error	Beta		
1	(Constant)	1,737	,417		4,170	,000
	Informational Content - Phase 3	,842	,114	,793	7,373	,000
2	(Constant)	1,493	,400		3,731	,001
	Informational Content - Phase 3	,675	,127	,635	5,329	,000
	Support - Phase 3	,302	,124	,291	2,442	,021

a. Dependent Variable: Network Size in Phase 3

Excluded Variables[c]

Model		Beta In	t	Sig.	Partial Correlation	Collinearity Statistics Tolerance
1	Provision of Resources - Phase 3	,066[a]	,367	,716	,066	,369
	Support - Phase 3	,291[a]	2,442	,021	,402	,705
2	Provision of Resources - Phase 3	-,037[b]	-,213	,832	-,039	,346

a. Predictors in the Model: (Constant), Informational Content - Phase 3
b. Predictors in the Model: (Constant), Informational Content - Phase 3, Support - Phase 3
c. Dependent Variable: Network Size in Phase 3

Table 5-50: Multiple Regression Analysis Phase 3: Network Size with Relational Content

Regarding the prevalent type of content determining the size of the personal network, following the results of the correlation analysis after Spearman of Chapter 5.2.2, all three possible content types, i.e. information, resources, and support, are put to regression analysis. Below the line, as for the previous two phases, *Provision of Resources* is excluded from the regression equation due to its lack of significance. Remaining variables are *Informational Content* and *Support*, with *Informational Content* featuring by a much higher influence, i.e. *Beta* of 0.635 as opposed to a *Beta* value of 0.291 for *Support*. In this respect, the difference in influence between those two variables has increased compared to the second phase, yet, without reaching the large gap of the first phase.

5.2.3.3.2 Informational Content

With a view on the different sources of information in the last phase according to Table 5-51, out of the types of contact submitted to regression analysis following the results of the previous correlation analysis after Spearman, all except variable except from the Number of Females are kept in the regression equation. This means that in the last phase information is primarily obtained from friends (*Beta* 0.583) and funding partners (*Beta* 0.551), followed far behind by academics (Beta 0.243).

Model Summary

Model	R	R Square	Adjusted R Square	Std. Error of the Estimate
1	,582[a]	,338	,318	1,73759
2	,791[b]	,625	,601	1,32859
3	,825[c]	,680	,648	1,24828

a. Predictors: (Constant), N° of Friends - Phase 3
b. Predictors: (Constant), N° of Friends - Phase 3, N° of Funding Partners - Phase 3
c. Predictors: (Constant), N° of Friends - Phase 3, N° of Funding Partners - Phase 3, N° of Academics - Phase 3

ANOVA[d]

Model		Sum of Squares	df	Mean Square	F	Sig.
1	Regression	49,385	1	49,385	16,357	,000[a]
	Residual	96,615	32	3,019		
	Total	146,000	33			
2	Regression	91,280	2	45,640	25,856	,000[b]
	Residual	54,720	31	1,765		
	Total	146,000	33			
3	Regression	99,254	3	33,085	21,233	,000[c]
	Residual	46,746	30	1,558		
	Total	146,000	33			

a. Predictors: (Constant), N° of Friends - Phase 3
b. Predictors: (Constant), N° of Friends - Phase 3, N° of Funding Partners - Phase 3
c. Predictors: (Constant), N° of Friends - Phase 3, N° of Funding Partners - Phase 3, N° of Academics - Phase 3
d. Dependent Variable: Informational Content - Phase 3

Coefficients[a]

Model		Unstandardized Coefficients		Standardized Coefficients	t	Sig.
		B	Std. Error	Beta		
1	(Constant)	1,725	,434		3,978	,000
	N° of Friends - Phase 3	1,008	,249	,582	4,044	,000
2	(Constant)	,513	,415		1,238	,225
	N° of Friends - Phase 3	1,122	,192	,648	5,847	,000
	N° of Funding Partners - Phase 3	1,344	,276	,540	4,872	,000
3	(Constant)	,309	,400		,773	,446
	N° of Friends - Phase 3	1,011	,187	,583	5,403	,000
	N° of Funding Partners - Phase 3	1,372	,259	,551	5,287	,000
	N° of Academics - Phase 3	,423	,187	,243	2,262	,031

a. Dependent Variable: Informational Content - Phase 3

Excluded Variables[d]

Model		Beta In	t	Sig.	Partial Correlation	Collinearity Statistics Tolerance
1	N° of Females - Phase 3	,388[a]	2,909	,007	,463	,945
	N° of Academics - Phase 3	,216[a]	1,472	,151	,256	,926
	N° of Funding Partners - Phase 3	,540[a]	4,872	,000	,659	,985
2	N° of Females - Phase 3	,177[b]	1,419	,166	,251	,756
	N° of Academics - Phase 3	,243[b]	2,262	,031	,382	,924
3	N° of Females - Phase 3	,198[c]	1,717	,097	,304	,752

a. Predictors in the Model: (Constant), N° of Friends - Phase 3
b. Predictors in the Model: (Constant), N° of Friends - Phase 3, N° of Funding Partners - Phase 3
c. Predictors in the Model: (Constant), N° of Friends - Phase 3, N° of Funding Partners - Phase 3, N° of Academics - Phase 3
d. Dependent Variable: Informational Content - Phase 3

Table 5-51: Multiple Regression Analysis Phase 3: Informational Content with Functional Roles

As Table 5-52 reveals, out of the types of information subjected to regression analysis for this phase of the spin-off process legal information no longer features a significant influence on the total number of informational ties in egos' personal networks at this stage. Yet, the most important impact in this respect can be observed again for *Other Information* with a standardised regression coefficient of 0.571 followed by *Economic Information* with a respective value of 0.416. More far behind comes Information about *Potential Customers* with a Beta value of 0.241.

Model Summary

Model	R	R Square	Adjusted R Square	Std. Error of the Estimate
1	,724[a]	,525	,510	1,47233
2	,884[b]	,782	,768	1,01281
3	,906[c]	,822	,804	,93155

a. Predictors: (Constant), Economic information - Phase 3
b. Predictors: (Constant), Economic information - Phase 3, Other Information - Phase 3
c. Predictors: (Constant), Economic information - Phase 3, Other Information - Phase 3, Information about Potential Customers - Phase 3

ANOVA[d]

Model		Sum of Squares	df	Mean Square	F	Sig.
1	Regression	76,632	1	76,632	35,351	,000[a]
	Residual	69,368	32	2,168		
	Total	146,000	33			
2	Regression	114,201	2	57,100	55,665	,000[b]
	Residual	31,799	31	1,026		
	Total	146,000	33			
3	Regression	119,966	3	39,989	46,081	,000[c]
	Residual	26,034	30	,868		
	Total	146,000	33			

a. Predictors: (Constant), Economic information - Phase 3
b. Predictors: (Constant), Economic information - Phase 3, Other Information - Phase 3
c. Predictors: (Constant), Economic information - Phase 3, Other Information - Phase 3, Information about Potential Customers - Phase 3
d. Dependent Variable: Informational Content - Phase 3

Coefficients[a]

Model		Unstandardized Coefficients B	Unstandardized Coefficients Std. Error	Standardized Coefficients Beta	t	Sig.
1	(Constant)	1,374	,372		3,691	,001
	Economic information - Phase 3	,970	,163	,724	5,946	,000
2	(Constant)	,947	,266		3,566	,001
	Economic information - Phase 3	,748	,118	,558	6,331	,000
	Other Information - Phase 3	,777	,128	,534	6,052	,000
3	(Constant)	,723	,259		2,790	,009
	Economic information - Phase 3	,557	,131	,416	4,237	,000
	Other Information - Phase 3	,831	,120	,571	6,929	,000
	Information about Potential Customers - Phase 3	,405	,157	,241	2,578	,015

a. Dependent Variable: Informational Content - Phase 3

Excluded Variables[d]

Model		Beta In	t	Sig.	Partial Correlation	Collinearity Statistics Tolerance
1	Legal Information - Phase 3	,076[a]	,404	,689	,072	,434
	Information about Potential Customers - Phase 3	,128[a]	,875	,388	,155	,704
	Other Information - Phase 3	,534[a]	6,052	,000	,736	,903
2	Legal Information - Phase 3	,174[b]	1,376	,179	,244	,427
	Information about Potential Customers - Phase 3	,241[b]	2,578	,015	,426	,682
3	Legal Information - Phase 3	,123[c]	1,028	,312	,188	,413

a. Predictors in the Model: (Constant), Economic information - Phase 3
b. Predictors in the Model: (Constant), Economic information - Phase 3, Other Information - Phase 3
c. Predictors in the Model: (Constant), Economic information - Phase 3, Other Information - Phase 3, Information about Potential Customers - Phase 3
d. Dependent Variable: Informational Content - Phase 3

Table 5-52: Multiple Regression Analysis Phase 3: Informational Content with Information Types

Taking a closer look at the nature of the informational ties, Spearman correlations suggest significant interrelations between *Informational Content* and *Strong Ties*, *Closure*, and *Multiplexity*. Subjected to multiple regression analysis as predictors, all three parameters featured significant regression coefficients and are thus kept in the equation of the following regression model as depicted in Table 5-53. While *Strong Ties* feature the most important influence on *Informational Content* with a *B* value of 1.142 and *Beta* value of 0.989, again as in the previous phase, the influence of network closure on the provision of information is a negative one (*B* of -4.927 and *Beta* of -0.839). This means the closer the network the smaller the number of informational ties it contains. *Multiplexity* again has positive, even though smaller (*Beta* of 0.638) influence on *Informational Content* than *Strong Ties*.

Model Summary

Model	R	R Square	Adjusted R Square	Std. Error of the Estimate
1	,806^a	,650	,640	1,26290
2	,832^b	,693	,673	1,20310
3	,875^c	,765	,741	1,06990

a. Predictors: (Constant), Strong Ties - Phase 3
b. Predictors: (Constant), Strong Ties - Phase 3, Closure - Phase 3
c. Predictors: (Constant), Strong Ties - Phase 3, Closure - Phase 3, Multiplexity - Phase 3

ANOVA^d

Model		Sum of Squares	df	Mean Square	F	Sig.
1	Regression	94,963	1	94,963	59,541	,000^a
	Residual	51,037	32	1,595		
	Total	146,000	33			
2	Regression	101,129	2	50,564	34,933	,000^b
	Residual	44,871	31	1,447		
	Total	146,000	33			
3	Regression	111,660	3	37,220	32,516	,000^c
	Residual	34,340	30	1,145		
	Total	146,000	33			

a. Predictors: (Constant), Strong Ties - Phase 3
b. Predictors: (Constant), Strong Ties - Phase 3, Closure - Phase 3
c. Predictors: (Constant), Strong Ties - Phase 3, Closure - Phase 3, Multiplexity - Phase 3
d. Dependent Variable: Informational Content - Phase 3

Coefficients^a

Model		Unstandardized Coefficients B	Std. Error	Standardized Coefficients Beta	t	Sig.
1	(Constant)	1,330	,306		4,342	,000
	Strong Ties - Phase 3	,931	,121	,806	7,716	,000
2	(Constant)	1,523	,306		4,971	,000
	Strong Ties - Phase 3	1,300	,212	1,126	6,117	,000
	Closure - Phase 3	-2,232	1,081	-,380	-2,064	,047
3	(Constant)	1,316	,281		4,687	,000
	Strong Ties - Phase 3	1,142	,196	,989	5,825	,000
	Closure - Phase 3	-4,927	1,309	-,839	-3,763	,001
	Multiplexity - Phase 3	3,116	1,027	,638	3,033	,005

a. Dependent Variable: Informational Content - Phase 3

Excluded Variables^c

Model		Beta In	t	Sig.	Partial Correlation	Collinearity Statistics Tolerance
1	Multiplexity - Phase 3	,101^a	,547	,588	,098	,329
	Closure - Phase 3	-,380^a	-2,064	,047	-,348	,293
2	Multiplexity - Phase 3	,638^b	3,033	,005	,484	,177

a. Predictors in the Model: (Constant), Strong Ties - Phase 3
b. Predictors in the Model: (Constant), Strong Ties - Phase 3, Closure - Phase 3
c. Dependent Variable: Informational Content - Phase 3

Table 5-53: Multiple Regression Analysis Phase 3: Informational Content with Structural Parameters

As the regression analysis of Table 5-52 discloses, in the last phase relevant information types exchanged in the personal networks of academic entrepreneurs are economic know-how, other information, and information about potential customers. Consequently, in the following all three information categories are subjected to regression analysis with possible sources of information positively associated with them according to the correlation results of Chapter 5.2.2.2.

Model Summary

Model	R	R Square	Adjusted R Square	Std. Error of the Estimate
1	,523[a]	,274	,251	1,35970

a. Predictors: (Constant), N° of Friends - Phase 3

ANOVA[b]

Model		Sum of Squares	df	Mean Square	F	Sig.
1	Regression	22,280	1	22,280	12,051	,002[a]
	Residual	59,162	32	1,849		
	Total	81,441	33			

a. Predictors: (Constant), N° of Friends - Phase 3
b. Dependent Variable: Economic information - Phase 3

Coefficients[a]

Model		Unstandardized Coefficients		Standardized Coefficients	t	Sig.
		B	Std. Error	Beta		
1	(Constant)	,820	,339		2,417	,022
	N° of Friends - Phase 3	,677	,195	,523	3,471	,002

a. Dependent Variable: Economic information - Phase 3

Excluded Variables[b]

Model		Beta In	t	Sig.	Partial Correlation	Collinearity Statistics Tolerance
1	N° of Funding Partners - Phase 3	,268[a]	1,825	,078	,312	,985

a. Predictors in the Model: (Constant), N° of Friends - Phase 3
b. Dependent Variable: Economic information - Phase 3

Table 5-54: Regression Analysis Phase 3: Sources of Economic Information

As indicated by the Spearman correlations, business know-how is significantly correlated with the *Number of Friends* and the *Number of Funding Partners*. When subjected to regression analysis as depicted in Table 5-45, however, only the *Number of Friends* remains as significant predictor of the amount of economic know-how provided in the regression model.

Model Summary

Model	R	R Square	Adjusted R Square	Std. Error of the Estimate
1	,642[a]	,412	,394	,97311

a. Predictors: (Constant), N° of Friends - Phase 3

ANOVA[b]

Model		Sum of Squares	df	Mean Square	F	Sig.
1	Regression	21,256	1	21,256	22,447	,000[a]
	Residual	30,302	32	,947		
	Total	51,559	33			

a. Predictors: (Constant), N° of Friends - Phase 3
b. Dependent Variable: Information about Potential Customers - Phase 3

Coefficients[a]

Model		Unstandardized Coefficients		Standardized Coefficients	T	Sig.
		B	Std. Error	Beta		
1	(Constant)	,370	,243		1,522	,138
	N° of Friends - Phase 3	,661	,140	,642	4,738	,000

a. Dependent Variable: Information about Potential Customers - Phase 3

Table 5-55: Regression Analysis Phase 3: Sources of Information about Potential Customers

Information about possible customers is mainly associated with the number of friends in the network. This conclusion of the Spearman correlation analysis is also confirmed by the subsequent regression analysis of Table 5-55. *Number of Friends* features a positive influence on the amount of customer-related information with a regression coefficient B of 0.661.

Model Summary

Model	R	R Square	Adjusted R Square	Std. Error of the Estimate
1	,501[a]	,251	,228	1,27034

a. Predictors: (Constant), N° of Funding Partners - Phase 3

ANOVA[b]

Model		Sum of Squares	df	Mean Square	F	Sig.
1	Regression	17,330	1	17,330	10,739	,003[a]
	Residual	51,640	32	1,614		
	Total	68,971	33			

a. Predictors: (Constant), N° of Funding Partners - Phase 3
b. Dependent Variable: Other Information – Phase 3

Coefficients[a]

Model		Unstandardized Coefficients		Standardized Coefficients	t	Sig.
		B	Std. Error	Beta		
1	(Constant)	,348	,301		1,157	,256
	N° of Funding Partners - Phase 3	,858	,262	,501	3,277	,003

a. Dependent Variable: Other Information - Phase 3

Table 5-56: Regression Analysis Phase 3: Sources of Other Information

Other information is positively correlated with the number of funding partners in the network. Again this indication of the correlation analysis of Chapter 5.2.2.2 is also reflected by the results of the regression analysis summarised in Table 5-56 hereafter. With a regression coefficient of 0.858, the *Number of Funding Partners* features even a rather strong positive impact on the amount of *Other Information*.

5.2.3.3.3 Resources

As regards the main providers of resources in the last phase of the spin-off process, friends, funding partners, academics and women, all having featured significant Spearman correlations in Chapter 5.2.2.3, are subjected to the subsequent regression analysis of Table 5-57. Apart from the *Number of Females* all other predictors for the *Provision of Resources* are kept in the final regression equation. The largest impact on the number of resource-providing ties is effected by the *Number of Friends* with a standardised regression coefficient of 0.533 followed by the *Number of Funding Partners* with 0.383 and the *Number of Academics* with 0.376.

Model Summary

Model	R	R Square	Adjusted R Square	Std. Error of the Estimate
1	,588[a]	,346	,325	1,42986
2	,691[b]	,478	,444	1,29802
3	,780[c]	,608	,569	1,14285

a. Predictors: (Constant), N° of Friends - Phase 3
b. Predictors: (Constant), N° of Friends - Phase 3, N° of Funding Partners - Phase 3
c. Predictors: (Constant), N° of Friends - Phase 3, N° of Funding Partners - Phase 3, N° of Academics - Phase 3

ANOVA[d]

Model		Sum of Squares	df	Mean Square	F	Sig.
1	Regression	34,576	1	34,576	16,912	,000[a]
	Residual	65,424	32	2,045		
	Total	100,000	33			
2	Regression	47,769	2	23,885	14,176	,000[b]
	Residual	52,231	31	1,685		
	Total	100,000	33			
3	Regression	60,817	3	20,272	15,521	,000[c]
	Residual	39,183	30	1,306		
	Total	100,000	33			

a. Predictors: (Constant), N° of Friends - Phase 3
b. Predictors: (Constant), N° of Friends - Phase 3, N° of Funding Partners - Phase 3
c. Predictors: (Constant), N° of Friends - Phase 3, N° of Funding Partners - Phase 3, N° of Academics - Phase 3
d. Dependent Variable: Provision of Resources - Phase 3

Coefficients[a]

Model		Unstandardized Coefficients		Standardized Coefficients	t	Sig.
		B	Std. Error	Beta		
1	(Constant)	,933	,357		2,615	,013
	N° of Friends - Phase 3	,843	,205	,588	4,112	,000
2	(Constant)	,253	,405		,625	,536
	N° of Friends - Phase 3	,908	,188	,633	4,839	,000
	N° of Funding Partners - Phase 3	,754	,269	,366	2,798	,009
3	(Constant)	-,008	,366		-,022	,982
	N° of Friends - Phase 3	,765	,171	,533	4,465	,000
	N° of Funding Partners - Phase 3	,790	,238	,383	3,325	,002
	N° of Academics - Phase 3	,542	,171	,376	3,161	,004

a. Dependent Variable: Provision of Resources - Phase 3

Excluded Variables[d]

Model		Beta In	t	Sig.	Partial Correlation	Collinearity Statistics Tolerance
1	N° of Females - Phase 3	,277[a]	1,967	,058	,333	,945
	N° of Academics - Phase 3	,357[a]	2,612	,014	,425	,926
	N° of Funding Partners - Phase 3	,366[a]	2,798	,009	,449	,985
2	N° of Females - Phase 3	,138[b]	,920	,365	,166	,756
	N° of Academics - Phase 3	,376[b]	3,161	,004	,500	,924
3	N° of Females - Phase 3	,171[c]	1,309	,201	,236	,752

a. Predictors in the Model: (Constant), N° of Friends - Phase 3
b. Predictors in the Model: (Constant), N° of Friends - Phase 3, N° of Funding Partners - Phase 3
c. Predictors in the Model: (Constant), N° of Friends - Phase 3, N° of Funding Partners - Phase 3, N° of Academics - Phase 3
d. Dependent Variable: Provision of Resources - Phase 3

Table 5-57: Multiple Regression Analysis Phase 3: Resources with Functional Roles

Coming to the distribution of resources in the course of establishing on the market, *Financial Resources* are no longer of significant influence on the amount of resources provided. The most important resource at this point as regards the results of the regression analysis is personnel with a standardised regression coefficient of 0.456. At some distance follow infrastructure with a *Beta* of 0.329 and material resources with a *Beta* of 0.286.

Model Summary

Model	R	R Square	Adjusted R Square	Std. Error of the Estimate
1	,699[a]	,488	,472	1,26441
2	,834[b]	,696	,676	,99044
3	,858[c]	,736	,709	,93857

a. Predictors: (Constant), Provision of Infrastructure - Phase 3
b. Predictors: (Constant), Provision of Infrastructure - Phase 3, Provision of Personnel - Phase 3
c. Predictors: (Constant), Provision of Infrastructure - Phase 3, Provision of Personnel - Phase 3, Material Resources - Phase 3

ANOVA[d]

Model		Sum of Squares	df	Mean Square	F	Sig.
1	Regression	48,840	1	48,840	30,549	,000[a]
	Residual	51,160	32	1,599		
	Total	100,000	33			
2	Regression	69,590	2	34,795	35,470	,000[b]
	Residual	30,410	31	,981		
	Total	100,000	33			
3	Regression	73,572	3	24,524	27,839	,000[c]
	Residual	26,428	30	,881		
	Total	100,000	33			

a. Predictors: (Constant), Provision of Infrastructure - Phase 3
b. Predictors: (Constant), Provision of Infrastructure - Phase 3, Provision of Personnel - Phase 3
c. Predictors: (Constant), Provision of Infrastructure - Phase 3, Provision of Personnel - Phase 3, Material Resources - Phase 3
d. Dependent Variable: Provision of Resources - Phase 3

Coefficients[a]

Model		Unstandardized Coefficients		Standardized Coefficients	t	Sig.
		B	Std. Error	Beta		
1	(Constant)	1,111	,270		4,114	,000
	Provision of Infrastructure - Phase 3	1,163	,210	,699	5,527	,000
2	(Constant)	,686	,231		2,974	,006
	Provision of Infrastructure - Phase 3	,865	,177	,520	4,884	,000
	Provision of Personnel - Phase 3	,693	,151	,489	4,599	,000
3	(Constant)	,672	,219		3,074	,004
	Provision of Infrastructure - Phase 3	,548	,225	,329	2,439	,021
	Provision of Personnel - Phase 3	,646	,144	,456	4,471	,000
	Material Resources - Phase 3	,445	,209	,286	2,126	,042

a. Dependent Variable: Provision of Resources - Phase 3

Excluded Variables[d]

Model		Beta In	T	Sig.	Partial Correlation	Collinearity Statistics
						Tolerance
1	Material Resources - Phase 3	,378[a]	2,239	,032	,373	,498
	Financial Resources - Phase 3	,343[a]	2,194	,036	,367	,584
	Provision of Personnel - Phase 3	,489[a]	4,599	,000	,637	,866
2	Material Resources - Phase 3	,286[b]	2,126	,042	,362	,486
	Financial Resources - Phase 3	,255[b]	2,036	,051	,348	,569
3	Financial Resources - Phase 3	,202[c]	1,611	,118	,287	,534

a. Predictors in the Model: (Constant), Provision of Infrastructure - Phase 3
b. Predictors in the Model: (Constant), Provision of Infrastructure - Phase 3, Provision of Personnel - Phase 3
c. Predictors in the Model: (Constant), Provision of Infrastructure - Phase 3, Provision of Personnel - Phase 3, Material Resources - Phase 3
d. Dependent Variable: Provision of Resources - Phase 3

Table 5-58: Multiple Regression Analysis Phase 3: Resources with Resource Types

In the last phase, resource-offering ties are not only characterised by high strength as in the previous phases, but also by decreasing network closure. This is reflected by the regression coefficient of Table 5-59. While *Strong Ties* feature a considerable relation to *Provision of Resources* with a *Beta* of 1.178, *Closure* exhibits a reverse effect. Its *Beta* value of -0.367 indicates that the number of resource-related ties decreases with increasing degree of closure.

Model Summary

Model	R	R Square	Adjusted R Square	Std. Error of the Estimate
1	,869[a]	,756	,748	,87348
2	,892[b]	,795	,782	,81264

a. Predictors: (Constant), Strong Ties - Phase 3
b. Predictors: (Constant), Strong Ties - Phase 3, Closure - Phase 3

ANOVA[c]

Model		Sum of Squares	df	Mean Square	F	Sig.
1	Regression	75,585	1	75,585	99,067	,000[a]
	Residual	24,415	32	,763		
	Total	100,000	33			
2	Regression	79,528	2	39,764	60,213	,000[b]
	Residual	20,472	31	,660		
	Total	100,000	33			

a. Predictors: (Constant), Strong Ties - Phase 3
b. Predictors: (Constant), Strong Ties - Phase 3, Closure - Phase 3
c. Dependent Variable: Provision of Resources - Phase 3

Coefficients[a]

Model		Unstandardized Coefficients		Standardized Coefficients	t	Sig.
		B	Std. Error	Beta		
1	(Constant)	,510	,212		2,407	,022
	Strong Ties - Phase 3	,831	,083	,869	9,953	,000
2	(Constant)	,665	,207		3,211	,003
	Strong Ties - Phase 3	1,126	,144	1,178	7,842	,000
	Closure - Phase 3	-1,785	,730	-,367	-2,443	,020

a. Dependent Variable: Provision of Resources - Phase 3

Excluded Variables[c]

Model		Beta In	t	Sig.	Partial Correlation	Collinearity Statistics Tolerance
1	Multiplexity - Phase 3	-,058[a]	-,375	,710	-,067	,329
	Closure - Phase 3	-,367[a]	-2,443	,020	-,402	,293
2	Multiplexity - Phase 3	,328[b]	1,758	,089	,306	,177

a. Predictors in the Model: (Constant), Strong Ties – Phase 3
b. Predictors in the Model: (Constant), Strong Ties – Phase 3, Closure - Phase 3
c. Dependent Variable: Provision of Resources - Phase 3

Table 5-59: Multiple Regression Analysis Phase 3: Resources with Structural Parameters

Taken from the results of the regression analysis as depicted in Table 5-58, physical assets mainly requested in the last phase of new venture creation are money, infrastructure, and most importantly personnel. *Number of Females* and *Number of Friends* are significantly correlated (see Chapter 5.2.2.3) with the brokerage of employees, and thus subjected to a corresponding regression analysis. As highlighted thereinafter in Table 5-60 only female contacts feature a significant impact with a regression coefficient B of 0.368 on the amount of contacts providing the required staff for the young start-up.

Model Summary

Model	R	R Square	Adjusted R Square	Std. Error of the Estimate
1	,352[a]	,124	,096	1,16872

a. Predictors: (Constant), N° of Females - Phase 3

ANOVA[b]

Model		Sum of Squares	Df	Mean Square	F	Sig.
1	Regression	6,173	1	6,173	4,519	,041[a]
	Residual	43,709	32	1,366		
	Total	49,882	33			

a. Predictors: (Constant), N° of Females - Phase 3
b. Dependent Variable: Provision of Personnel - Phase 3

Coefficients[a]

Model		Unstandardized Coefficients		Standardized Coefficients	t	Sig.
		B	Std. Error	Beta		
1	(Constant)	,616	,252		2,445	,020
	N° of Females - Phase 3	,368	,173	,352	2,126	,041

a. Dependent Variable: Provision of Personnel - Phase 3

Excluded Variables[b]

Model		Beta In	t	Sig.	Partial Correlation	Collinearity Statistics
						Tolerance
1	N° of Friends - Phase 3	,182[a]	1,070	,293	,189	,945

a. Predictors in the Model: (Constant), N° of Females - Phase 3
b. Dependent Variable: Provision of Personnel - Phase 3

Table 5-60: Regression Analysis Phase 3: Sources of Personnel

In line with the results of the Spearman correlation analysis of Chapter 5.2.2.3, the provision of infrastructure is related to the number of friends, academics, females, and industry contacts in entrepreneurs' personal networks in the course of the following multivariate regression analysis. Respective results are presented in Table 5-61. In the end, only the *Number of Friends* and the *Number of Academics* are kept in the regression equation with amicable contacts exhibiting the stronger impact on the amount of infrastructure provided with a *Beta* value of 0.550 compared to a *Beta* value of 0.357 for the number academic alters in the network.

Model Summary

Model	R	R Square	Adjusted R Square	Std. Error of the Estimate
1	,647[a]	,419	,401	,80994
2	,733[b]	,537	,507	,73440

a. Predictors: (Constant), N° of Friends - Phase 3
b. Predictors: (Constant), N° of Friends - Phase 3, N° of Academics - Phase 3

ANOVA[c]

Model		Sum of Squares	Df	Mean Square	F	Sig.
1	Regression	15,126	1	15,126	23,057	,000[a]
	Residual	20,992	32	,656		
	Total	36,118	33			
2	Regression	19,398	2	9,699	17,983	,000[b]
	Residual	16,720	31	,539		
	Total	36,118	33			

a. Predictors: (Constant), N° of Friends - Phase 3
b. Predictors: (Constant), N° of Friends - Phase 3, N° of Academics - Phase 3
c. Dependent Variable: Provision of Infrastructure - Phase 3

Coefficients[a]

Model		Unstandardized Coefficients		Standardized Coefficients	t	Sig.
		B	Std. Error	Beta		
1	(Constant)	,059	,202		,293	,771
	N° of Friends - Phase 3	,558	,116	,647	4,802	,000
2	(Constant)	-,072	,189		-,380	,707
	N° of Friends - Phase 3	,474	,109	,550	4,334	,000
	N° of Academics - Phase 3	,310	,110	,357	2,815	,008

a. Dependent Variable: Provision of Infrastructure - Phase 3

Excluded Variables[c]

Model		Beta In	T	Sig.	Partial Correlation	Collinearity Statistics Tolerance
1	N° of Females - Phase 3	-,055[a]	-,392	,698	-,070	,945
	N° of Academics - Phase 3	,357[a]	2,815	,008	,451	,926
	N° of Industry Partners - Phase 3	,072[a]	,494	,624	,088	,889
2	N° of Females - Phase 3	-,024[b]	-,184	,856	-,033	,937
	N° of Industry Partners - Phase 3	,101[b]	,772	,446	,140	,883

a. Predictors in the Model: (Constant), N° of Friends - Phase 3
b. Predictors in the Model: (Constant), N° of Friends - Phase 3, N° of Academics - Phase 3
c. Dependent Variable: Provision of Infrastructure - Phase 3

Table 5-61: Regression Analysis Phase 3: Sources of Infrastructure

Finally, according the correlation analysis after Spearman of Chapter 5.2.2.3 the provision of material resources was positively associated with the amount of friends, women and industry contacts in the network. Now, when entered in the multivariate regression model only the *Number of Friends* remains as significant influence factor on the number of contacts supplying material assets with a positive regression coefficient of 0.698.

Model Summary

Model	R	R Square	Adjusted R Square	Std. Error of the Estimate
1	,756[a]	,571	,557	,74555

a. Predictors: (Constant), N° of Friends - Phase 3

ANOVA[b]

Model		Sum of Squares	df	Mean Square	F	Sig.
1	Regression	23,654	1	23,654	42,555	,000[a]
	Residual	17,787	32	,556		
	Total	41,441	33			

a. Predictors: (Constant), N° of Friends - Phase 3
b. Dependent Variable: Material Resources - Phase 3

Coefficients[a]

Model		Unstandardized Coefficients		Standardized Coefficients	t	Sig.
		B	Std. Error	Beta		
1	(Constant)	-,206	,186		-1,105	,277
	N° of Friends - Phase 3	,698	,107	,756	6,523	,000

a. Dependent Variable: Material Resources - Phase 3

Excluded Variables[b]

Model		Beta In	t	Sig.	Partial Correlation	Collinearity Statistics
						Tolerance
1	N° of Females - Phase 3	,219[a]	1,914	,065	,325	,945
	N° of Industry Partners - Phase 3	,010[a]	,080	,936	,014	,889

a. Predictors in the Model: (Constant), N° of Friends - Phase 3
b. Dependent Variable: Material Resources – Phase 3

Table 5-62: Regression Analysis Phase 3: Sources of Material Resources

5.2.3.3.4 Support

Out of the two sources of support submitted to regression analysis based on their significant Spearman correlations, i.e. *Number of Funding* and *Industry Partners*, only *Number of Industry Partners* is kept in the regression equation (see Table 5-63). This can be interpreted as such that in particular partners out of the industry provide support, presumably in the form of institutional support, in the last phase of the spin-off process.

Model Summary

Model	R	R Square	Adjusted R Square	Std. Error of the Estimate
1	,376[a]	,142	,115	2,02230

a. Predictors: (Constant), N° of Industry Partners - Phase 3

ANOVA[b]

Model		Sum of Squares	df	Mean Square	F	Sig.
1	Regression	21,600	1	21,600	5,282	,028[a]
	Residual	130,871	32	4,090		
	Total	152,471	33			

a. Predictors: (Constant), N° of Industry Partners - Phase 3
b. Dependent Variable: Support - Phase 3

Coefficients[a]

Model		Unstandardized Coefficients		Standardized Coefficients	t	Sig.
		B	Std. Error	Beta		
1	(Constant)	1,871	,434		4,309	,000
	N° of Industry Partners - Phase 3	,600	,261	,376	2,298	,028

a. Dependent Variable: Support - Phase 3

Excluded Variables[b]

Model		Beta In	t	Sig.	Partial Correlation	Collinearity Statistics
						Tolerance
1	N° of Funding Partners - Phase 3	,279[a]	1,756	,089	,301	,999

a. Predictors in the Model: (Constant), N° of Industry Partners - Phase 3
b. Dependent Variable: Support - Phase 3

Table 5-63: Multiple Regression Analysis Phase 3: Support with Functional Roles

Also in the last phase both types of support determine the overall amount of support. Again *Institutional Support* features the larger influence on the overall number of supportive ties with a standardised regression coefficient of 0.59 followed by *Emotional Support* with a respective value of 0.5 (see Table 5-64).

Model Summary

Model	R	R Square	Adjusted R Square	Std. Error of the Estimate
1	,861[a]	,741	,733	1,11019
2	,958[b]	,918	,913	,63538

a. Predictors: (Constant), Institutional Support - Phase 3
b. Predictors: (Constant), Institutional Support - Phase 3, Emotional Support - Phase 3

ANOVA[c]

Model		Sum of Squares	df	Mean Square	F	Sig.
1	Regression	113,030	1	113,030	91,707	,000[a]
	Residual	39,441	32	1,233		
	Total	152,471	33			
2	Regression	139,956	2	69,978	173,336	,000[b]
	Residual	12,515	31	,404		
	Total	152,471	33			

a. Predictors: (Constant), Institutional Support - Phase 3
b. Predictors: (Constant), Institutional Support - Phase 3, Emotional Support - Phase 3
c. Dependent Variable: Support - Phase 3

Coefficients[a]

Model		Unstandardized Coefficients		Standardized Coefficients	t	Sig.
		B	Std. Error	Beta		
1	(Constant)	,681	,267		2,551	,016
	Institutional Support - Phase 3	,998	,104	,861	9,576	,000
2	(Constant)	,141	,166		,848	,403
	Institutional Support - Phase 3	,684	,071	,590	9,634	,000
	Emotional Support - Phase 3	,735	,090	,500	8,167	,000

a. Dependent Variable: Support - Phase 3

Excluded Variables[b]

Model		Beta In	t	Sig.	Partial Correlation	Collinearity Statistics Tolerance
1	Emotional Support - Phase 3	,500[a]	8,167	,000	,826	,706

a. Predictors in the Model: (Constant), Institutional Support - Phase 3
b. Dependent Variable: Support - Phase 3

Table 5-64: Multiple Regression Analysis Phase 3: Support with Support Types

Finally, also the nature of supportive ties is looked at. Following the correlation analysis after Spearman as carried out in Chapter 5.2.2, not only *Strong* but also *Weak Ties*, followed by *Multiplexity* and *Closure* are subjected to multiple regression analysis with Support as dependent variable. In the end, only *Multiplexity* and *Weak Ties* feature significant regression coefficients, i.e. B of 2.872 and 0.47, respectively, and are thus kept in the regression model.

Model Summary

Model	R	R Square	Adjusted R Square	Std. Error of the Estimate
1	,568[a]	,322	,301	1,79721
2	,707[b]	,499	,467	1,56900

a. Predictors: (Constant), Multiplexity - Phase 3
b. Predictors: (Constant), Multiplexity - Phase 3, Weak Ties - Phase 3

ANOVA[c]

Model		Sum of Squares	df	Mean Square	F	Sig.
1	Regression	49,111	1	49,111	15,205	,000[a]
	Residual	103,359	32	3,230		
	Total	152,471	33			
2	Regression	76,156	2	38,078	15,468	,000[b]
	Residual	76,315	31	2,462		
	Total	152,471	33			

a. Predictors: (Constant), Multiplexity - Phase 3
b. Predictors: (Constant), Multiplexity - Phase 3, Weak Ties - Phase 3
c. Dependent Variable: Support - Phase 3

Coefficients[a]

Model		Unstandardized Coefficients		Standardized Coefficients	t	Sig.
		B	Std. Error	Beta		
1	(Constant)	1,086	,470		2,309	,028
	Multiplexity - Phase 3	2,832	,726	,568	3,899	,000
2	(Constant)	,334	,469		,711	,482
	Multiplexity - Phase 3	2,872	,634	,575	4,528	,000
	Weak Ties - Phase 3	,470	,142	,421	3,314	,002

a. Dependent Variable: Support - Phase 3

Excluded Variables[c]

Model		Beta In	t	Sig.	Partial Correlation	Collinearity Statistics	
						Tolerance	
1	Strong Ties - Phase 3	,240[a]	,943	,353	,167	,329	
	Closure - Phase 3	-,821[a]	-2,692	,011	-,435	,191	
	Weak Ties - Phase 3	,421[a]	3,314	,002	,512		1,000
2	Strong Ties - Phase 3	,350[b]	1,601	,120	,281	,322	
	Closure - Phase 3	-,366[b]	-,991	,330	-,178	,118	

a. Predictors in the Model: (Constant), Multiplexity - Phase 3
b. Predictors in the Model: (Constant), Multiplexity - Phase 3, Weak Ties - Phase 3
c. Dependent Variable: Support - Phase 3

Table 5-65: Multiple Regression Analysis Phase 3: Support with Structural Parameters

As indicated by the correlation analysis of Chapter 5.2.2.4 possible sources of institutional support during the establishment phase of the young spin-off were industry and funding partners. Yet, when subjected to regression analysis with *Institutional Support*, both according variables featured no significant regression coefficients and thus no respective regression model could be set up.

Regarding emotional backing potential supporters in the last phase according to the Spearman correlations turned out to be friends and funding partners. However, only for the first alter type the indicated association with Emotional Support could be confirmed by regression analysis (see Table 5-66). While the Number of Funding Partners had to be excluded from the regression equation, Number of Friends exhibits a positive impact with a regression coefficient of 0.586.

Model Summary

Model	R	R Square	Adjusted R Square	Std. Error of the Estimate
1	,487[a]	,237	,213	1,29655

a. Predictors: (Constant), N° of Friends - Phase 3

ANOVA[b]

Model		Sum of Squares	df	Mean Square	F	Sig.
1	Regression	16,707	1	16,707	9,938	,004[a]
	Residual	53,793	32	1,681		
	Total	70,500	33			

a. Predictors: (Constant), N° of Friends - Phase 3
b. Dependent Variable: Emotional Support - Phase 3

Coefficients[a]

Model		Unstandardized Coefficients		Standardized Coefficients	t	Sig.
		B	Std. Error	Beta		
1	(Constant)	,759	,324		2,344	,025
	N° of Friends - Phase 3	,586	,186	,487	3,153	,004

a. Dependent Variable: Emotional Support - Phase 3

Excluded Variables[b]

Model		Beta In	t	Sig.	Partial Correlation	Collinearity Statistics
						Tolerance
1	N° of Funding Partners - Phase 3	,297[a]	1,998	,055	,338	,985

a. Predictors in the Model: (Constant), N° of Friends - Phase 3
b. Dependent Variable: Emotional Support - Phase 3

Table 5-66: Regression Analysis Phase 3: Sources of Emotional Support

5.2.3.4 Summary

To provide an overview about the development of the interrelations identified in the previous three sections over time, in the following the standardised and unstandardised regression coefficients of all explanatory variables are confronted in one table per independent variable.

At first, as depicted in Table 5-67, regression coefficients of the multiple regressions with *Network Size* as dependent variable are illustrated for all three phases of the spin-off process. Following trends over time can be detected: first the *Number of Funding Partners* and the *Number of Friends* are those predictors for *Network Size* that feature significant coefficients for all three phases. While *Funding Partners* has the larger explanatory power, both variables display a peak in the second phase as regards their relative importance for determining the size of egos' personal networks. *Industry Partners* on the other hand are only of relevance in the first and in the last phase, with a slump in significance for the second phase. While the *Number of Academics* exhibits a significant influence on *Network Size* only for the first phase, the same holds true for the *Number of Others* in the last phase.

As to the influence of the number of informational and support providing ties on the size of the personal network, both related variables feature significant regression coefficients for all three phases of the academic spin-off process. While the explanatory power of *Informational Content* slightly drops in the second phase, *Support* exhibits its maximal influence on entrepreneurs' network size at this stage of the process.

Dependent Variable: Network Size	Unstandardised Regression Coefficients (B)			Standardised Regression Coefficients (Beta)		
Phase	1	2	3	1	2	3
N° of Funding Partners	0,991	1,033	0,839	0,472	0,477	0,317
N° of Friends	0,609	0,798	0,752	0,383	0,707	0,409
N° of Industry Partners	0,568		0,597	0,367		0,361
N° of Academics	0,542			0,302		
N° of Others			0,801			0,551
Phase	1	2	3	1	2	3
Informational Content	0,756	0,541	0,675	0,707	0,568	0,635
Support	0,273	0,345	0,302	0,259	0,301	0,291

Table 5-67: Dependent Variable: Network Size – Overview Regression Coefficients

Going into the details of the determinants of *Informational Content* over time as depicted in Table 5-68, regarding the sources of information friends are the ones that feature a significant influence in all three phases of the spin-off process, as in the second phase, where they are the only providers of information. The *Number of Academics* and *Funding Partners* both exhibit significant regression coefficients in the first and the last stage the first with decreasing, the last with increasing explanatory power. Industry partners are only significantly related to the number of information-providing ties in the first phase and then loose their relevance in this regard.

Looking at the evolution of informational content over time, those two types of information that stretch over all three spin-off phases are economic knowledge and other information in the form of scientific and technical know-how. While the former features a peak in the middle of the process, the later in turn encounters a loss at this point in time. *Legal Information* and *Information about Potential Employees* are only relevant at the beginning. *Information about Potential Customers* in contrast is only requested at the end, when the firm strives for establishment on the market.

As to the nature of informational ties, in all three phases they are partly characterised by high tie strength. In the first phase, however, the relevance of weak ties preponderates. Yet, *Weak Ties* feature no more significant relations with *Informational Content* in the last two phases of the spin-off process. *Multiplexity* in turn is only significantly associated with *Informational Content* in the last phase. As mentioned, *Closure*, as sole variable is characterised by a negative influence on the number of informational ties, i.e. the denser the network, the fewer the amount of information-providing contacts.

Dependent Variable: Informational Content	Unstandardised Regression Coefficients (B)			Standardised Regression Coefficients (Beta)		
Phase	1	2	3	1	2	3
N° of Industry Partners	0.589			0.407		
N° of Academics	0.695		0.423	0.414		0.243
N° of Funding Partners	0.637		1.372	0.325		0.551
N° of Friends	0.457	0.694	1.011	0.307	0.585	0.583
Phase	1	2	3	1	2	3
Economic Information	0.627	0.817	0.557	0.425	0.726	0.416
Legal Information	0.500			0.247		
Information about Potential Employees	0.589			0.189		
Information about Potential Customers			0.405			0.241
Other Information	0.912	0.735	0.831	0.631	0.465	0.571
Phase	1	2	3	1	2	3
Weak Ties	0.73			0.619		
Strong Ties	0.723	1.629	1.142	0.469	1.291	0.989
Closure		-3.523	-4.927		-0.656	-0.839
Multiplexity			3.116			0.638

Table 5-68: Dependent Variable: Informational Content – Overview Regression Coefficients

The development of the main predictors of *Resources* is highlighted in the subsequent Table 5-69. As to the different types of resource suppliers, in the first phase only *Funding Partners* feature a significant influence on the amount of resources provided. For the second phase, no significant relations could be obtained at all. In the last phase, however, besides the *Number of Funding Partners* also the *Number of Friends* and *Academics* positively influences the amount of resource-providing ties.

Personnel and infrastructure are required throughout the whole process thus significantly contributing to the number of resource-delivering ties in the personal network in all three phases. *Financial Resources* in turn are primarily relevant during the first two phases, while *Material Resources* only feature a significant relation to the overall amount of resource provision at the final stage.

As regards the nature of resource-offering contacts they are characterised by *Strong Ties* that are of particular importance in the last phase. *Closure* in contrast features a negative impact on resource provision, i.e. the higher the degree of closure of an entrepreneur's personal network, the lower the amount of resource-relevant contacts.

Dependent Variable: Resources	Unstandardised Regression Coefficients (B)			Standardised Regression Coefficients (Beta)		
Phase	1	2	3	1	2	3
N° of Funding Partners	0.889		0.790	0.580		0.383
N° of Friends			0.765			0.533
N° of Academics			0.542			0.376
Phase	1	2	3	1	2	3
Financial Resources	0.759	0.548		0.480	0.504	
Provision of Personnel	1.071	0.804	0.646	0.528	0.662	0.456
Provision of Infrastructure	0.571	0.404	0.548	0.267	0.266	0.329
Material Resources			0.445			0.286
Phase	1	2	3	1	2	3
Strong Ties	0.597	0.452	1.126	0.496	0.553	1.178
Closure			-1.785			-0.367

Table 5-69: Dependent Variable: Resources – Overview Regression Coefficients

Regarding the determining factors of support and their development in the course of the spin-off process, a recapitulatory overview is provided by Table 5-70. While female contacts feature a positive impact on the amount of support provided in entrepreneurs' networks of the first two phases, yet with decreasing order of magnitude, funding partners, friends and industry partners alternate in their influence on *Support*, the first significantly contributing in the first, the second in the second, and third in the last phase.

Both types of support, *Institutional Support* and *Emotional Support*, feature a continuous impact on the overall amount of support rendered across all three phases of the academic spin-off process. Yet, while *Institutional Support* encounters a peak in the second phase, *Emotional Support* suffers a slight reduction of its predictive power at this point in time.

In all three phases supportive relations are multiplex as the subsequent regression coefficients indicate. The largest influence of *Multiplexity* on *Support* can be observed for the first phase. It decreases in the next phase to regain in importance in the last stage of the process, yet, without reaching the maximum level of the first phase. *Closure* and *Strong Ties* are only of relevance for support in the first phase. *Weak Ties* interestingly features a significant regression coefficient for the last phase.

Dependent Variable: Support	Unstandardised Regression Coefficients (B)			Standardised Regression Coefficients (Beta)		
Phase	1	2	3	1	2	3
N° of Females	1.143	0.458		0.519	0.288	
N° of Funding Partners	0.744			0.373		
N° of Friends		0.626			0.635	
N° of Industry Partners			0.6			0.376
Phase	1	2	3	1	2	3
Institutional Support	0.704	0.778	0.684	0.588	0.636	0.590
Emotional Support	0.642	0.618	0.735	0.450	0.489	0.500
Phase	1	2	3	1	2	3
Strong Ties	1.313			0.839		
Weak Ties			0.47			0.421
Multiplexity	3.221	2.199	2.872	0.615	0.525	0.575
Closure	4.501			0.719		

Table 5-70: Dependent Variable: Support – Overview Regression Coefficients

Finally, Table 5-71 provides an overview of the relationship of different relational contents as dependent variables with respect to potential sources of supply.

Dependent Variable	Source	Unstandardised Regression Coefficients (B)			Standardised Regression Coefficients (Beta)		
		Phase					
		1	2	3	1	2	3
Economic Information	N° of Industry Partners	0.559			0.569		
	N° of Friends		0.549	0.677		0.522	0.523
	N° of Funding Partners		0.304			0.398	
Legal Information		x			x		
Information about Potential Employees	N° of Industry Partners	0.228			0.492		
Information about Potential Customers	N° of Friends			0.661			0.642
Other Information	N° of Academics	0.743	x		0.640	x	
	N° of Funding Partners		x	0.858		x	0.501
Financial Resources	N° of Funding Partners	0.635	0.570		0.654	0.421	
Infrastructure	N° of Friends	0.262	0.292	0.474	0.483	0.577	0.550
	N° of Funding Partners	0.236			0.329		
	N° of Academics			0.310			0.357
Personnel	N° of Females	x	x	0.368	x	x	0.352
Material Resources	N° of Friends			0.698			0.756
Emotional Support	N° of Females	0.785	0.506		0.509	0.402	
	N° of Friends	0.343	0.308	0.586	0.324	0.486	0.487
Institutional Support	N° of Females	1.054		x	0.573		x
	N° of Friends		0.501	x		0.621	x

Table 5-71: Sources of Different Content Types – Overview Regression Coefficients

Starting with business-related information, in the first phase industry partners feature a unique position as sources of *economic know-how* which disappears in the course of the spin-off process in favour of friends and funding partners. While the former keep their relevance regarding the provision of business-related issues till the last phase, the latter exclusively fulfil this function in the second stage. *Legal information*, yet, an important type of information in the first phase, does not relate to a particular source. This is indicated by the x in the first columns of the regression coefficients. *Information about* qualified *personnel* for the new venture is also primarily sought in the first phase. Here again, industry partners function as important source of information. On the contrary, the search for potential customers becomes pressing only in the last phase. In this context, friends serve as primary source of relevant information. As mentioned, *other information* in the beginning mainly refers to topics of scientific and / or technical nature that primarily relate to the business opportunity of the planned venture. Later on, respective emphasis shifts to reflective feedback from others and to information about funding agencies. Yet, these topics are exchanged in the course of the whole spin-off process, however, with changing sources of information. In the beginning, the academic community is consulted for respective inputs, in the last phase funding partners take over this function. During the formation phase, no explicit source of other information could be identified.

Coming to the different types of resources supplied, of prevalent importance, at least in the first two phases, is assuring the required *financing*. As to be expected, funding partners are the main source of monetary capital. The structure of the sources of *infrastructure* in turn is more complex and distributed over three different types of alters. Friends play a paramount role in this regard. Not only do they keep their function as suppliers of infrastructure throughout the whole process, but also with dominant impact compared to the share of the two other types of alter. While funding partners add to infrastructure in the first phase, academic colleagues take over this function in the last phase. *Personnel* are required during the whole start-up process. Yet, a particular source addressed to satisfy this need can only be observed for the last phase, where female contacts in the entrepreneurs' networks assume this function. *Material resources* on contrary are primarily requested in the last phase, where they are provided by egos' friends.

Both types of support are rendered throughout the whole spin-off process. As to *emotional support*, moral backing is primarily provided by women and friends during the first two phases and then only by friends in the last phase. *Institutional support* in turn is issued by female contacts at the beginning, and then handed over to friends in the second stage. For the last phase no particular source of institutional backup could be identified.

5.2.4 Testing of Hypotheses

Detected relational structures are related to the hypotheses of this research in view to corroborate or discard the underlying assumptions. To this end the hypotheses of Chapter 5.1.2.4 that were revised according to the results of the exploratory qualitative analysis are opposed to the results of the quantitative part of the previous chapter. As managed hitherto in the course of this thesis, the phaseal trichotomy of the spin-off process is applied to structure the comparison.

5.2.4.1 Opportunity Recognition

> *H1.A: The opportunity recognition phase of the entrepreneurial process is characterised by information networks, i.e. the personal social network of the academic entrepreneur consists predominantly of information-related interactions.*

When looking at the influence of the different types of content exchanged on the size of entrepreneurs' personal networks in the first phase of new venture formation as depicted in Table 5-67 of the previous chapter, it can be clearly stated that informational content features the largest contribution to the growth of the network with a regression coefficient of 0.707. Support lies clearly behind with a coefficient of only 0.259, while resources exhibit no significant influence on the size of the network at all.

Consequently, it can be confirmed that at the first stage of spinning-off the largest and thus dominant part of business founders' personal networks consists of contacts with alters providing some sort of information. Yet, hypothesis **H1.A is corroborated**.

> *H1.B: The opportunity recognition phase of the entrepreneurial process is characterised by sparse information networks of academic entrepreneurs that are rich in structural holes.*

Already in the course of data exploration of Chapter 5.2.2.1 no significant Spearman correlations for structural holes and network sparsity could be detected. Yet, both variables were not included in the subsequent regression analysis of Chapter 5.2.3.1.1. This means that hypothesis **H1.B** is not confirmed and thus **discarded**. Personal networks at the first phase of the spin-off process do not necessarily have to be sparse and rich in structural holes to result in a successful realisation of the academic venture.

> H1.C: *The opportunity recognition phase of the entrepreneurial process is characterised by information networks dominated by weak ties.*

Following the indications of the Spearman correlation analysis between the number of informational ties and tie strength of Chapter 5.2.2.1 both weak and strong ties were included in the regression analysis with informational content as dependent variable and both retained their significant influence. Having a look at the overview provided in Table 5-68, one may find that the relative influence of weak ties on the amount of information-providing contacts is by far larger, i.e. Beta value of 0.619, than the one of strong ties with a Beta value of 0.469. As a result hypothesis **H1.C** is not rejected and can be deemed as **corroborated**.

> H1.D: *During the opportunity recognition phase of the spin-off process the information networks of academic entrepreneurs are dominated by contacts to academic actors and industry.*

Looking at the Beta values of Table 5-68 expressing the relative importance of different types of alters for the provision of informational content, clear evidence is provided in favour of this hypothesis. Academic contacts feature closely followed by the number of industry partners exhibit the largest regression coefficient in explaining increases in the number of informational ties. On this account hypothesis **H1.D** is **confirmed**.

> H1.E: *It is usually one special contact that provides the main impetus for opportunity recognition.*

The last hypothesis does not really require a statistical analysis of significance. A simple look at the descriptive statistics offers overwhelming proof of the above made statement: all respondents, i.e. 100%, stated that one person had been responsible for providing the main impetus to spinning of an own business (see Table 4-11 in Chapter 4.3.8.2). Yet, also the last hypothesis **H1.E** can be regarded as **confirmed**.

5.2.4.2 Formation

> H2.A: *The formation phase of the entrepreneurial process is characterised by both information and resource networks, i.e. the personal network of the academic entrepreneur contains information- and resource-related interactions.*

As the results of the regression analysis summarised in Table 5-67 reveal, at the second stage of the spin-off process the size of egos' personal networks is about half, i.e. regression coefficient of about 0.54, determined by informational ties, and to one-third, i.e. regression coefficient of about 0.35, coined by support-rendering contacts. Yet, hypothesis **H2.A** cannot be completely confirmed and thus has to be **rejected**.

> H2.B: *The information networks of the formation phase are characterised by sparse social networks of academic entrepreneurs that are rich in structural holes.*

No significant relations between sparsity or number of structural holes and the size of information networks could be detected in the course of the regression analysis of Chapter 5.2.3.2.2. Hence, hypothesis **H2.B** is **discarded**.

H2.C: The resource networks of the formation phase are characterised by dense social networks of academic entrepreneurs.

Again as for the previous hypothesis no significant interaction between resource networks and density could be identified. As a result hypothesis **H2.C is rejected**.

H2.D: The information network of the formation phase is characterised by weak ties.

Also with respect to hypothesis H2.D no positive evidence could be obtained in the course of regression analysis. The number of weak ties features no significant influence on the amount of resource provision at this stage of the spin-off process. Consequently, hypothesis **H2.D** has to be **discarded**.

H2.E: The resource networks of the formation phase are characterised by strong ties.

As the results of the regression analysis captured in Table 5-69 reveal, a positive relation between resource-providing ties and increase tie strength exist. Yet, hypothesis **H2.E** is not rejected and can be deemed as **confirmed** at this stage.

H2.F: The information networks of the formation phase are characterised by contacts to academic actors.

In the second phase only contacts to friends exhibit a positive relation to the number of information-related ties. Following Table 5-68 *Number of Friends* is the only variable with a significant regression coefficient (0.694). However, this means that contacts to academics feature no influence on information provision at this point in time. Hypothesis **H2.F** is thus **discarded**.

H2.G: The resource networks of the formation phase are dominated by exchanging financial resources.

As Table 5-69 reveals financial resources have an important share (*Beta* of 0.504) in the overall amount of resources provided in the second phase, however, the provision of personnel seems to be even more important with a *Beta* value of 0.662. Yet, hypothesis **H2.G** has to be **rejected**.

H2.H: In the formation phase infrastructural resources are provided by the university in the form of shared assets.

According to Table 5-69 also infrastructure has a positive share in the total of resources obtained at this point in time with regression coefficient of about 0.4 and a *Beta* value of 0.266. Yet, when looking at the distribution of cases, where university offered sharing infrastructure and the cases, where this was not the case, as indicated by the descriptive statistics of Chapter 4.3.8.2, one will see that no clear tendency can be observed. 50% of the time infrastructure was shared between new venture and university, and 50% not. This seems to be too little evidence to support the hypothesis. Moreover, as Table 5-71 reveals, the primary source of infrastructure in the second phase are entrepreneur's friends and not the academic environment. Hypothesis **H2.H** is therefore **discarded**.

H2.I: In the formation phase the home department features a positive attitude; while the home university features a more neutral to positive attitude towards the spin-off intention.

Again this hypothesis can be best checked by having a look at the descriptive statistics of Chapter 4.3.8.2 and here at the results highlighted in Table 4-16. As the distribution of cases indicates entrepreneurs' home departments exhibit a tendency towards a more positive attitude towards their researchers' spin-off intentions, while the university as such adopts a rather neutral attitude in this respect. Yet, hypothesis **H2.I** is deemed without contradictive evidence and thus is seen as **confirmed**.

5.2.4.3 Establishment

> *H3.A: The establishment phase of the entrepreneurial process is characterised by information, resource and support networks, i.e. the personal network of the academic entrepreneur comprises of interactions transferring information as well as material and symbolic support.*

As the regression coefficients of Table 5-50 indicate only information and support are relevant contents exchanged in the personal networks of the last phase. Resources the same as in the previous phases do not feature any significant influence on the size of the overall network. Yet, hypothesis **H3.A** has to be **discarded**.

> *H3.B: The information networks of the establishment phase are characterised by a dense and interconnected structure.*

According to the regression analysis of Chapter 5.2.3.3.2 no significant relations could be found between the number of informational ties and the network density in the third stage of the spin-off process. As a result, hypothesis **H3.B** has to be **rejected** as well.

> *H3.C: The information networks of the establishment phase are characterised by strong ties.*

However, a significant regression coefficient was specified for the impact of strong ties on the number of information-related contacts (see Table 5-68). Hence, hypothesis **H3.C** can be deemed as **corroborated** so far.

> *H3.D: The resource networks of the establishment phase are characterised by a dense and interconnected structure.*

Again no significant interconnections could be detected for the influence of networks density on the amount of resource-providing contacts. Consequently, hypothesis **H3.D** is **discarded**.

> *H3.E: The resource networks of the establishment phase are characterised by strong ties.*

However, the number of alters supplying resources to ego in the third phase is positively influenced by the number of strong ties in the network (see Table 5-69). As a result, hypothesis **H3.E** can be regarded as **confirmed**.

> *H3.F: The support networks of the establishment phase are characterised by strong ties.*

As the regression analysis of Chapter 5.2.3.3.4 indicates, no significant influence of the number of strong ties on the amount of support rendered could be observed for the last phase. Hypothesis **H3.F** is thus **rejected**.

> *H3.G: Emotional support is predominantly obtained from members of the entrepreneur's academic network.*

Already the correlation analysis of Chapter 5.2.2.4 revealed no significant Spearman correlation coefficients between emotional support and the number of academic alters in egos' networks. Yet, no further testing in the form of a regression analysis was carried out. Hypothesis **H3.G** has to be **discarded** for lack of empirical evidence.

H3.H: Institutional support is predominantly obtained from members of the entrepreneur's business network.

While in this case significant correlations after Spearman could be ascertained, when subjecting *Number of Industry Partners* to regression analysis with *Institutional Support* as independent variable no valid regression equation could be set up due to lack of predictor's significance. This is also reflected by the results of Table 5-71. Yet, hypothesis **H3.H** has to be **rejected** as well.

H3.I: The resource networks of the establishment phase are characterised by the provision of financial resources from academic contacts.

Having a look at Table 5-69, no significant regression coefficient could be specified for the influence of financial resources on the overall amount of resources obtained at this stage of the process. Hence, hypothesis **H3.I** has to be **discarded**.

H3.J: The resource networks of the establishment phase are characterised by the provision of personnel resources from academic contacts.

As regards hypothesis **H3.J** it can only be **partly confirmed**. As can be learned from Table 5-69 the amount of personnel placing contacts features a significant positive impact with a regression coefficient of 0.646 on the overall amount of resources made available at this stage. Relatively speaking, placement of staff amounts for the largest part of resources provided at this stage, with a standardised Beta value of 0.456 over 0.329 for the provision of infrastructure and 0.286 for the supply of material resources. However, when looking at the relationship between the provision of personnel and the number of academic contacts in entrepreneurs' personal networks, already the Spearman correlation analysis as depicted in Table 5-12 and consequently also the results of Table 5-71 show no significant association between those two network parameters. This means that the first part of the hypothesis can be regarded as confirmed, while when determining the origin of personnel the assumed influence of academic alters has to be rejected.

H3.K: During the establishment phase reputation on the market is gained from the academic career and from industry contacts.

To assess the validity of the last hypothesis the descriptive statistics of Chapter 4.3.8.2 have to be drawn upon. As Table 4-17 illustrates according to the respondents 38% of the reputation on the market could be ascribed to industry contacts, 31% to the academic career. Therefore, hypothesis **H3.K** is regarded as **confirmed**.

5.2.4.4 Summary

Table 5-72 gives a recapitulatory overview of the hypotheses and their status after the quantitative statistical analysis of the previous chapter.

As can be stated on first sight, only a limited number of hypotheses could be confirmed by the quantitative network results, in particular regarding the later phases of the spin-off process.

In the first phase as assumed information is the predominant form of content exchanged between academic founders and their alters, who in this context are primarily colleagues of the scientific community and industry contacts. These information networks are characterised by weak ties, yet, they are not as expected sparsely knit and thus rich in structural holes. However, all founders were stimulated by a particular person to spin-off their venture.

Phase	Hypothesis	Final	Corroborated
Opportunity Recognition	H1.A	Predominance of informational content	Yes
	H1.B	Sparse information networks rich in structural holes	No
	H1.C	Information network: weak ties > strong ties	Yes
	H1.D	Information network: informational content from academic actors and industry contacts	Yes
	H1.E	One person as main impetus for spin-off	Yes
Formation	H2.A	Informational and resource content	Information – Yes
			Resources - No
	H2.B	Information networks sparse rich in structural holes	No
	H2.C	Resource networks dense	No
	H2.D	Information network: weak ties > strong ties	No
	H2.E	Resource network: strong ties > weak ties	Yes
	H2.F	Information network: informational content from academic actors	No
	H2.G	Resource network: dominant exchange of financial resources	No
	H2.H	Resources: shared usage of university infrastructure	No
	H2.I	Reaction department -> positive Reaction university -> neutral to positive	Yes

Establish-ment	H3.A	Informational, resource and support content	No (only information and support)
	H3.B	Information network dense	No
	H3.C	Information network: strong ties > weak ties	Yes
	H3.D	Resource network dense	No
	H3.E	Resource network: strong ties > weak ties	Yes
	H3.F	Support network: strong ties > weak ties	No
	H3.G	Support network: emotional support from academics / co-founders	No
	H3.H	Support network: institutional support from business contacts	No
	H3.I	Resource network: financial resources from academic contacts	No
	H3.J	Resource network: human capital from academic contacts	Yes – Dominance of human resources No – Source academic contacts
	H3.K	Reputation on the market due to academic career and industry contacts	Yes

Table 5-72: Final Status of Hypotheses – Overview

Regarding the second phase, only two hypotheses can be completely confirmed. According to them, resource networks are dominated by strong ties, and the reaction of the home department to the spin-off intention of its researcher is for the most part positive, while the university as such exhibits a rather neutral attitude. An other hypothesis is at least partly confirmed, which means that information networks, as stipulated, maintain their relevance, while resource networks, however, do not gain in significance, as originally supposed. As to the rest of the assumptions, they could not be corroborated: information networks are not sparse dominated by weak ties with academic actors as main sources of content, and resource networks are not dense characterised by the transfer of funds and the shared usage of university infrastructure.

In the last phase, while the provision of support has also attained relevance, resource-providing contacts still play no overly important role in terms of their contribution to personal network size, which means that the first hypothesis cannot be completely confirmed. Hypothesis H3.J on resource networks is also only partly confirmed. While as already hypothesized the acquisition of manpower earmarks this phase, it is not predominantly brokered, as presumed, by academic contacts. However, three other hypotheses are corroborated: information networks of the last phase feature strong ties, the same holds true for resource networks, and market recognition is supported by the entrepreneurs' academic background and their contacts to industry. In contrast, resource and information networks of the last phase are not, as assumed, dense, and resources exchanged do not primarily relate to funds provided by colleagues out of the scientific community. Moreover, support is not as expected related to strong ties with emotional support being provided by academic colleagues and institutional support obtained from industry contacts.

Despite these rather disillusioning results of the statistical examination of the established hypotheses, the quantitative network analysis has revealed a series of other interesting findings that are depicted in the next chapter.

5.2.5 Additional Findings

While the testing of hypotheses did reveal rather disappointing results, looking at the outcome of the regression analysis as summarised in Chapter 0, one can deduct a series of other fruitful conclusions instead. To this end, the Beta values put together in the tables of Chapter 0 are taken as basis of interpretation with Beta allowing for the comparison of parameters of differing scales. Starting with general findings applying to all types of content exchanged, in the following a distinction is made between the three network types identified: information, resource, and support network.

5.2.5.1 General

As regards entrepreneurs' social environment in general, friends and funding partners dominate egos' networks throughout the whole spin-off process with friends becoming considerably important in the second phase. Moreover, a basis level of support can be observed, which forms part of the entrepreneurs' networks during the whole start-up process.

5.2.5.2 Information Networks

Having a closer look at the nature of information exchange, in the first and last phase sources of information are quite diverse including almost all types of contacts. Interestingly, industry partners are no longer consulted in the last phase. In the second phase in return, the only source of information are friends. Consequently friends are the only ones to furnish information throughout the whole process of new venture creation. Academics and funding partners regain importance in the provision of information in the last phase.

As to the content of information, economic know-how and other information, predominantly in the form of technical and scientific knowledge, are the determining subject matters of information exchange throughout the whole spin-off process. Altogether both types of information are equally important. Yet, while business-related topics become in particular relevant in the second phase of spin-off creation, when the new venture is actually set-up, development-related topics of scientific and technical nature exhibit higher relevance in the first and the last phase, when the initial idea is shaped and later on when the prototype is advanced into a marketable product. In the beginning, economic know-how is primarily obtained in the interaction with industry partners; a function that is taken over by friends from the second stage on, and to some extent also provided by funding partners, but only temporarily during the formation phase. In the first part of the spin-off process, when other information is characterised to scientific and technical know-how, academic colleagues serve as main source of information. Later on, as reflective feedback from others and information about funding agencies dominate, emphasis is shifted towards funding partners as primary source of this type of information.

Legal information is only relevant at the beginning prior to the actual set-up of the firm. Information about customers in turn prevails in the end of the start-up process with friends as dominant source of supply. Both of them, legal know-how and customer-related information, are required to a far lesser extent than economic and technical know-how. Interestingly, information about qualified staff is only sought in the beginning, where it is primarily found from industry contacts.

Other than originally expected, the provision of information is only associated with weak ties during the first phase. Strong ties in turn remain influential all over the time, yet, particularly in the last two phases taking over the effect of weak ties displayed in the first phase. In the last phase the exchange of information becomes increasingly complex and is linked to the transfer of other contents as indicated by the positive impact of multiplexity on the number of informational ties.

5.2.5.3 Resource Networks

In general, as depicted in Chapter 5.2.5.1, entrepreneurs' networks are dominated by informational and supportive relations. However, if resources are sought this takes primarily place in the last phase, where particularly friends and to a lesser extent also funding partners and academic colleagues serve as sources of supply. In addition, funding partners are also addressed in the first phase.

Types of resources gathered vary. Financial resources are only requested in the first two phases, where they are primarily obtained from funding relations. Personnel in contrary are sought all the time, but particularly in the second phase, when the new venture is actually spun-off. The provision of infrastructure is only of minor relevance, but still important during all phases with a slightly stronger effect during the last stage. Continuous sources of infrastructure are friends that, particularly in the end, provide respective frameworks. To a lesser extent funding partners are also addressed in this regard, however, only in the first and the last phase. Material resources are only needed in the last phase and are mainly provided by entrepreneurs' friends.

5.2.5.4 Support Networks

The provision of support forms part of entrepreneurs' social relations throughout the whole start-up process with a stronger emphasis on the last two phases. Both types, institutional and emotional support, are sought, institutional support even a bit more. As to the provision of support in general, during the first phase, women and funding partners are addressed. In the second phase, again women, but then friends constitute the supportive relations. In the last phase finally, industry partners are the only ones to provide support. While the amount of institutional support remains rather steady, the extent of emotional support in turn increases over time.

Providers of support alter during the process, but are basically the same two for both types of support. Emotional backing is rendered by friends and women in entrepreneurs' networks. While the former keep this function throughout the whole start-up process, women cease to fulfil this role in the last phase completely handing it over to entrepreneur's friends. Coming to institutional support, interestingly in the first phase women play a strong role in this respect that is taken over by friends for the second stage. In the end, not particular source of institutional support could be identified.

Surprisingly, whereas in the case of resources, as expected, strong ties play a significant role throughout the whole spin-off process, in the case of support, other than originally assumed, strong ties are only influential in the first phase. Later on in the last phase, weak ties take over. Support is always related to the provision of other types of content with the number support-providing relations being strongly related to multiplex ties.

6 Conclusions

The following chapter provides a reflective review of the results of the previous analysis part. What can be learnt from the empirical results and what insights on the academic spin-off process do they offer? How can the statistical values and outcomes be interpreted? These and other questions are dealt with in the Section , while the subsequent Section outlines the boundaries of these interpretations and illuminates the cautions, with which the results have to be treated.

Despite their limitations the results gathered in the course of the empirical enquiry furnish worthwhile contributions to the theoretical base in this field, in particular as regards the intersection of social capital and entrepreneurship research in the analysis of academic spin-off formation. Taking these contributions, as depicted in Section 6.3.1, and confronting them with the major limitations presented in the previous sub-chapter, recommendable future research efforts to overcome these drawbacks and blind spots are drafted in the last section 6.4 of this chapter. If pursued, they may valuably add to the stock of knowledge developed in the course of this research.

6.1 Discussion of Empirical Results

Summing up what are the central findings of this research. First of all we have to bear in mind the central questions posed at the beginning of the work. As formulated, the research project aimed at shedding more light at the social environment of academic spin-off founders and its development in the course of the new venture creation process. So possible answers should relate to such topics as, whether there are certain particular characteristics in entrepreneurs' social networks that facilitate the spin-off process, and whether these characteristic significantly differ between the different phases of new venture creation, or with respect to the different contents exchanged. As regards the last aspect, as mentioned, a distinction has been made between information, resource, and support networks; a differentiation that is also drawn upon in the course of the following interpretation of empirical results.

6.1.1 General

Looking at the overall composition of entrepreneurs' personal networks in the course of time, friends and funding partners are those contacts that predominantly accompany them throughout the whole start-up process. The number of friends even increases in the end. The first phase of opportunity recognition, where the initial idea to spin-off matures, is characterised by a rather diverse network. Besides friends and funding partners also industry contacts and academic colleagues are addressed. This supports the findings of the qualitative phase indicating that previous contract research and the interaction with the academic environment play a decisive role in identifying and formulating the business opportunity. It is this combination of already existing first contacts to the market-driven business world paired with the innovative climate of the scientific environment at the university that offers the right medium for emerging successful business ideas. Later on as this initial idea becomes more focussed and concrete, trust comes into play and emphasis thus is shifted towards more amicable contacts that can be confided increasingly confidential business information. The same holds true for funding partners that in the beginning are primarily addressed for financial and then later on for consulting services as to the right approach to the market.

Entrepreneurs' personal networks are characterised by informational and support-rendering relations throughout the whole process of spinning-off. This means that interactions are primarily directed towards alters providing some sort of information or support. Of course resources, and here in particular money and personnel, are also required, however, they are not so strongly associated with networking activities than gathering information and seeking support. This can be interpreted such that the amount of contacts in the network to gather the requested level of information and support has to be much higher than to obtain necessary resources. Given the right information about where to get them, resources can be rather easily tapped by a small number of interactions. This fact is reflected in the low share of resource-related ties in the personal networks of the academic entrepreneurs. Yet, the key is getting the right information. Consequently number of informational ties preponderates in the personal networks through all phases, particular in the beginning, when the opportunity is spotted and formed, and in the end when the market is entered. Type and composition of information exchanged will be dealt with in more detail in the subsequent Section 6.1.2. Support, too, requires a lot of interaction. However, it does not seem as pressing as information for the success of the start-up process. Interestingly, in the second phase, when the new firm is actually legally established, the amount of support-providing interactions reaches its peak. Having a look at the type of support concerned in this context as carried out in Section 6.1.3, i.e. emotional or institutional support will shed more light on the underlying reasons of this phenomenon.

6.1.2 Information

Having a closer look at the actual type of information exchanged, it can be found that in particular two types of information, i.e. economic know-how and other information, are requested throughout the whole spin-off process.

Regarding the former, in the beginning it is basically provided by industry contacts established in the course of contract research carried out the home department. Later on till the end of the process friends serve as main source of business-related inputs. Funding partners temporarily add to this in the second stage, where the legal set-up of the spin-off is effected. It is also at this stage that the largest amount of information is required. Summarising it can be said that for opportunity recognition business experiences of industry partners serve as stimulating factors for the motivation to spin-off. When this initial motivation or idea advances into a more and more elaborated business concept, as mentioned above, trusting relationships as provided by friends come to the fore with increasing relevance the closer the market entry is.

Other information as specified by the respondents basically is of two-fold nature. On the one hand it refers to project-related technical and scientific knowledge, on the other hand to critical feedback and consulting services. While scientific and technical inputs prevail through all phases of the process, feedback and above all consulting services are named primarily for the last phase of establishment on the market. Yet, also the sources of other information vary over the spin-off process. At first, academic colleagues are consulted for scientific and technological know-how in order to further develop the initial idea into a viable opportunity. Later on, as other information is increasingly enriched with advisory services, funding partners take over the role as main sources of information. It is also at these later stages when the original project idea materialises that the need for and possibility of public support is increasingly realised and thus funding partners are approached.

Surprisingly legal information is only gathered to a larger extent during the opportunity recognition phase. It can be assumed that this is done in view of the increasingly form taking intention to establish an own company in the near future. Then when the time of actual business set-up arises, enough information has already been collected to complete this task without any further interactions in this respect. Moreover, legal information is now longer actively sought by the academic founder him- or herself, but increasingly substituted in the form of appropriately informed contacts that just exercise their knowledge without necessarily transferring it to the founders.

Information about adequate employees is gathered early on right from the beginning as one of the most important pillars for a young research-intensive start-up. Other than originally expected it is not the academic environment that serves as main source of respective information, but the contacts out of the industry projects carried out during the scientific career of the entrepreneur. One reasonable argumentation in favour of this phenomenon may be found in the fact that the entrepreneurial researcher is looking for complementary qualifications that add to his / her scientific background and thus can be most likely found with partners experienced in business-related issues.

Finally, in the last phase of establishment on the market the need for first customers becomes more and more urgent. This reflected by the increasing amount of ties informing about potential clients in the third phase. The most important source in this context is the entrepreneur's circle of friends. These are the one with the least competitive attitude towards the spinning-off research and possibly are thus more eager to disclose respective information.

As to the intensity of interaction accompanying the information exchange throughout the spin-off process, only in the beginning weak ties prevail and dominate over strong ties that also have their share at this stage. It is at this stage that obviously the most diverse information has to be collected in order to be brought into the position to recognise and shape a highly innovative opportunity as basis for a successful scientific spin-off. Then, as required information becomes more and more exclusive and confidential, trusting relationships, as represented by strong ties characterised by high intensity of interaction, gain importance. In particular in the second phase, when the new venture is actually spun off and thus the spin-off intention is officially disclosed to the public, bonding relationships seem to offer most relevant and trusted information.

Too many interactions impair the amount of information exchanged. This is reflected by the impact increasing network closure has on the number of informational contacts. The larger the proportion of strong ties in the network the fewer the number of information providing contacts. This may be explained by the fact that the increased networking effort larger numbers of time-intensive relationships entail, at a certain point has a negative effect on the informative capacity of the relation. The entrepreneur is too busy to maintain all his or her strong ties that he or she has no time left to direct them towards a more informational nature or to adequately absorb and process the information provided.

The information exchange of the last phase, when the newly founded spin-off enters the market, is characterised by a high degree of multiplexity. This means that provision of information is connected to the transfer of other contents in the same relation. Information is not transferred as content as such, but related to other types of content exchanged. More and more contacts at this point of the process serve more than one purpose.

6.1.3 Resources

While resource-collecting activities do not have an overly important share in the total composition of entrepreneurs' networks as opposed to information or support in the course of the whole spin-off process, they are of course still necessary with changing emphasis of types of resources and supplying alters.

Interestingly interactions to broker personnel and not to raise capital consume to most amount of social activity with respect to procuring resources. Attracting appropriate staff preoccupies the successful entrepreneur throughout the whole process from the time the first rough business idea emerges until the entry in and establishment on the market. This may be explained as follows: While sufficient funding is certainly important, it cannot be effectively transformed into valuable business activities without the performance of respectively qualified employees. They represent the transforming capacity turning money into successful competitive business results. For the brokerage of personnel no specific alter type can be identified as primary source. Only in the last phase, female contacts take feature a positive impact. One possible explanation may lay in the more pronounced social behaviour of women entailing a larger pool of potential contacts and thus of potential staff members.

Furthermore, adequate financing has to be safeguarded during the first two phases to ensure a substantiated development of the young venture. This is reflected by the absence of a significant regression coefficient for this type of resource on the totality of resource-providing relations in the last phase. Only the first two phases are characterised by a distinct proportion of capital-related contacts in the overall amount of resource-seeking activities. In this regard, funding partner serve as main source of capital. A fact that is also supported by the high usage of public support as depicted in Figure 4-20 of Chapter 4.3.8.2.

On contrary, material resources are not needed until the end of the process. It is at this phase that the start-up has already been set-up and starts requiring physical assets to expand its production process to persist on the market. Main sources of material are friends that seem to dispose of sufficiently developed relations with ego to timely anticipate their needs in this regard and to provide for the required trust to advance respective assets.

Infrastructure is sought for right from the beginning, yet, at a lower level of activity. Respective efforts intensify a little bit during the last phase; possibly as a result of the already assumed enlargement of production capacities in the course of establishing on the market. Infrastructure is obtained from a variety of resources. Friends serve as continuous broker in this respect. Funding partners in turn are primarily addressed in the beginning prior to the actual start-up when the project particularly qualifies for public support. Later on when approaching the market and thus being faced with a reduced eligibility for public assistance due to competitive reasons, academic colleagues are tapped instead.

As already presumed in previous chapters of this research, the actual handing over of resources whatever type to the young entrepreneur represents a rather immediate action requiring a high level of trust between ego and alter. Consequently, respective interactions are characterised by strong ties that even increase in importance in the course of time as relationships further advance and intensify. Hence, to obtain the required resources the academic entrepreneur has to build up trusting relationships and carefully cultivate them over time. This is also reflected by the increasing importance of friends in the provision of resources in the course of the spin-off process. However, as already observed for informational content (see 6.1.2), too many strong relationships also exhibit a negative influence on the amount of resources obtained; in particular in the last phase, when networking activities seem to culminate.

6.1.4 Support

Both types of support, emotional backing and institutional anchorage, pervade the whole spin-off process. As reminder, institutional support refers to the transfer of positive image and trust from the person of alter on the academic entrepreneur and / or to the improvement of visibility of the new venture on the market. In this context, institutional backing is even slightly more dominant in the network than emotional support. Further interpretations, especially with respect to the sources of institutional support, are difficult due to the blurry perception respondents had of the concept. Answers suggest that to some extent institutional backing is mistreated for emotional assistance. This may explain the dominance of female ties in the first phase and of friends in the second.

To this end and in view of more reasonable results, in the following when interpreting its different sources, support is treated in its meta-dimension without distinguishing between the institutional and emotional component. In the first two phases, but in particular during opportunity recognition, female contacts play a predominant role. Accounting for the fact that the majority of respondents are male (almost 80% see Table 4-11), female contacts in this regard possibly stand for entrepreneurs life partners, who are consulted in the beginning to check for principle accordance with the researcher's plans to spin-off. Later on, as this accordance has already been obtained, friends and industry partners come into fore. While friends encourage the entrepreneur shortly prior to and during actual spin-off, industry partners replace them in the last phase, when increased visibility on the market is required. It can be assumed and is also indicted by the Spearman correlations of 5.2.2.4, yet, not confirmed by regression analysis, that industry partners rather refer to institutional than emotional support.

Moreover, when checking for the viability of the initial business idea during opportunity recognition, also funding partners assume some sort of supportive function with their expertise and approving feedback.

Regarding the nature of supportive ties, in the first phase they are for the most part marked by high tie strength. This seems to reinforce the assumed role of female life partners as important source of emotional support. In the end, when it comes to the establishment of the young start-up on the market, the weak character of the ties takes over. This is in line with the increased dominance of industry partners observed for the last phase that was attributed to the fact that respective contacts are sought for market visibility and strategic image-building. Opposed to the contact to life partners and friends, relations to industry contacts are usually rather infrequent, loose and of distant nature, i.e. weak.

Finally, support as such is multiplex, i.e. it does not come alone, but is always provided in connection with other types of content transferred. High degrees of multiplexity prevailing in all three phases confirm this assumption.

6.1.5 Synthesis

To provide an integrative picture of the social and environmental factors influencing the academic spin-off process, both the results of the social network analysis in the form of social capital indicators and the results of traditional questions on the surrounding factors are summarised in the subsequent Figure 6-1. Regarding the structure of social capital, the horizontal layers represent the three different sub-network types of entrepreneurs' personal networks. The vertical columns in turn represent the different stages of the spin-off process. The descending order of the bubbles on the right embodying the different contents exchanged corresponds to their declining share in the respective network type. For example, having a look at the resource networks of the first phase, one can find that personnel are the most frequent resource exchanged at this time, followed by financial means and infrastructure. The arrows express the relationship between content provider and type of content. Coming to our example of the resource networks at stage one, funding partners furnish money and infrastructure, while friends provide infrastructure alone. The italicised terms at the bottom line of each cell, stand for the prevailing structural parameters in the network type of the respective phase. Yet, again in our example case strong ties dominate the exchange of resources during the opportunity recognition phase. Finally, the box at the bottom highlights the most distinct realisations of environmental framework conditions.

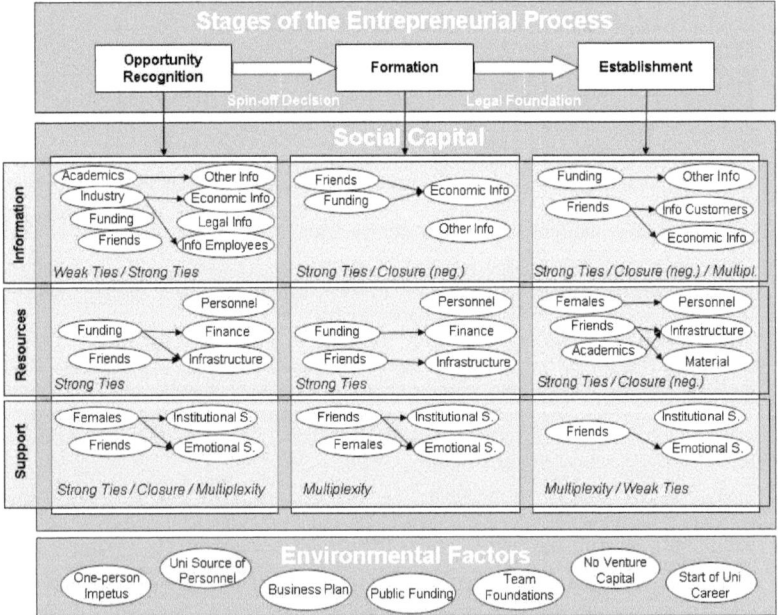

Figure 6-1: Final Model of the Social and Environmental Factors of the Academic Spin-off Process

As the properties of social capital have already been described in detail in the previous sections, only the most prominent parameter values are again briefly highlighted at this point. A new aspect included is providing for the relation to the most important environmental conditions. At the beginning of the spin-off process, when the business opportunity is deliberated about, sources of information are rather diverse. The relevance of academics and industry partners clearly shows the connex to and importance of the entrepreneurs' prior activities for the first phase of the start-up. Contract research and the exchange with colleagues serve as vital informational input to opportunity recognition. The novelty of information is guaranteed via weak ties, while strong ties provide for more confidential insider information. Moreover, according to Table 4-11 of Chapter 4.3.8.1 all respondents referred to one decisive person that provided the main impetus to the researchers' motivation to go into business for themselves. Later on two main sources of information emerge, i.e. friends and funding partners. Contacts to both of them are characterised by a high frequency of interaction as the share of strong ties in the network indicate. The role of the latter as important source of information reflects the dominance of public funding as environmental factor. As Table 4-15 of Chapter 4.3.8.2 indicates more than 88% of the interviewed spin-off founders reported to have received public funding. This can be regarded as essential supporting factor for the success of the actual establishment of the spin-off in particular against the light of the low usage of venture capital. Only 8.8% entrepreneurs quoted to have been financed by venture capital (see Table 4-15 in section 4.3.8.2). Interestingly legal information and information about personnel is already sought early on, but looses importance in the subsequent phases. As the market is approached data about potential customers becomes increasingly important. Yet, the right mixture of business- and content-related technical know-how accompanies the whole process. As to economic knowledge of course the increased dealing with this subject resulting from the predominant building of business plans – according to Table 4-11 in Chapter 4.3.8.1 more than 94% of the respondents reported to have set-up a business plan – reflects the added value of this exercise. It has to be noted at this point that the existence of a business plan represents an indispensable requirement for most funding schemes and thus can be also attributed to the large prevalence of public funding.

The provision of resources in turn is characterised by a similar structure as to type of content and its providers with respect to the first two phases prior to actual legal start-up. The most important resource is personnel and according to the results of a direct question on this aspect mainly derived from the academic professional background of the entrepreneur. More than 72% of the responding entrepreneurs quoted the university or their department as main supplier of manpower (see Table 4-18 in Chapter 4.3.8.2). Moreover, financial capital and necessary infrastructure are important assets aspired during this time. While money is primarily obtained from funding partners, infrastructure is mainly provided by friends. In the end when establishing on the market, financial aspects take a back seat in favour of material assets that are obtained from friends. Interestingly, academics also come into play at this point of time assuming the role of infrastructure providers. This may be attributed to the partly joint usage of university structures as depicted in Table 4-15 of Chapter 4.3.8.2. Summarising what is needed the most is money, people and office space. Respective relations are characterised by high levels of frequency, i.e. strong ties, indicating that the provision of these assets comes along with repeated meetings and a trustful climate.

Coming to the provision of support, both types emotional backing and institutional support in the sense of projecting positive image and reputation on the entrepreneur, are almost equally prevalent during all three phases of the spin-off process. Also the type of contacts providing for those forms of support remains nearly unchanged, yet, with different associations to the two contents of support. In the first phase of opportunity recognition, women are the ones that back the aspiring entrepreneurs both emotionally and institutionally, friends only emotionally. This is mirrored reversely for the second stage, when the actual start-up is prepared. Here, friends render both types of support, while women only cater for moral support. In the last phase, when entering the market, it is only friends that furnish the young company with a trusting image due to their association to the entrepreneur. As to the nature of relations they are in the first instance multiplex, which means that they do not come alone but primarily in connection with other types of content transferred.

Finally, some remarks on further environmental aspects. First of all, in the predominant part of the cases, i.e. following Figure 4-15 in Chapter 4.3.8.1 more than 85%, researchers did not start their spin-off alone, but together with at least one partner. Secondly, when looking at the position entrepreneurial researchers held at the university prior to setting up their own firm, one may observe that in the majority of cases researchers at the beginning of their career ranging from research associate to university assistant exhibited entrepreneurial spirits, while established university members such as professors rarely undertook this effort and risk (see Table 4-13 in Chapter 4.3.8.1). Moreover, according to Table 4-11 almost 60% of them left university after having spun off.

6.2 Limitations

Every theoretical model is based on a simplification of reality. And human behaviour and social interactions are in particular complex realities. Yet, their simplification in the form of a research model that is to be tested in the course of a statistical analysis by nature entails a couple of limitations that have to be taken into account. Moreover, in order not to induce a method-related deformation of the research object, the limitations of the applied method, i.e. social network analysis, were identified and taken into account accordingly to ensure suitability of the method for the object of investigation (Polya 1995 p. 75).

In the following, the research results presented above have to be considered against the background of these limitations. Only then they can be put in the right perspective and encountered boundaries loose their acuity. In the context of this research, three major limitations have to be highlighted and are depicted in more detail in the next sections.

6.2.1 Sampling

First of all, strictly speaking the sampling procedure was not really carried out in a random manner. Rather all identifiable targets meeting the specified sampling criteria were addressed. Actual response determined the final composition of the sample and thus as such runs the risk of selection biases. Due to the lack of documentation as regards the specific target population of academic spin-offs in Austria, the primary database applied were the websites of the AplusB centres, publicly supported academic spin-off incubators. This adds an additional bias in the form of a couple of rather homogenous properties resulting from the specific funding criteria applied to AplusB spin-offs. In particular, when it comes to the usage of public funding in the course of the spin-off process a certain dominance of affirmative answers in this regard can be observed. On the other hand, it has to be noted that in Austria the majority of university spin-offs has to be attributed to the favourable framework conditions provided by the AplusB programme thus partly reflecting reality.

Moreover, as a result of the small size of the total population, the actual size of the sample, too, is rather small, particularly after having eliminated not appropriate data sets. Even with a comparably high response rate of 27% only 34 valid data sets remain for statistical processing. While this represents a rather high number of respondents for a social network analysis, most statistical procedures and tests of SPSS are designed for larger data volumes.

Altogether these factors compromise the assumed normal distribution of the sample. While the Spearman correlations of Chapter 5.2.2 account for this fact, the results of the subsequent regression analysis of Chapter 5.2.3 are explicitly based on the assumption of normal distribution. However, as mentioned, for that reason the correlation analysis after Spearman was prepended to make sure that only statistically relevant associations are subjected to regression analysis.

Nevertheless, the lack of randomised sampling and the small number of data sets render the transfer of results from the sample to the underlying population difficult and thus have to be treated with caution.

6.2.2 Lack of Control Group

Another serious shortcoming of the research project is the sole concentration on established firms without a control group of failed business founders.

According to Davidsson and Honig (Davidsson and Honig 2003), studies based on samples of established companies dealing with questions related to the earliest stages of development, such as start-up motivations or how resources for the planned business were acquired, encounter a couple of shortcomings. Firstly, it has been estimated that only half of all aspiring business founders succeed in creating new organizations that are ever recorded in public records (Aldrich 1999). Therefore, samples composed only of successful firms are subject to success bias. The results are based only on those cases that successfully completed and survived the start-up process. We learn nothing about those that drop out early, and it cannot be excluded that presumed success factors among survivors are equally characteristic for entrepreneurs that quit in earlier stages (Davidsson and Honig 2003).

Nevertheless, this approach was chosen primarily for feasibility reasons. A control group would have had to be sampled for each phase in terms of would-be entrepreneurs that fail and quit at the end of the respective phase. Identification of drop-out cases turned to be nearly impossible. In particular for the first phase, when the first idea to spin-off slowly emerges and moves towards recognising a substantial opportunity, no way to identify those researchers the quit before coming to the actual decision to spin-off seemed to be available.

As a result this research project has to be regarded more as exploratory contribution to empiricism that offers insights of descriptive rather than confirmatory nature. The network characteristics presented are deemed to offer a picture of the spin-off process against the background of different assumed network properties and parameters. They cannot be validly judged as decisive success factors against the properties of more luckless entrepreneurial researchers. However, they may serve public decision-makers and aspiring spin-off founders as orientation frame for their actions (see 6.3.2).

6.2.3 Cross-sectional Analysis

Finally, while referring to a dynamic process that describes a phaseal development over time, due to practical reasons only a cross-sectional analysis was carried out in the course of this research. This means that data was only collected at one particular point in time, although relating to different periods that depending on the respondent and the history of his or her firm dated back varying times. Needless to say that this approach entails a limited comparability of results as surrounding context factors at that time are difficult to grasp and control for.

Moreover, research asking respondents to report about their history suffers from potential bias due to memory decay and hindsight bias, or rationalization after the fact. It is possible that outcomes are attributed to factors that were not truly present at the relevant time. In particular, singular or sporadic social encounters of the earliest phase dating back long time ago are difficult to commemorate for the respondents. The research design implies that only the point of view of the entrepreneur is captured lacking any guarantee that this perception really coincides with reality. Yet, as previously mentioned above, all that counts in this context is the subjective feeling of the entrepreneurs, because it is their perceptions that determines their actual behaviour and as such has factual importance.

In any way, a longitudinal approach accompanying academic entrepreneurs over time from the initial idea to the actual spin-off and establishment on the market asking questions at the respective phase would have provided a more profound and valid picture of the founding researchers particular social environment at that time. However, given the long duration of the start-up process lasting up to seven to ten years this would have gone beyond the scope of an individual dissertation project. The cross-sectional alternative of asking entrepreneurs at different states of the spin-off process about their relations at the actual phase turned out of no reasonable option either, since in the case of earlier phases no information would have been available whether the respective researcher in the end really succeeded in spinning off and thus could be regarded as desirable case.

6.3 Contributions for Theory and Practice

After having summarised and interpreted the major results of this research and depicting their boundaries, what can be learnt from these findings and how can they contribute to the theoretical work in this field? And most importantly, how can they support practitioners in their work, in particular aspiring founders to successfully spin-off and public decision-makers to provide for the supporting framework conditions? These questions are dealt with in the subsequent sections, starting with a brief estimation on the epistemic value of the results.

6.3.1 Theory

The thesis provides detailed insights in the social structure of academic business founders and the nature and the dynamics of their relations over time. As such it closes the gap of analyses solely concentrating on individual spin-off phases, i.e. points in time, or very specific properties and thus rather narrow sections of the founding process. Instead a dynamic approach is pursued assuming a resource-based understanding of firms with social capital as both resource and access to resources. In order to dispose of social capital, an individual must have external relationships of a certain quality. Following Davidsson and Honig (2003), in this research social capital was operationalised in terms of social exchange, to examine the effects of exchange ties on performance. An important aspect in this context is the consideration of information as important resource in particular during early phases of the spin-off process. Above all it is information that is exchanged via social interactions of all type. The knowledge base of a firm is thus to a large extent coined by its local social and industrial context (Meso and Smith 2000). Networking is considered as important way of closing the resource gap between available and needed resources.

As such this thesis aims at closing a couple of gaps the emerging research on social networks or social capital in entrepreneurship in general (Aldrich and Zimmer 2001; Birley 1985; Christensen et al. 2000; Dubini and Aldrich 1991; Elfring and Hulsink 2001; Greve and Salaff 2003; Hulsink and Elfring 2003; Larson and Starr 1993; Singh et al. 2000) still faces. As depicted in Chapter 1.3, these limitations have been best summarised by Hoang and Antoncic (2003) as follows:

First of all, according to them (Hoang and Antoncic 2003), there are only few process-oriented studies accounting for the dynamic nature of networks and taking them as dependent variables to answer the question how entrepreneurial processes and outcomes influence network development over time. Many network approaches lack to be specific about the context and the timing of the role of network relations. This missing information is accounted for by the phaseal approach to the spin-off process applied in the scope of this research. By doing so the underlying work contributes important empirical confirmation for a theory of network development. In particular, deeper insights on the variation of network content, structure and composition at the different stages of the academic spin-off formation process were generated.

Moreover, according to Hoang and Antoncic (2003), current work on entrepreneurial success is limited by considerable conceptual vagueness regarding the resources vital to success, and how to measure the networks that provide those resources. There is little specification of the various dimensions of networks may exhibit and their impact on the early development of a venture. Mapping networks of general information or resource flows seems not fine-grained enough to shed sufficient light on the determining factors of business establishment. Currently, a standard question to gather network data asks the entrepreneur to whom he or she would turn for advice or information (Burt 1984). This may not allow revealing meaningful differences in network structure, and if differences are observed, there is little insight into the nuances of the entrepreneurial process that would explain them. In contrast, this research differentiated not only between three different content types, i.e. information, resources, and support, but also within each type of content further subcategories allowing for a very sophisticated and extensive picture of the social factors determining the spin-off process and their constituting dimensions.

As mentioned, despite the general notion of the importance of social networks and social capital for the entrepreneurial process, little research has been performed so far transferring these concepts to the analysis of the academic spin-off process. Even already existing studies (Johansson et al. 2005; Nicolaou and Birley 2003a; Nicolaou and Birley 2003b) remain limited to investigating the relationship between an academic's personal network structure and the outcome of the spin-off efforts in terms of the researcher's involvement in the newly founded firm and his affiliation with the mother university. Again a rather static perspective is chosen, taking social networks as independent variable and presuming that the network structure remains unchanged in the course of the spin-off process. Taking this research gap into account, the research project exclusively focused on analysing the process and the conditions of spin-offs from academia in terms of the founding academic's social network and its development over time.

Finally, the research project adds to the rare studies dealing with academic spin-offs in the context of the specific Austrian framework. Given the major differences with respect to funding system, legal background, attitude towards entrepreneurship and academic spin-offs, or market developments, existing research dealing with the spin-off phenomenon in other countries may feature limited transferability. To overcome this deficiency this thesis explicitly focuses on the conditions and features academic start-ups have to face in Austria. In doing so, it aims at providing Austrian policy makers and the scientific establishment with useful insights on the Austrian situation as basis for targeted measures to promote spin-off creations.

6.3.2 Practice

The effective combination of research, economy, and required finance is unanimously regarded as decisive driver for technology-oriented business foundations by entrepreneurship researchers. Statistics show that successful entrepreneurs considerably benefit from scientific education and professional experience. A Europe-wide study of the European Innovation Monitoring Service (EIMS 1997) confirms that most founders of new technology-based firms (NTBF) possess several years of business experience. The primarily male founders are usually between 30 and 50 years old and in most of the cases dispose of higher education, frequently a PhD. Ideally, idea and technological basis result from economically oriented university research. The EIMS-study even infers from the present development that in the upcoming decade most NTBFs will arise from scientific and engineering doctoral projects. These predicted developments emphasise the importance of academic spin-offs for the entrepreneurial dynamics of economic systems and explain the increased political attention of this field. In this context, following main barriers for a spin-off-oriented innovation policy were identified (Malek and Ibach 2004):

- Finance bottlenecks and access to venture capital
- Lack of qualified personnel
- Complex authorisation procedures and laws

- Science and management consulting in operative business
- Smooth transfer between scientific activity and business foundation
- Co-operative networks

With respect to these quoted barriers, this thesis intends to support public decision-makers in furnishing supportive framework conditions for academic spin-off formations. The results provided on the nature of entrepreneurial researchers social networks and their development over time, contribute to efficiently concentrate public funding efforts to effective areas of action. In particular increasingly popular measures to foster networking and cooperation among beneficiaries of public funding can be better judged as to their actual necessity and content in the light of the presented social structures of the spin-off process.

Yet, on the other hand the insights of this research intend to support academic would-be entrepreneurs in providing for the optimal composition of their social environments and interactions to facilitate the spin-off process over time. To this end, it illuminates the typical social structure of entrepreneurial researchers of various scientific backgrounds depending the respective phase of the start-up process. Given the limitations cited above, young academic entrepreneurs may take these research results as orientation and guiding principles when it comes to their networking activities.

Starting with the first phase of recognising and developing the underlying business opportunity, in the scope of the social network analysis of this research it turned out that the diversity of contacts and information in the beginning of the process serve as important medium for innovative business opportunities. Thus, on the one hand aspiring entrepreneurs should strive for a respectively stratified social environment of different contacts from industry, funding agencies, academia, and friends, and on the other hand public authorities should envisage respective support measures to realise apt framework conditions for "multi-social" environments. In this context, the importance of prior industry contacts, i.e. contract research for private clients, in the course of the entrepreneur's academic career has to be highlighted again. It seems to be the possibility were the researcher gets into contact with the business world and makes his or her first experiences with marketability of his or her research results. This not only provides him or her with the respective professional know-how, but also frequently represents the catalytic process that brings the researcher to consider spinning-off an own firm. Besides the economic know-how from industry contacts, scientific and technical knowledge from academic colleagues play an equally important role in the first phase. This development is also supported by the relevance of team foundations with interdisciplinary teams. The success of the newly established venture seems to strongly depend on the right mixture of founding members combining complementary types of know-how and experiences. As can be learnt from Chapter 4.3.8.1 only about 14% of founding researchers acted as solo entrepreneurs. The majority of them set up the new business together with at least one or two partners. This suggests identifying and establishing respective public support measures that promote the brokerage of adequate founding teams with members of complementary backgrounds.

Having a look at pre-seed, seed and early stage financing, it are funding agencies that are quoted as main sources in particular during the first two phases of the spin-off process, i.e. from the initial idea to the actual start-up. One reason of this increased importance of public funding may be derived from the low usage of venture capital. As depicted in Chapter 4.3.8.2, only 8.8% of respondents indicated venture capital as source of early stage finance. This reflects the lack of early stage venture capital in Austria that is primarily compensated by public support. However, it is not the right quantity of venture capital that is missing, but its dedication to risk-prone undertakings such as knowledge- and technology-intensive ventures in their early life-cycles. Austrian venture capitalists are characterised by risk-aversion and lack of appropriate funds management know-how in the form of industry knowledge in research-intensive sectors that usually are the primary origin of the academic spin-offs. Thus, public funding agencies later on in the last phase close to the market assume another important role. They provide consulting services in the form of the required market know-how for a successful market entry and information about necessary further steps for a proper commercialisation of the researcher's development results.

Physical assets obviously require stronger relationships characterised by mutual trust and understanding. Hence, friendship ties are the ones drawn upon the most when it comes to the provision of infrastructure and material resources. The usage of joint infrastructure at the mother institution does not really seem to be an issue. Rather respective framework conditions are additionally sought at funding institutions.

Lack of qualified personnel as cited as one of the main barriers for a spin-off-oriented innovation policy is overcome right at the beginning of the start-up process as a result of previous contract research with industry. In the course of common projects with industry partners entrepreneurial researchers gather relevant information about potential future personnel that comes into effect in later stages, when personnel is employed.

In general, the analysis of this research considerably contributes illuminating the relationship between university and spin-off founders in the light of the social interactions involved. A purely positive relationship between the young company and the incubating institution, however, is not necessarily required, as long as no obstacles are issued by the home department or the mother university in general. The benefits for both, home department and university, associated with spinning off a venture are nevertheless primarily recognised on the part of the researcher's department being the one that immediately capitalises on the close co-operation with the new company in terms of contract research, third-party-funds, and common theses and dissertation projects. The university itself focuses more on the drain of intellectual capital and property rights, which entails a more reserved attitude towards the nascent entrepreneurs. In general, the majority of entrepreneurs subject to this research did not encounter a hostile reaction from their incubator. This asks for respective public measures ensuring at least a neutral attitude towards young spin-offs from their mother universities and home departments by emphasising the benefits all parties may realise in particular when the future relationship is characterised by co-operative social interactions.

6.4 Research Agenda

In the light of the major limitations of this thesis as depicted in Chapter 6.2, following possible subjects of continuative research on this topic are depicted below.

First, a major contribution to knowledge gain can be achieved by establishing a consistent and more or less complete data base of all Austrian academic spin-off companies. In that case, the target population can be addressed in its totality without any biases resulting from facilitated access to certain subpopulations that are better documented such as publicly funded companies. Thus, specific insights on the financing structure and the relation to public agencies of primarily privately financed research-based start-ups are generated that at the current stage of research are by trend slightly underrepresented due to the restricted data base available at this point. A larger data base may in turn also produce a higher response rate and thus a larger sample that allows for more sophisticated statistical procedures than those possible in the scope of this thesis. As a result confirmatory output is generated that contributes to a secured understanding of the actual nature of the process.

Yet, another important aspect consists in generating a control group of unsuccessful entrepreneurs for differential analytics. To this end, future research may take the results of this project as starting point to be complemented by asking prospective spin-off founders that finally did not succeed in realising their venture about their social networks. It has to be noted that due to underlying phase model of this research different control groups for each phase are required. Otherwise it cannot be excluded that presumed success factors among survivors are equally characteristic for entrepreneurs that quit in the respective stage.

This in turn leads to another critical aspect: the phaseal nature of the analysed phenomenon. While the spin-off process investigated is assumed to evolve along three stages, still a cross-sectional approach was applied to this thesis for pragmatic reasons. A research project with a long-term dedication to this matter could accompany a specified sample of young academic entrepreneurs throughout the whole spin-off process starting from the initial intention to spin off until the final establishment of the newly founded company on the market. Against the background of the typical average duration of the start-up process an appropriate study has to cover a horizon of up to seven to ten years. This approach redounds to a more profound and valid picture of the founding researchers' particular social environment at that time. Besides, entrepreneuring academics dropping out of the spin-off process at a certain stage may serve as negative control group for the respective phase. As a result the problem of different control groups for the different stages of the process is also overcome. Moreover, the danger of biases due to memory decay, hindsight, or rationalization after the fact is reduced by asking respondents to report about their social interactions within a shorter period of time. In particular, singular or sporadic social encounters are better captured.

The thesis is focused on the establishment of new ventures out of a particular academic background. It is assumed that this type of start-up is characterised by particular social network characteristics that are attributed to the scientific nature of the incubating entity. Yet, it may be of interest to test these assumptions by comparing the results with those of a comparable social network analysis with "normal" knowledge-based start-ups that feature no particular academic origin, i.e. addressing entrepreneurs that prior to start-up formation did not pursue an academic profession. Another promising extension to this thesis may consist in distinguishing between different scientific disciplines or industries to determine whether thus categorised entrepreneuring individuals are characterised by certain particular network structures that can be assigned to the very specific nature of the respective scientific field or the market characteristics and other context factors of the particular industry.

And finally, a look over the borders in the form of a cross-country analysis with other European or Western nations that feature comparable innovation systems may serve as promising subject of a subsequent research project. Respective results may be appropriate to reveal specific socio-cultural, legal or economic context factors influencing entrepreneuring academics social structure and most notably resulting from the particular national university or research system. Such a comparison may provide valuable insights to political decision-makers about the framework parameters to be applied as correcting variables to support the entrepreneurial dynamics with respect to the creation of innovative knowledge- und technology-intensive companies in Austria.

7 References

Adler, Paul and Seok-Woo Kwon (2002), "Social capital: Prospects for a New Concept," Academy of Management Review, 27 (1), 17-40.

Agresti, Alan and Barbara Agresti (1978), "Statistical Analysis of Qualitative Variation," in Sociological Methodology, Karl F. Schuessler, Ed. San Francisco: Jossey-Bass.

Aldrich, Howard (1999), Organizations Evolving. Newbury Park, CA: Sage Publications.

Aldrich, Howard and Catherine Zimmer (2001), "Entrepreneurship Through Social Networks," in The Art and Science of Entrepreneurship, Donald L. Sexton and Raymond W. Smilor, Eds. Cambridge: Ballinger Publishing Company.

Allcott, Hunt, Dean Karlan, Markus M. Möbius, Tanya Rosenblat, and Adam Szeidl (2007), "Community Size and Network Closure." Cambridge, MA: Harvard University and National Bureau of Economic Research.

Alvarez, Sharon A. and Lowell W. Busenitz (2001), "The entrepreneurship of resource-based theory," Journal of Management, 27, 755–75.

Amit, Raphael and Paul J.H. Schoemaker (1993), "Strategic Assets and Organizational Rent," Strategic Management Journal 14 (1), 33-46.

Anderson, Alistair R. and Sarah L. Jack (2002), "The Articulation of Social Capital in Entrepreneurial Networks: A Glue or a Lubricant?," Entrepreneurship and Regional Development, 14, 193-210.

Ardichvili, Alexander, Richard Cardozo, and Sourav Ray (2003), "Theory of Entrepreneurial Opportunity Identification and Development," Journal of Business Venturing, 18, 105-23.

Arenius, Pia Maria (2002), "Creation of Firm-Level Social Capital, Its Exploitation, and the Process of Early Internationalisation," Dissertation, Helsinki University of Technology.

Atteslander, Peter (1993), Methoden der empirischen Sozialforschung (7th ed.). Berlin: de Gruyter.

Barney, Jay (1991), "Firm Resources and Sustained Competitve Advantage," Journal of Management, 17 (1), 99-120.

Bastians, Frauke (2001), "Die Bedeutung sozialer Netzwerke für die Integration rußlanddeutscher Spätaussiedler in der Bundesrepublik Deutschland," Dissertation, Universität Osnabrück.

Bazeley, Pat (2004), "Issues in Mixing Qualitative and Quantitative Approaches to Research," in Applying Qualitative Methods to Marketing Management, Renate Buber and Johannes Gadner and Lyn Richards, Eds. New York: Palgrave Macmillan.

Beer, Heike (2000), Hochschul-Spin-Offs im High-Tech-Wettbewerb. Hamburg: Verlag Dr. Kovac.

Bellini, Emilio (1999), "Innovation Strategies in Small Firms: A Competence-Based Approach," in 44th ICBS World Conference, ICSB189 (Ed.). Naples.

Bellini, Emilio and Giuseppe Zollo (1997), "Academic Spin-off and Regional Development: Empirical Evidences in Southern Italy," in ERSA 97 - European Regional Science Association, 37th European Congress. Rome, Italy.

Bergmann Lichtenstein, Benyamin M. and Candida G. Brush (2001), "How Do 'Resource Bundles' Develop and Change in New Ventures? A Dynamic Model and Longitudinal Exploration," Entrepreneurship Theory and Practice, 25 (3), 37-58.

Bernard, H. Russell, Peter D. Killworth, David Kronenfeld, and Lee Sailer (1984), "On the Validity of Retrospective Data. The Problem of Informant Accuracy," Annual Review of Anthropology, 13, 495-517.

Bernard, H. Russell, Peter Killworth, and Lee Sailer (1982), "Informant Accuracy in Social-network Data V. An Experimental Attempt To Predict Actual Communication from Recall Data," Social Science Research, 11 (1), 30-66.

Bhave, Mahesh P. (1994), "A Process Model of Entrepreneurial Venture Creation," Journal of Business Venturing, 9 (3), 223-42.

Birley, Sue (1985), "The Role of Networks in the Entrepreneurial Process," Journal of Business Venturing, 1 (1), 107-17.

Blalock, Hubert M. (1985), Causal Models in the Social Sciences (2nd ed.). New York: AldineTransaction.

Bloodgood, James M., Harry J. Sapienza, and Alan Carsrud (1995), "The dynamics of new business startups: person, context, and process," in Advances in Entrepreneurship, Firm Emergence, and Growth, J.A. Katz and R.H. Brockhaus, Eds. Vol. 2. Greenwich, CN: JAI Press.

Boari, Christina and Manuela Presutti (2004), "Social Capital and Entrepreneurship Inside an Italian Cluster: Empirical Investigation." Uppsala: Uppsala University, Department of Business Studies.

Böhner, Ingo (2007), "Network, Network Position and the Deal Flow of Venture Capital Firms," Dissertation, Rhenish-Westphalian Technical University.

Bortz, Jürgen and Nicola Döring (1995), Forschungsmethoden und Evaluation (2nd ed.). Berlin / Heidelberg: Springer-Verlag.

Bourdieu, Pierre (1985), "The Forms of Capital," in Handbook of Theory and Research for the Sociology of Education, John G. Richardson, Ed. New York: Greenwood.

Bourdieu, Pierre and Loïc Wacquant (1992), An Invitation to Reflexive Sociology. Chicago: University of Chicago Press.

Bouty, Isabelle (2000), "Interpersonal and Interaction Influences On Informal Resource Exchanges Between R&D Researchers Across Organisational Boundaries," Academy of Management Journal, 43 (1), 50-65.

Bower, D. Jane (2003), "Business Model Fashion and the Academic Spinout Firm," R & D Management, 33 (2), 97-106.

Brüderl, Josef and Peter Preisendörfer (1998), "Network Support and the Success of Newly Founded Businesses," Small Business Economics, 10, 213–25.

Brush, Candida G., Patricia G. Greene, and Myra M. Hart (2001), "From Initial Idea To Unique Advantage: The Entrepreneurial Challenge of Constructing A Resource Base," Academy of Management Executive, 15 (1), 64-78.

Bührer, Susanne (2002), "Network Analysis," in RTD Evaluation Toolbox - Assessing the Socio-Economic Impact of RTD-Policies - Strata Project HPV 1 CT 1999-00005, Gustavo Fahrenkrog and Wolfgang Polt and Jaime Rojo and Alexander Tübke and Klaus Zinöcker,

Eds. Seville: EUROPEAN COMMISSION Joint Research Centre - Institute for Prospective Technological Studies (IPTS).

Burgelman, R. A., Thomas J. Kosnik, and M. Van den Poel (1987), "The innovative capabilities audit framework," in Strategic Management of Technology and Innovation, R. A. Burgelman and M. Maidique, Ed. Homewood: Richard D. Irwin.

Burt, Ronald S. (1984), "Network Items and the General Social Survey," Social Networks, 6, 293-339.

---- (2000), "The Network Structure of Social Capital," in Research in Organizational Behavior, Robert I. Sutton and Barry M. Staw, Eds. Vol. 22. Greenwich: JAI Press.

---- (1997), "A Note on Social Capital and Network Content," Social Networks, 19, 355-73.

---- (1983), "Range," in Applied Network Analysis: A Methodological Introduction, Ronald S. Burt and M.J. Minor, Eds. Beverly Hills/London/New Delhi.

---- (2001), "Structural Holes versus Network Closure as Social Capital," in Social Capital: Theory and Research, Nan Lin and Karen S. Cook and Ronald S. Burt, Eds. New York: Aldine de Gruyter.

---- (1992), Structural Holes: The Social Structure of Competition. Cambridge: Harvard Business Press.

Bygrave, William D. (1997), "The Entrepreneurial Process," in The Portable MBA In Entrepreneurship, William D. Bygrave, Ed. 2nd ed. New York: John Wiley & Sons, Inc.

Bygrave, William D. and Charles W. Hofer (1991), "Theorizing About Entrepreneurship," Entrepreneurship Theory and Practice, 16, 13-22.

Callan, Benedicte (2001), "Generating Spin-offs: Evidence from Across the OECD," STI Review, 26 (Special Issue on Fostering High-tech Spin-offs: A Public Strategy for Innovation), 13-55.

Carayannis, Elias G., Everett M. Rogers, Kazuo Kurihara, and Marcel M. Allbritton (1998), "High-technology Spin-offs from Government R&D Laboratories and Research Universities," Technovation, 18 (1), 1-11.

Chandler, Gaylen N. and Steven H. Hanks (1998), "An Examination of the Substitutability of Founders Human and Financial Capital In Emerging Business Ventures.," Journal of Business Venturing, 13 (5), 353-70.

---- (1994), "Market Attractiveness, Resource-Based Capabilities, Venture Strategies, and Venture Performance," Journal of Business Venturing 9(4), 331-49.

Chiesa, Vittorio and Andrea Piccaluga (2000), "Exploitation and Diffusion of Public Research: The Case of Academic Spin-off Companies in Italy," R & D Management, 30 (4), 329-39.

Christensen, Patrizia V., John P. Ulhoi, and Henning Madsen (2000), "The Entrepreneurial Process In a Dynamic Network Perspective: A Review and Future Directions for Research." Copenhagen: LOK Center.

Christensen, Patrizia V., John P. Ulhoi, and Helle Neergaard (2001), "The Entrepreneurial Process In High-Tech and Knowledge-Based Sectors In Denmark." Copenhagen: LOK Research Center, Department of Management, Politics & Philosophy.

Clarysse, Bart, Nathalie Moray, and Ans Heirman (2002), "Transferring Technology by Spinning Off Ventures: Towards an Empirically Based Understanding of the Spin Off Process." Gent: Gent University, Faculty of Economics and Business Administration.

Clarysse, Bart, Mike Wright, Andy Lockett, Els Van de Veldea, and Ajay Vohorab (2005), "Spinning out new ventures: a typology of incubation strategies from European research institutions," Journal of Business Venturing, 20, 183–216.

Cohen, Wesley, M. and Daniel A. Levinthal (1990), "Absorptive Capacity: A New Perspective on Learning and Innovation," Administrative Science Quarterly, 35, 128-52.

Coleman, James S. (1990), Foundations of Social Theory. Cambridge: Harvard University Press.

Collis, David J. and Cynthia A. Montgomery (1995), "Competing on resources: Strategy in the 1990s," Harvard Business Review, 73 118-28.

Conner, Kathleen R. (1991), "A Historical Comparison of Resource-based Theory and Five Schools of Thought Within Industrial Economics: Do We Have a New Theory of the Firm?," Journal of Management, 17, 121-54.

Conner, Kathleen R. and C.K. Prahalad (1996), "A Resource-based Theory of the Firm: Knowledge Versus Opportunism," Organization Science, 7 (5), 477-501.

Cortright, Joseph and Heike Mayer Eds. (2001), High Tech Specialization: A Comparison of High Technology Centers: Center on Urban & Metropolitan Policy.

Dahlstrand, Asa Lindholm (1997), "Entrepreneurial spin-off enterprises in Göteborg, Sweden," European Planning Studies, 5 (5), 659-73.

Dasgupta, Parthaa and Paul A. David (1994), "Toward a New Economics of Science," Research Policy 23 (5), 487-521.

Davidsson, Per and Benson Honig (2003), "The Role of Social and Human Capital Among Nascent Entrepreneurs," Journal of Business Venturing, 18, 301-31.

De Coster, Rebecca and Clive Butler (2005), "Assessment of proposals for new technology ventures in the UK: characteristics of university spin-off companies," Technovation, 25 (5), 535-43.

deBono, Edward (1978), "When Opportunity Knocks," Management Today, 102-05.

Degenne, Alain and Michel Forsé (1999), Introducing Social Networks. London / Thousand Oaks / New Delhi: Sage Publications Ltd.

Degroof, Jean-Jacques and Edward B. Roberts (2004), "Overcoming Weak Entrepreneurial Infrastructures for Academic Spin-Off Ventures," Journal of Technology Transfer, 29, 327-52.

Diekmann, Andreas (2002), Empirische Sozialforschung - Grundlagen, Methoden, Anwendungen (9 ed.). Reinbek bei Hamburg: Rowohlt Taschenbuch Verlag GmbH.

Dierickx, Ingemar and Karel Cool (1989), "Asset Stock Accumulation and Sustainability of Competitive Advantage," Management Science, 35 (12), 1504-11.

Dorf, Richard C. and Thomas H. Byers (2005), Technology Ventures: From Idea to Enterprise. New York: McGraw-Hill.

Drucker, Peter (1985), Innovation and Entrepreneurship: Practice and Principles. New York: Harper & Row.

Druilhe, Céline and Elizabeth Garnsey (2004), "Do Academic Spin-Outs Differ and Does it Matter?," Journal of Technology Transfer, 29, 269–85.

Dubini, Paola and Howard Aldrich (1991), "Personal and Extended Networks Are Central to the Entrepreneurial Process," Journal of Business Venturing, 6 (5), 305-73.

Egeln, Jürgen, Helmut Fryges, Sandra Gottschalk, and Christian Rammer (2006), "Dynamik von Spinoff-Gründungen in Österreich: Performance und Erfolgsfaktoren." Mannheim: Zentrum für Europäische Wirtschaftsforschung.

Egeln, Jürgen, Sandra Gottschalk, Christian Rammer, and Alfred Spielkamp (2002), "Spin-off Gründungen aus der öffentlichen Forschung in Deutschland. ," Federal Ministry for Education and Research (Ed.). Mannheim: Zentrum für Europäische Wirtschaftsforschung (ZEW).

Egeln, Jürgen, Sandra Gottschalk, Georg Metzger, Christian Rammer, Helmut Gassler, and Nikolaus Gretzmacher (2003), "Akademische Spin-off Gründungen in Österreich," ZEW / Joanneum Research.

EIMS (1997), "Profiling Entrepreneurs in New Technology Based Firms," European Innovation Monitoring Service.

Elfring, Tom and Willem Hulsink (2001), "Networks in Entrepreneurship: The Case of High-Technology Firms." Rotterdam: Erasmus Research Institute of Management (ERIM), Rotterdam School of Management.

Etzkowitz, Henry, E. Schuler, and Magnus Gulbrandsen (2000), "The Evolution of the Entrepreneurial University," in The Future of Knowledge Production in the Academy, Merle Jacob and Tomas Hellström, Eds. Buckingham: The Society for Research into Higher Education & Open University Press.

Federal Ministry of Science and Research, Innovation and Technology Federal Ministry of Transport, and Federal Ministry of Economics and Labour (2007), "Austrian Research and Technology Report 2007." Vienna.

Flick, Uwe (1998), An Introduction to Qualitative Research. London / Thousand Oaks / New Delhi: Sage Publications.

Fontes, Margarida (2005a), "The process of transformation of scientific and technological knowledge into economic value conducted by biotechnology spin-offs," Technovation, 25 (4), 339-47.

Fontes, Margarida (2005b), "The Process of Rransformation of Scientific and Technological Knowledge into Economic Value Conducted by Biotechnology Spin-offs," Technovation 25 339–47.

Foss, Nicolai and Bo Eriksen (1995), "Competitive advantage and industry capabilities," in Resource-based and evolutionary theories of the firm: Towards a synthesis, Cynthia Montgomery, Ed. Boston: Kluwer Academic Publishers.

Foss, Nicolai J. (1999), "Networks, Capabilities, and Competitive Advantage," Scandinavian Journal of Management, 15 1-15.

Fountain, Jane E. (1998), "Social Capital: Its Relationship to Innovation in Science and Technology.," Science and Public Policy, 25 (2), 103-15.

Franke, Sandra (2005), "Measurement of Social Capital: Reference Document for Public Policy Research, Development, and Evaluation," in PRI Project: Social Capital as a Public Policy Tool. Canada.

Freeman, Linton C. (1979), "Centrality In Social Networks: I. Conceptual Clarification," Social Networks, 1, 215–39.

Froschauer, Ulrike and Manfred Lueger (1998), Das qualitative Interview zur Analyse sozialer Systeme (2nd ed.). Wien: WUV-Universitätsverlag.

Fuchs, Susanne (2005), "Organizational Adoption Models for Early ASP Technology Stages - Adoption and Diffusion of Application Service Providing (ASP) in the Electric Utility Sector," Doctoral Thesis, Vienna University for Economics and Business Administration.

Fukuyama, Francis (1995), Trust. The Social Virtues and the Creation of Prosperity. New York: Free Press.

Galbraith, Jay (1982), "The Stages of Growth," Journal of Business Strategy, 3 (1), 70-79.

Gardner, David M., Frank Johnson, Moonkyu Lee, and Ian Wilkinson (2000), "A Contingency Approach to Marketing High Technology Products," European Journal of Marketing, 34 (9/10), 1053-77.

Gargiulo, Martin and Mario Benassi (2000), "Trapped in Your Own Net? Network Cohesion, Structural Holes, and the Adaptation of Social Capital," Organization Science, 11 (2), 183-96.

Gartner, William B. (1985), "A Conceptual Framework for Describing the Phenomenon of New Venture Creation," Academy of Management Review, 10 (4), 696-706.

Gibson, David V. and Christopher E. Stiles (2000), "Technopoleis, Technology Transfer, and Globally Networked Entrepreneurship," in Science, Technology, and Innovation Policy, Pedro Conceicao and David V. Gibson and Manuel V. Heitor and Syed Shariq, Eds. Westport, CT: Quorum Books.

Granovetter, Mark S. (1985), "Economic Action, Social Structure, and Embeddedness," American Journal of Sociology, 91, 481-510.

---- (1973), "The Strength of Weak Ties," American Journal of Sociology, 78, 1360-80.

---- (1982), "The Strength of Weak Ties: A Network Theory Revisited," in Social Structure and Network Analysis, P.V. Marsden and N. Lin, Eds. Beverly Hills: Sage.

Granstrand, Ove (1998), "Towards a Theory of the Technology-Based Firm," Research Policy 27, 465–89.

Grant, Robert M. (1991), "The Resource-Based Theory of Competitve Advantage: Implications for Strategy Formulation," California Management Review, 33 (3).

Grebel, Thomas, Andreas Pyka, and Horst Hanusch (2003), "An Evolutionary Approach to the Theory of Entrepreneurship," Industry and Innovation, 10 (4), 493-514.

Greve, Arent and Janet W. Salaff (2003), "Social Networks and Entrepreneurship," Entrepreneurship Theory and Practice, 28 (1), 1-22.

Hall, Alan and Barry Wellman (1985), "Social Networks and Social Support," in Social Support and Health, Sheldon Cohen and S. Leonard Syme, Eds. Orlando: Academic Press.

Hall, Richard (1993), "A Framework Linking Intangible Resources and Capabilities To Sustainable Competitive Advantage," Strategic Management Journal, 14 (8), 607-19.

Hammer, Muriel (1985), "Implications of Behavioral and Cognitive Reciprocity in Social Network Data," Social Networks 7, 189-201.

Hanneman, Robert A. (2001), "Introduction to Social Network Methods." Riverside: Department of Sociology, University of California.

Hansen, Eric L. and Barbara J. Bird (2001), "The Stages Model of High-Tech Venture Founding: Tried But True?," Entrepreneurship Theory and Practice, 22 (2), 111-22.

Harary, Frank (1969), Graph Theory. Reading: Addison-Wesley.

Harwood, E (1982), "The Sociology of Entrepreneurship," in Encyclopedia of Entrepreneurship, C. Kent and D. Sexton and K. Vesper, Eds. Englewood Cliffs, NJ: Prentice Hall.

Hasenauer, Rainer, Fritz Scheuch, Walter Aigner, Michael Schreiber, and Rudolf Sinkovics (1994), "High Tech Marketing zur Effizienzsteigerung der Technologiepolitik." Wien: Wirtschaftsuniversität Wien, Institut für Absatzwirtschaft, Ordinariat Absatzlehre.

Haß, Wolfgang (2002), "Soziale Unterstützungsnetzwerke von Menschen mit chronischer Polyarthritis. Eine explorative, netzwerkanalytische Studie," Dissertation, University of Cologne.

Hayek, Friedrich (1945), "The Use of Knowledge in Society," American Economic Review, 35 (4), 519-30.

Heinze, Thomas (1995), Qualitative Sozialforschung: Erfahrungen, Probleme und Perspektiven (3rd ed.). Opladen: Westdeutscher Verlag GmbH.

Heirman, Ans, Bart Clarysse, and Vicky van den Haute (2003), "Starting Resource Configurations of Research-Based Start-Ups and the Interaction With Technology, Institutional Background, and Industrial Dynamics," in Vlerick Leuven Gent Working Paper Series. Gent: Vlerick Leuven Gent Management School.

Hempel, Carl Gustav and Paul Oppenheim (1948), "Studies in the Logic of Explanation," Philosophy of Science, 15, 135-75.

Hills, Gerald E., G.T. Lumpkin, and Robert P. Singh (1997), "Opportunity Recognition: Perceptions and Behaviors of Entrepreneurs," Frontiers of Entrepreneurship Research, 1997 Edition.

Hite, Julie M. and William S. Hesterly (2001), "The Evolution of Firm Networks: From Emergence to Early Growth of the Firm," Strategic Management Journal, 22 (3), 275-86.

Hitt, M. A. and R. D. Ireland (1985), "Corporate Distinctive Competence, Strategy, Industry and Performance," Strategic Management Journal, 6, 273-93.

Hoang, Ha and Bostjan Antoncic (2003), "Network-Based Research in Entrepreneurship: A Critical Review," Journal of Business Venturing, 18 (2), 165-87.

Hodson, Randy and Robert E. Parker Eds. (1988), Work in High-Technology Settings: A Review of the Empirical Literature. Greenwich, CN: JAI.

Hoepfl, Marie C. (1997), "Choosing Qualitative Research: A Primer for Technology Education Researchers," Journal of Technology Education, 9 (1), 47-63.

Hofer, Charles W. and Dan Schendel (1978), Strategy Formulation: Analytical Concepts. New York: West.

Hoffmeyer-Zlotnik, Jürgen (1987), "Egozentrierte Netzwerke in Massenumfragen: Ein ZUMA Methodenforschungsprojekt," ZUMA-Nachrichten, 20, 37-43.

Hölzl, Erik (1994), "Qualitatives Interview," in Verführung zum Qualitativen Forschen: Eine Methodenauswahl, Chorherr et al., Ed. Wien: WUV-Universitätsverlag.

Hoopes, David G., Tammy L. Madsen, and Gordon Walker (2003), "Guest Editors' Introduction to the Special Issue: Why Is There a Resource-based View? Toward a Theory of Competitive Heterogeneity," Strategic Management Journal, 24 (10), 889-902.

Hulbert, Bev, Reva Berman Brown, and Sophie Adams (1997), "Towards an Understanding of 'Opportunity'," Marketing Education Review, 7 (3).

Hulsink, Willem and Tom Elfring (2003), "Networks Effects On Entrepreneurial Processes: Start-Ups in the Dutch ICT Industry 1990-2000." Rotterdam: Erasmus Research Institute of Management (ERIM), Rotterdam School of Economics.

Hunt, Shelby Dean (2000), A General Theory of Competition: Resources, Competences, Productivity, Economic Growth (Marketing for a New Century). Thousand Oaks / London / New Dehli: Sage Publications.

Jackson, Robert Max, Claude S. Fischer, and Lynn M. Jones (1977), "The Dimensions of Social Networks," in Networks and Places: Social Relations in the Urban Setting, Claude S. Fischer and Robert Max Jackson and C. Ann Stueve and Kathleen Gerson and Lynne McCallister Jones and Mark Baldassare, Eds. New York: Free Press.

Jacobs, Jane (1965), The Death and Life of Great American Cities. New York: Penguine Books.

Jansen, Dorothea (2003), Einführung in die Netzwerkanalyse: Grundlagen, Methoden, Forschungsbeispiele (2 ed.). Opladen: Leske + Buderich.

---- (2000), "Netzwerke und soziales Kapital: Methoden zur Analyse struktureller Einbettung," in Soziale Netzwerke: Konzepte und Methoden der sozialwissenschaftlichen Netzwerkforschung, Johannes Weyer, Ed. München: Oldenbourger Wissenschaftsverlag GmbH.

---- (2001), "Soziales Kapital von Unternehmensgründern: Theoretische Überlegungen und erste empirische Ergebnisse." TU Berlin: Lehrstuhl für Soziologie der Organisation, DHV / FÖV Speyer.

Johansson, Mattias, Merle Jacob, and Tomas Hellström (2005), "The Strength of Strong Ties: University Spin-offs and the Significance of Historical Relations," The Journal of Technology Transfer, 30 (3), 271-86.

Kalish, Yuval and Garry Robins (2006), "Psychological Predispositions and Network Structure: The Relationship Between Individual Predispositions, Structural Holes and Network Closure," Social Networks, 28 (1), 56-84.

Kaplan, B. and J.A. Maxwell (1994), "Qualitative Research Methods for Evaluating Computer Information Systems," in Evaluating Health Care Information Systems: Methods and Applications, J.G. Anderson and C.E. Aydin and S.J. Jay, Eds. Thousand Oaks: Sage.

Kazanjian, Robert K. and Robert Drazin (1990), "A stage-contingent model of design and growth for technology based new ventures," Journal of Business Venturing, 5 (3), 137-50.

Kepper, Gaby (1994), Qualitative Marktforschung. Wiesbaden: Deutscher Universitätsverlag.

Kirzner, Israel, M. (1973), Competition and Entrepreneurship. Chicago: University of Chicago Press.

Kirzner, Israel M. (1997), "Entrepreneurial Discovery and the Competitive Market Process: An Austrian Perspective," Journal of Economic Literature, 25, 60-85.

---- (1979), Perception, Opportunity and Profit. Chicago, IL: University of Chicago Press.

Klofsten, M. and D. Jones-Evans (2000), "Comparing Academic Entrepreneurship in Europe - The Case of Sweden and Ireland," Small Business Economics, 14 (4), 299-309.

Kohli, Martin (1978), ""Offenes" und "geschlossenes" Interview: Neue Argumente zu einer alten Kontroverse," Soziale Welt: Zeitschrift für Sozialwissenschaftliche Forschung und Praxis, 1 (78), 1-25.

Kostopoulos, Konstantinos C., Yiannis E. Spanos, and Gregory P. Prastacos (2002), "The Resource-Based View of the Firm and Innovation: Identification of Critical Linkages," in 2nd European Academy of Management Conference. Stockholm.

Kromrey, Helmut (1994), Empirische Sozialforschung: Modelle und Methoden der Datenerhebung und Datenauswertung (6th ed.). Opladen: Leske Verlag + Buderich GmbH.

Kurtako, Donald F. and Richard M. Hodgetts (1994), Entrepreneurship: A Contemporary Approach. New York: The Dryden Press.

Kurz, Andrea, Constanze Stockhammer, Susanne Fuchs, and Dieter Meinhard (2007), "Das problemzentrierte Interview," in Qualitative Marktforschung: Konzepte - Methoden - Analysen, Renate Buber and Hartmut H. Holzmüller, Eds. Wiesbaden: Betriebswirtschaftlicher Verlag Dr. Th. Gabler.

Lamnek, Siegfried (1989), Qualitative Sozialforschung, Methoden und Techniken. München: Psychologie Verlags Union.

Landry, Réjean, Nabil Amara, and Moktar Lamari (2000), "Does Social Capital Determine Innovation? To What Extent?," in 4th International Conference on Technology Policy and Innovation. Curitiba, Brazil.

Larson, Andrea (1992), "Network Dyads in Entrepreneurial Settings: A Study of the Governance of Exchange Relationships," Administrative Science Quarterly, 37 (1), 76-104.

Larson, Andrea and Jennifer A. Starr (1993), "A Network Model of Organization Formation," Entrepreneurship Theory and Practice, 17 (2), 5-15.

Leana, Carrie and Harry J. Van Buren III (1999), "Organizational Social Capital and Employment Practices," Academy of Management Review, 24 (3), 538-55.

Liao, Jianwen and Harold P. Welsch (2003), "Social Capital and Entrepreneurial Growth Aspiration: a Comparison of Technology- and Non-technology-based Nascent Entrepreneurs," Journal of High Technology Management Research, 95, 1-22.

---- (2001), "Social Capital and Growth Intention: The Role of Entrepreneurial Networks," in Technology-Based New Ventures, Frontiers of Entrepreneurship Research.

Lichtenstein Bergmann, Benyamin M. and Candida G. Brush (2001), "How Do "Resource Bundles" Develop and Change in New Ventures? A Dynamic Model and Longitudinal Exploration," Entrepreneurship Theory and Practice, Spring.

Lin, Bou-Wen , Po-Chien Li, and Ja-Shen Chen (2006), "Social capital, capabilities, and entrepreneurial strategies: a study of Taiwanese high-tech new ventures," Technological Forecasting & Social Change, 73 (2).

Lin, Nan (1999), "Building a Network Theory of Social Capital," CONNECTIONS, 22 (1), 28-51.

Lindholm Dahlstrand, Asa (1997), "Growth and Inventiveness in Technology-Based Spin-Off Firms," Research Policy, 26 (3), 331-44.

Lippman, S. A. and Richard P. Rumelt (1982), "Uncertain Imitability: An Analysis of Interfirm Differences in Efficiency under Competition," Bell Journal of Economics, 13, 418–38.

Lockett, Andy, Mike Wright, and Stephen Franklin (2003), "Technology Transfer and Universities' Spin-Out Strategies," Small Business Economics, 20, 185–200.

Loury, Glenn C. (1977), "A Dynamic Theory of Racial Income Differences," in Woman, Minorities, and Employment Discrimination, Phyllis.A. Wallace and Annette Le Mund, Eds. Lexington, MA: Lexingtion Books.

---- (1987), "Why Should We Care About Group Inequality?," Social Philosophy and Policy, 5 (1), 249-71.

Lueger, Manfred and Christof Schmitz (1984), Das offene Interview: Theorie - Erhebung - Rekonstruktion latenter Strukturen. Wien: Service Fachverlag der Wirtschaftsuniversität Wien.

Mahoney, Joseph T. and Steven Michael (2004), "A Subjectivist Theory of Entrepreneurship," Working Paper, University of Illinois at Urbana - Champaign.

Mahoney, Joseph T. and J. Rajendran Pandian (1992), "The Resource-Based View Within the Conversation of Strategic Management," Strategic Management Journal, 13 (5), 363-80.

Makadok, Richard (2001), "Toward a Synthesis of the Resource-based and Dynamic-capability Views of Rent Creation," Vol. 22.

Mäkinen, Helena (2001), "Acquiring New Resources In New Startup Companies: The Case of Pharmaceuticals In Finland," Turku School of Economics and Business Administration.

Malek, Miroslaw and Peter K. Ibach (2004), Entrepreneurship: Prinzipien, Ideen und Geschäftsmodelle zur Unternehmensgründung im Informationszeitalter: dpunkt.verlag GmbH.

Malhotra, Naresh K. and David F. Birks (2003), Marketing Research: An Applied Approach (Second European ed.). Essex: Pearson Education Limited.

Marsden, Peter V. (2002), "Egocentric and Sociocentric Measures of Network Centrality," Social Networks, 24 407–22.

---- (1990), "Network Data and Measurement," Annual Review of Sociology, 16, 435-63.

---- (2005), "Recent Developments in Network Measurement," in Models and Methods of Social Network Analysis, Peter J. Carrington and John Scott and Stanley Wasserman, Eds. Cambridge: Cambridge University Press.

---- (1993), "The Reliability of Network Density And Composition Measures," Social Networks, 15, 399-421.

Mayring, Philipp (1990), Einführung in die qualitative Sozialforschung. München: Psychologie Verlags Union.

---- (2000), "Qualitative Inhaltsanalyse," in Forum: Qualitative Sozialforschung / Forum: Qualitative Social Reseach.

---- (1993), Qualitative Inhaltsanalyse: Grundlagen und Techniken (4th ed.). Weinheim: Deutscher Studien Verlag.

McCallister, Lynne and Claude S. Fischer (1978), "A Procedure for Surveying Personal Networks," Sociological Methods and Research, 7, 131-48.

McCarty, Christopher (2003), "Social Network Analysis," Vol. 2004: Bureau of Economic & Business Research.

McKenna, Regis (1985), The Regis touch: million-dollar advice from America's topmarketing consultant. Reading, Mass.: Addison-Wesley.

Medjedovic, Irena and Andreas Witzel (2005), "Sekundäranalyse qualitativer Interviews. Verwendung von Kodierungen der Primärstudie am Beispiel einer Untersuchung des Arbeitsprozesswissens junger Facharbeiter," in Forum: Qualitative Sozialforschung / Forum: Qualitative Social Reseach.

Merton, Robert K. and Patricia L. Kendall (1946), "The Focused Interview," American Journal of Sociology, 51, 541-57.

Meso, Peter and Robert Smith (2000), "A Resource-Based View of Organisational Knowledge Management Systems," Journal of Knowledge Management, 4 (3), 224-34.

Miles, Matthew B. and A. Michael Huberman (1994), Qualitative Data Analysis: An Expanded Sourcebook (2nd ed.). Thousand Oaks / London / New Delhi: Sage Publications.

Miller, Danny and Jamal Shamsie (1996), "The Resource-Based View of the Firm in Two Environments: The Hollywood Film Studios from 1936 to 1965 " The Academy of Management Journal, 39 (3), 519-43.

Moore, Geoffrey A. (1999), Crossing the Chasm: Marketing and Selling High-Tech Products to Mainstream Customers New York: HarperCollins Publishers.

Moreno, Jacob L. (1954), Die Grundlagen der Soziometrie. Opladen: Westdeutscher Verlag.

---- (1934), How Shall Survive? New York: Beacon Press.

Moriarty, Rowland. T. and Thomas J. Kosnik (1989), "High-Tech Marketing: Concepts, Continuity, and Change," Sloan Management Review, 30 (4), 7-17.

Morris, Michael H. (1998), Entrepreneurial Intensity: Sustainable Advantages for Individuals, Organizations and Societies. Westport: Quorum Books.

Morris, Michael H., Ramon A. Avila, and Jeffrey Allen (1993), "Individualism and the modern corporation: implications for innovation
and entrepreneurship, ," Journal of Management 19 (3), 595–612.

Mosakowski, Elaine (1993), "A Resource-Based Perspective on the Dynamic Strategy-Performance Relationship: An Empirical Examination of the Focus and Differentiation Strategies in Entrepreneurial Firms " Journal of Management, 19 (4), 819-39.

Mueller, John H. and Karl F. Schuessler (1961), Statistical Reasoning in Sociology. Boston: Houghton Miffin.

Müller, Christoph, Barry Wellman, and Alexandra Marin (1999), "How to Use SPSS to Study Ego-centered Networks," Bulletin de Méthodologies Sociologiques BMS, 64, 63-76.

Nahapiet, Janine and Sumantra Ghoshal (1998), "Social Capital, Intellectual Capital and the Organisational Advantage," Academy of Management Review, 23 (2), 242-66.

Ndonzuau, Frédéric N., Fabrice Pirnay, and Bernard Surlemont (2002), "A Stage Model of Academic Spin-Off Creation," Technovation, 22 (5), 281-89.

Nelson, Richard R. and Sidney G. Winter (1982), An Evolutionary Theory of Economic Change. Cambridge, MA: Harvard University Press.

Nicolaou, Nicos and Sue Birley (2003a), "Academic Networks In a Trichotomous Categorisation of University Spinouts," Journal of Business Venturing, 18, 333-59.

---- (2003b), "Social Networks in Organizational Emergence: The University Spinout Phenomenon," Management Science, 49 (12), 1702-25.

Nonaka, Ikujiro and Hirotaka Takeuchi (1995), The Knowledge-Creating Company: How Japanese Companies Create the Dynamics of Innovation. New York / Oxford: Oxford University Press.

OECD (2000), "Special Issue on "Fostering High-tech Spin-offs: A Public Strategy for Innovation"," STI Review, 26 (1).

Pappi, Franz U. (1987), "Die Netzwerkanalyse aus soziologischer Perspektive," in Methoden der Netzwerkanalyse, Franz U. Pappi, Ed. München: Oldenbourg.

Pappi, Franz Urban and Gunter Wolf (1984), "Wahrnehmung und Realität sozialer Netzwerke: Zuverlässigkeit und Gültigkeit der Angaben über beste Freunde im Interview," in Soziale Realität im Interview, Heiner Meulemann and Karl-Heinz Reuband, Eds. Frankfurt am Main: Campus.

Park, John S. (2005), "Opportunity recognition and product innovation in entrepreneurial hi-tech start-ups: a new perspective and supporting case study," Technovation, 25, 739–52.

Patton, Michael Quinn (1990), Qualitative Evaluation and Research Methods (2nd ed.). Newbury Park, CA: Sage Publications, Inc.

---- (2001), Qualitative Research & Evaluation Methods (3rd ed.). Newbury Park: Sage Publications Inc

Penrose, Edith T. (1959), The Theory of Growth of the Firm. New York: John Wiley & Son.

Peteraf, Margaret (1993), "The Cornerstones of Competitive Advantage: A Resource-Based View," Strategic Management Journal, 14 (3), 179-91.

Pfennig, Astrid, Uwe Pfennig, and Peter P. Mohler (1991), "Zur Reliabilität von egozentrierten Netzwerken in Massenumfragen," ZUMA-Nachrichten, 28, 92-108.

Pfenning, Astrid, Uwe Pfenning, and Peter Mohler (1991), "Zur Reliabilität von egozentrierten Netzwerken in Massenumfragen," ZUMA-Nachrichten, 28, 92-108.

Pfenning, Uwe (1996), Soziale Netzwerke in der Forschungspraxis: zur theoretischen Perspektive, Vergleichbarkeit und Standardisierung von Erhebungsverfahren sozialer Netzwerke - Zur Validität und Reliabilität von egozentrierten Netz- und Namensgeneratoren. Darmstadt: Dissertations Druck Darmstadt GmbH Druck und Verlag.

---- (1995), Soziale Netzwerke in der Forschungspraxis: zur theoretischen Perspektive, Vergleichbarkeit und Standardisierung von Erhebungsverfahren sozialer Netzwerke - Zur Validität und Reliabilität von egozentrierten Netz- und Namensgeneratoren. Darmstadt: Dissertations Druck Darmstadt GmbH Druck und Verlag.

Philipsen, Kristian (1998), "Entrepreneurship as Organizing - A Literature Study of Entrepreneurship," in DRUID Summer Conference. Bornholm, Denmark.

Polya, Georg (1995), Schule des Denkes. Vom Lösen mathematischer Probleme. (4th ed.). Tübinge / Basel.

Porter, Michael (1980), Competitive Strategy. New York: Free Press.

Portes, Alejandro (1998), "Social Capital: Its Origins and Applications In Modern Sociology " Annual Review of Sociology, 24, 1-24.

Prahalad, C. K. and Gary Hamel (1990), "The Core Competence of the Corporation," Harvard Business Review, 68 (3), 79-91.

Putnam, Robert D. (2000), Bowling Alone - The Collapse and Revival of American Community. New York: Simon & Schuster.

---- (1993), Making Democracy Work. Princeton, NJ: Princeton University Press.

Radosevich, Raymond (1995), "A Model for Entrepreneurial Spin-offs from Public Technology Sources," International Journal of Technology Management, 10 (7-8), 879-93.

Ramachandran, k. and Sougata Ray (2006), "Networking and New Venture Resource Strategies: A Study of Information Technology Start-ups," Journal of Entrepreneurship, 15 (2), 145-68.

Rasmussen, Einar and Odd Jarl Borch (2004), "University Resources Facilitating Strategic Entrepreneurship," University of Twente.

Reed, Richard and Robert J. Defillippi (1990), "Causal Ambiguity, Barriers to Imitation, and Sustainable Competitive Advantage," The Academy of Management Review, 15 (1), 88-102.

Riche, Richard W., Daniel E. Hecker, and John U. Burgan (1983), "High technology today and tomorrow: a small slice of the employment pie," Monthly Labor Review, 50-58.

Roberts, Edward B. (1991), Entrepreneurs In High Technology: Lessons Learned from MIT and Beyond. New York: Oxford University Press.

Roberts, Edward B. and Denis E. Malone (1996), "Policies and Structures for Spinning Off New Companies From Research and Development Organizations," R & D Management, 26 (1), 17-48.

Roessner, David, Alan Porter, Xiao-Yin Jin, Nils Newman, and Elmer Yglesias (2001), "Indicators of technology-based competitiveness: incorporating recent changes in the concept, "high-technology," and in "data availability"," in Final Report to the National Science Foundation. Atlanta, Georgia, USA: Georgia Institute of Technology.

Rogers, Everett M. (1995), Diffusion of Innovations (4th ed.). New York: The Free Press.

Rogers, Everett M., Shiro Takegami, and Jing Yin (2001), "Lessons Learned About Technology Transfer," Technovation, 21, 253-61.

Roth, Erwin and Klaus Heidenreich (1993), Sozialwissenschaftliche Methoden: Lehr- und Handbuch für Forschung und Praxis (3rd ed.). München / Wien: R. Oldenbourg Verlag GmbH.

Rowley, Timothy J. and Joel A.C. Baum (2004), "Sophistication of Interfirm Network Strategies In the Canadian Investment Banking Industry," Scandinavian Journal of Management, 20 103–24.

Rumelt, Richard P. (1984), "Towards a Strategic Theory of the Firm," in Competitive Strategic Management, Robert B. Lamb, Ed. Engelwood Cliffs: Prentice-Hall.

Saffu, Kojo and Takyiwaa Manu (2004), "Strategic Capabilities of Ghanaian Female Business Owners and the Performance of Their Ventures," Paper, Brock University.

Sale, Joanna E. M., Lynne H. Lohfeld, and Kevin Brazil (2002), "Revisiting the Quantitative-Qualitative Debate: Implications for Mixed-Methods Research," Quality and Quantity, 36 (1), 43-53.

Sandberger, Georg (1986), "Rechtliche Hemnisse und Schranken für den Technologietransfer in Deutschland," in Wissenstransfer zwischen Universität und Wirtschaft. Neue Formen der Kooperation in Westeuropa, Adolf Theis and Walter Graumann and Thomas Oppermann, Eds. Baden-Baden: Nomos Verlagsgesellschaft.

Sanders, Jimy M. and Victor Nee (1996), "Immigrant Self-Employment: The Family as Social Capital and the Value of Human Capital," American Sociological Review 61, 231–49.

Say, Jean Baptiste (1803), A Treatise on Political Economy (4th ed.). Paris: Deterville (reprinted by Lippincott, Grambo & Co., Philadelphia, 1855).

Schäfer, Jutta (1995), "Glossar qualitativer Verfahren." Berlin: Berliner Zentrum Public Health.

Schenk, Michael (1983), "Das Konzept des sozialen Netzwerks," in Gruppensoziologie. Perspektiven und Materialien. Kölner Zeitschrift für Soziologie und Sozialpsychologie., Friedhelm Neidhardt (Ed.) Vol. Sonderheft 25. Opladen: Westdeutscher Verlag.

---- (1984), Soziale Netzwerke und Kommunikation. Tübingen: J.C.B. Mohr (Paul Siebeck).

---- (1995), Soziale Netzwerke und Massenmedien: Untersuchungen zum Einfluss der persönlichen Kommunikation. Tübingen: Mohr.

Schnell, Rainer, Paul B. Hill, and Elke Esser (1999), Methoden der empirischen Sozialforschung (Ed. 6 ed.). München: R. Oldenbourg Verlag.

Schülein, Johann A. and Simon Reitze (2002), Wissenschaftstheorie für Einsteiger. Vienna: WUV Universitätsverlag.

Schumpeter, Joseph A. (1934), The Theory of Economic Development. Cambridge, MA: Harvard University Press.

Schütze, Fritz (1976), "Zur Hervorlockung und Analyse von Erzählungen thematisch relevanter Geschichten im Rahmen soziologischer Feldforschung – dargestellt an einem Projekt zur Erforschung von kommunalen Machtstrukturen," in Arbeitsgruppe Bielefelder Soziologen: Kommunikative Sozialforschung Vol. München. Munich.

Scott, John (2000), Social Network Analysis: A Handbook (2nd ed.). London / Thousand Oaks / New Delhi: Sage Publications.

Shane, Scott and S. Venkataraman (2000), "The Promise of Entrepreneurship as a Field of Research," Academy of Management Review, 25 (1), 217-26.

Shanklin, William L. and John K. Ryans (1984), "Organizing for high-tech marketing," Harvard Business Review, 62 (6), 164-71.

Siegel, Sidney and N. John Castellan Jr. (1988), Nonparametric Statistics for the Behavioral Sciences (2nd ed.). London: McGraw-Hill College.

Simmel, Georg (1908), Soziologie - Untersuchungen über die Formen der Vergesellschaftung (1st ed.). Berlin Duncker & Humblot.

Simon, Herbert A. (1976), Administrative Behavior. New York: Free Press.

Simpson, E.H. (1949), "Measurement of Diversity," Nature 163, 688.

Singh, Robert P., Gerald E. Hills, and Lumpkin G.T. "New Venture Ideas and Entrepreneurial Opportunities: Understanding the Process of Opportunity Recognition."

Singh, Robert P., Ralph C. Hybels, and Gerald E. Hills (2000), "Examining the Role of Social Network Size and Structural Holes," New England Journal of Entrepreneurship, 3 (2), 47-59.

Smilor, Raymond W. , David V. Gibson, and Glenn B. Dietrich (1990), "University Spin-out Companies: Technology Start-ups from UT Austin," Journal of Business Venturing, 5, 63-76.

Spratt, Peter (2003), "Methods of Social Research," in Sociology: Australian Connections, Ray Jureidini and Marilyn Poole, Eds. 3rd ed. Sydney: Allen & Unwin.

Starr, Jennifer A. and Ian C. MacMillan (1990), "Resource Cooptation Via Social Contracting: Resource Acquisition Strategies for New Ventures," Strategic Management Journal, 11 (5), 79-92.

Steffensen, Morten, Everett M. Rogers, and Kristen Speakman (2000), "Spin-offs From Research Centers At a Research University," Journal of Business Venturing, 15, 93-111.

Stevenson, Howard H. and David E. Gumpert (1985), "The Heart of Entrepreneurship," Harvard Business Review, Apr-May, 85-94.

Steward, Frances (1977), "Inappropriate Technology," in Technology and Underdevelopment. London: Macmillan Press.

Stinchcombe, Arthur L. (1965), "Social Structure and Organisations," in Handbook of Organizations, James D. March, Ed. Chicago: Rand McNally & Company.

Stone, Wendy (2001), "Measuring Social Capital." Melbourne: Australian Institute of Family Studies.

Stuart, Toby E., Ha Hoang, and Ralph C. Hybels (1999), "Interorganizational Endorsements and the Performance of Entrepreneurial Ventures," Administrative Science Quarterly, 44 (2), 315-49.

Suvanto, Tiina (2000), "Social Capital and Value Creation: A Theoretical Approach." Espoo: Institute of Strategy and International Business, Department of Industrial Engineering and Management, Helsinki University of Technology.

Sydow, Jörg and Arnold Windeler (1998), "Organizing and Evaluating Interfirm Networks: a Structurationist Perspective on Network Processes and Effectiveness," Organization Science, 9 (3), 265-84.

Teece, David J., Gary Pisano, and Amy Shuen (1997), "Dynamic Capabilities and Strategic Management," Strategic Management Journal, 18 (7), 509-33.

Thudium, Thomas (2005), Technologieorientiertes Strategisches Marketing: Die Entwicklung eines neuen Bezugsrahmens zur Generierung von Marketingstrategien für technologieorientierte Unternehmen. Wiesbaden: Deutscher Universitäts-Verlag.

Tsai, Wenpin and Sumantra Ghoshal (1998), "Social Capital and Value Creation: The Role of Intrafirm Networks," Academy of Management Journal, 41 (4), 464-76.

Tyson, Laura D'Andrea (1993), Who's bashing whom? Trade conflict in high-technology industries. Washington, D.C.: Institute for International Economics.

Ulhoi, John P. (2005), "The Social Dimensions of Entrepreneurship," Technovation, 25 (8), 939-46.

Uzzi, Brian (1996), "The Sources and Consequences of Embeddedness for the Economic Performance of Organizations: The Network Effect," American Sociological Review, 61, 674-98.

Van de Ven, Andrew H., Roger Hudson, and Dean M. Schroeder (1984), "Designing New Business Startups: Entrepreneurial, Organizational, and Ecological Considerations," Journal of Management, 10 (1), 87-107.

van der Gaag, Martin (2005), "Measurement of Individual Social Capital," Dissertation, Rijksuniversiteit Groningen.

van Geenhuizen, Marina (2003), "Fostering Academic Entrepreneurship: New Insights into Incubation from An Evolutionary Perspective," in ERSA 2003 Congress. University of Jyväskylä, Jyväskylä, Finland.

Viardot, Eric (2004), Successful Marketing Strategy for High-Tech Firms (3rd ed.). Boston / London: Artech House technology management library.

Vohora, Ajay, Mike Wright, and Andy Lockett (2004), "Critical Junctures in the Development of University High-Tech Spinout Companies," Research Policy, 22 (1), 147-75.

von Hippel, Eric (1994), ""Sticky Information"" and the Locus of Problem Solving: Implications for Innovation," Management Science, 40 (4), 429-39.

Wasserman, Stephen and Katherine Faust (1997), Social Network Analysis: Methods and Applications. Cambridge: Cambridge University Press.

Wernerfelt, Birger (1984), "A Resource-Based View of the Firm," Strategic Management Journal, 5 (2), 171-80.

West, G. Page and Julio O. De Castro (2001), "The Achilles Heel of Firm Strategy: Resource Weaknesses and Distinctive Inadequacies," Journal of Management Studies, 38, 417-42.

Wikipedia (2007a), "Sample (Statistics)," Wikimedia Foundation, Inc.

---- (2007b), "Social Capital," Wikimedia Foundation, Inc.

Wilken, Paul H. (1979), Entrepreneurship. A comparative and historical study. New Jersey: Ablex Publishing Corporation.

Williams, Fred and David V. Gibson (1990), Technology Transfer: A Communication Perspective. Newbury Park: Sage Publications.

Williamson, Oliver E. (1975), Markets and Hierarchies: Analysis and Antitrust Implications. New York: Free Press.

Witzel, Andreas (1985), "Das problemzentrierte Interview," in Qualitative Forschung in der Psychologie. Grundfragen, Verfahrensweisen, Anwendungsfelder G. Jüttemann, Ed. Beltz: Weinheim.

---- (1982), Verfahren der qualitativen Sozialforschung. Überblick und Alternativen. Frankfurt/New York: Campus.

World Bank, The (1999), "What is Social Capital," in PovertyNet

Wright, Mike, Sue Birley, and Simon Mosey (2004a), "Entrepreneurship and University Technology Transfer," Journal of Technology Transfer, 29, 235-46.

Wright, Mike, Ajay Vohora, and Andy Lockett (2004b), "The Formation of High-Tech University Spinouts: The Role of Joint Ventures and Venture Capital Investors," Journal of Technology Transfer, 29, 287-310.

Zahn, Erich, Andreas Koch, and Michaela Schaschke (2003), "Spin-offs als Gründungsform: Charakteristika, Entwicklungswege, Erfolg und Misserfolg," in Gründungen von Technologieunternehmen: Merkmale - Erfolg - empirische Ergebnisse, Claus Steinle and Katja Schumann, Eds. 1st ed. Wiesbaden: Betriebswirtschaftlicher Verlag Dr. Th. Gabler GmbH.

8 Appendices

8.1 Data Sources

8.1.1 Academia plus Business

AplusB[1] promotes the establishment of dedicated centres to consult and coach entrepreneurs from the academic sector. AplusB centres support new venture creations originating from universities, colleges of higher education and non-university research institutions.

Since 2002 a total of nine centres have been approved in the course of two respective calls that have already started operations (see Table 8-1):

Science Park	Website	Province
Accent	www.accent.at	Lower Austria
BCCS-Business Creation Center Salzburg	www.bccs.at	Salzburg
build!	www.build.or.at	Carinthia
CAST	www.cast-tyrol.com	Tyrol
INiTS	www.inits.at	Vienna
SPG-Science Park Graz	www.sciencepark.at	Styria
tech2b	www.tech2b.at	Upper Austria
v-start	www.v-start.at	Vorarlberg
ZAT - Zentrum für Angewandte Technologie	www.zat.co.at	Styria

Table 8-1: Overview Science Parks of AplusB

8.1.2 Life Science Austria

Life Science Austria Vienna Region[2] (LISA VR) is the central life sciences consultancy and coordination point in the Vienna region. LISA VR provides consultancy for business start-ups, preparation of business plans, and financing. It offers assistance in applying to national and regional funding programs and helps open up access to private financing. In addition to consultancy and financing, the centre's international activities, networking and training play a central role.

---

[1] http://www.ffg.at/aplusb
[2] http://www.lisavr.at

8.1.3 Universities

In Austria universities are financed by the State. To this end, the responsible minister concludes a performance agreement of three years with each university. The draft of this agreement, which is subject to negotiations, is submitted by the respective university. 20% of the budget is determined by indicators. 80% are allocated to each university on the basis of negotiations referring to the performance agreements. Criteria are need, demand, performance and social objectives. The three-year global budget of each university thus consists of the formula budget and the negotiated budget (basic budget).

Besides funding of the state, universities dispose of additional financial sources. Within the scope of their recently obtained total legal capacity, universities may acquire estate, carry out contractual research etc. and use these revenues to accomplish their tasks. Furthermore, a large amount of the funds is assigned by the Austrian Science Fund (FWF) that is fed by the federal budget. Finally, universities may use the tuition fees of their students as own revenues.

The University for Education Krems is financed by federal funds, the province of Lower Austria and by the fees levied for the university courses.

At present 22 public universities exist in Austria (see Table 8-2).

University	Website
University of Vienna	http://www.univie.ac.at
University of Graz	http://www.uni-graz.at
University of Innsbruck	http://www.uibk.ac.at
University of Salzburg	http://www.sbg.ac.at
Vienna University of Technology	http://www.tuwien.ac.at
Graz University of Technology	http://www.TUGraz.at
University of Leoben	http://www.unileoben.ac.at
University of Natural Resources and Applied Life Sciences, Vienna	http://www.boku.ac.at
University of Veterinary Medicine Vienna	http://www.vu-wien.ac.at
Vienna University of Economics and Business Administration	http://www.wu-wien.ac.at
University of Linz	http://www.jku.at
University of Klagenfurt	http://www.uni-klu.ac.at
Danube University Krems	http://www.donau-uni.ac.at
Medical University of Graz	http://www.meduni-graz.at/
Medical University of Innsbruck	http://www.i-med.ac.at

Medical University of Vienna   http://www.meduniwien.ac.at

University of Applied Arts Vienna   http://www.dieangewandte.at

University of Music and Dramatic Arts Vienna   http://www3.mdw.ac.at

University Mozarteum Salzburg   http://www.moz.ac.at

University of Music and Dramatic Arts Graz   http://www.kug.ac.at

Kunstuniversität Linz   http://www.ufg.ac.at

Academy of Fine Arts Vienna   http://www.akbild.ac.at

Table 8-2: Universities in Austria

With the University Accreditation Law (UniAkkG) of 1999, proceedings for acknowledging private universities were created. Currently ten private universities are accredited.

8.2 Data Sheets

8.2.1 Ego

Variable	Phase	Frage Nr.	Frage	Antwortmöglichkeiten				
ID Ego			ID Ego					
Haupt-motivation	1	1	Was war die zugrunde liegende Idee / Hauptmotivation für die spätere Spin-off Gründung?	Offen				
main	1	3	Gab es eine Person, die den Hauptanstoß zur Chancenerkennung / Ideenfindung gegeben hat / maßgeblich beigetragen hat?	1=Ja	0=Nein			
busp	2	10	Haben Sie einen Business Plan geschrieben?	1=Ja	0=Nein			
startD	2	11	Woher haben Sie das erforderliche Startkapital erhalten?	Offen				
vent	2	12	Haben Sie Venture Capital in Anspruch genommen?	1=Ja	0=Nein			
univ	2	13.1	Wie war die Reaktion auf den Ausgründungsgedanken seitens der Universität?	1=positiv	2=negativ	3=neutral		
inst	2	13.2	Wie war die Reaktion auf den Ausgründungs-gedanken seitens Ihres Instituts?	1=positiv	2=negativ	3=neutral		
reakD	2	13.3	Anmerkungen:	Offen				
infra	2	14	Konnten Sie die Infrastruktur der Universität mitnutzen?	1=Ja	0=Nein			
inrfD	2	14.1	Anmerkungen:	Offen				
part	3	21	Haben Sie das Unternehmen gemeinsam mit Partnern	1=Ja	0=Nein			

Variable	Phase	Frage Nr.	Frage	Antwortmöglichkeiten								
			gegründet?									
anza	3	21.1	Anzahl der Partner	Numerisch								
pins	3	22.1	Woher haben Sie primär das erforderliche Personal rekrutiert - Institut / Universität?	1=Ja	0=Nein							
frei	3	22.2	Woher haben Sie primär das erforderliche Personal rekrutiert - Freier Markt?	1=Ja	0=Nein							
beka	3	22.3	Woher haben Sie primär das erforderliche Personal rekrutiert - Bekanntenkreis?	1=Ja	0=Nein							
pson	3	22.4	Woher haben Sie primär das erforderliche Personal rekrutiert - Sonstiges?	1=Ja	0=Nein							
psonD	3	22.4.1	Spezifizieren Sie:	Offen								
funk	3	13	Haben Sie nach der Unternehmens-gründung Ihre Funktion an der Universität beibehalten?	1=Ja	0=Nein							
baka	3	24.1	Was hat Ihnen primär geholfen, Ihren Bekanntheitsgrad und Ruf am Markt aufzubauen - Akademische Laufbahn?	1=Ja	0=Nein							
bind	3	24.2	Was hat Ihnen primär geholfen, Ihren Bekanntheitsgrad und Ruf am Markt aufzubauen - Industriekontakte?	1=Ja	0=Nein							
bfre	3	24.3	Was hat Ihnen primär geholfen, Ihren Bekanntheitsgrad und Ruf am Markt aufzubauen - Beziehungen aus dem Freundeskreis?	1=Ja	0=Nein							
bfam	3	24.4	Was hat Ihnen primär geholfen,	1=Ja	0=Nein							

Variable	Phase	Frage Nr.	Frage	Antwortmöglichkeiten									
bson	3	24.5	Ihren Bekanntheitsgrad und Ruf am Markt aufzubauen - Familiärer Hintergrund?	1=Ja	0=Nein								
bsonD	3	24.5.1	Was hat Ihnen primär geholfen, Ihren Bekanntheitsgrad und Ruf am Markt aufzubauen - Sonstiges? Spezifizieren Sie:	Offen									
gesc	0	30	Welches Geschlecht haben Sie?	1=männlich	2=weiblich								
alte	0	31	Wie alt sind Sie?	Numerisch									
fachD	0	32	Von welcher wissenschaftlichen Fachrichtung kommen Sie?	Offen									
univD	0	33	Von welcher Universität kommen Sie?	Offen									
unipos	0	34	Welche Position haben Sie zum Zeitpunkt der Unternehmensgründung an der Universität bekleidet?	1=Professor	2=Dozent	3=Assistenter	4=wissenschaftlicher Mitarbeiter	5=Student	6=Sonstiges				
uniposD	0	34.1	Spezifizieren Sie:	Offen									
wann	0	35	Wann wurde Ihr Unternehmen gegründet?	Numerisch									
bran	0	36	Welcher Branche gehört Ihr Unternehmen an?	1=(Mikro-)Elektronik, IT, Telekommunikation	2=Fertigung, Transport	3=Industrielle Technologien	4=Energie	5=Wissenschaften (Mathematik, Physik, Chemie, Meteorologie, Erdwissen-	6= Medizin, Biologie / Biotechnologie	7= Mikro- & Nano-techno-logie	8 = Land - und Forstwirtschaft	9 = Lebens-mittel-industrie	10 = sonstige

Variable	Phase	Frage Nr.	Frage		Antwortmöglichkeiten	
						schaffen, etc.)
branD	0	36.1	Spezifizieren Sie:	Offen		
vorh	0	37	Haben Sie davor schon ein Unternehmen gegründet?	1=Ja	0=Nein	
unte	0	38	Wurden Sie bei der Unternehmens-gründung in irgendeiner Form durch öffentliche Fördergelder unterstützt?	1=Ja	0=Nein	
fundspecD	0	38.1	Warum handelte es sich dabei genau (bitte machen Sie nähere Angaben)?	Offen		
fundcom	0	38.2	Anmerkungen:	Offen		

8.2.2 Alter

Variable	Phase	Frage Nr.	Frage	Antwortmöglichkeiten	
EGOID			ID Ego		
ALTERID			ID Alter		
mainalt1	1	3	Hauptanstoß zur Chancenerkennung	0=nein	1=ja
sexa1	1	4	Geschlecht	1=männlich	2=weiblich
fam1	1	5.1	Familie	0=nein	1=ja
friend1	1	5.2	Freund	0=nein	1=ja
acad1	1	5.3	Akademia	0=nein	1=ja
disc1	1	5.3.1	Gleiche Disziplin?	0=nein	1=ja
discoth1	1	5.3.1.1	Andere Disziplin:	offen	
inst1	1	5.3.2	Eigenes Institut?	0=nein	1=ja
ind1	1	5.4	Industrie	0=nein	1=ja
funds1	1	5.5	Fördergeber	0=nein	1=ja
others1	1	5.6	Sonstiges	0=nein	1=ja

Variable	Phase	Frage Nr.	Frage	Antwortmöglichkeiten			
othspec1	1	5.6.1	Sonstiges spezifiziere:	Offen			
info1	1	6.1	Informationsaustausch/Diskussion	0=nein	1=ja		
infolaw1	1	6.1.1	Rechtliche Informationen	0=nein	1=ja		
infoeco1	1	6.1.2	Wirtschaftliches Know-how	0=nein	1=ja		
infocus1	1	6.1.3	Informationen zu / Vermittlung von potentiellen Kunden	0=nein	1=ja		
infoemp1	1	6.1.4	Informationen zu / Vermittlung von Personal	0=nein	1=ja		
infooth1	1	6.1.5	Sonstiges	0=nein	1=ja		
infotsp1	1	6.1.5.1	Bitte spezifizieren Sie:	Offen			
res1	1	6.2	Bereitstellung von Ressourcen	0=nein	1=ja		
resmat1	1	6.2.1	Material	0=nein	1=ja		
rescap1	1	6.2.2	Kapital	0=nein	1=ja		
resinfr1	1	6.2.3	Infrastruktur	0=nein	1=ja		
resemp1	1	6.2.4	Personal	0=nein	1=ja		
resoth1	1	6.2.5	Sonstiges	0=nein	1=ja		
resotsp1	1	6.2.5.1	Bitte spezifizieren Sie:	Offen			
sup1	1	6.3	Unterstützung	0=nein	1=ja		
supem1	1	6.3.1	Emotionale Unterstützung	0=nein	1=ja		
supins1	1	6.3.2	Institutionelle Unterstützung	0=nein	1=ja		
oth1	1	6.4	Sonstiges	0=nein	1=ja		
othspe1	1	6.4.1	Bitte spezifizieren Sie:	Offen			
frequ1	1	7	Wie intensiv war ihr Kontakt?	1=einmalig/ sporadisch (jährlich)	2=gelegentlich (monatlich)	3=häufig (wöchentlich)	4=ständig (fast täglich)
sexa2	2	15	Geschlecht	1=männlich	2=weiblich		
fam2	2	16.1	Familie	0=nein	1=ja		
friend2	2	16.2	Freund	0=nein	1=ja		
acad2	2	16.3	Akademia	0=nein	1=ja		
disc2	2	16.3.1	Gleiche Disziplin?	0=nein	1=ja		
discoth2	2	16.3.1.1	Andere Disziplin:	Offen			

Variable	Phase	Frage Nr.	Frage	Antwortmöglichkeiten			
inst2	2	16.3.2	Eigenes Institut?	0=nein	1=ja		
ind2	2	16.4	Industrie	0=nein	1=ja		
funds2	2	16.5	Fördergeber	0=nein	1=ja		
others2	2	16.6	Sonstiges	0=nein	1=ja		
othspec2	2	16.6.1	Sonstiges spezifizieren:	Offen			
info2	2	17.1	Informationsaustausch/ Diskussion	0=nein	1=ja		
infolaw2	2	17.1.1	Rechtliche Informationen	0=nein	1=ja		
infoeco2	2	17.1.2	Wirtschaftliches Know-how	0=nein	1=ja		
infocus2	2	17.1.3	Informationen zu / Vermittlung von potentiellen Kunden	0=nein	1=ja		
infoemp2	2	17.1.4	Informationen zu / Vermittlung von Personal	0=nein	1=ja		
infooth2	2	17.1.5	Sonstiges	0=nein	1=ja		
infotsp2	2	17.1.5.1	Bitte spezifizieren Sie:	Offen			
res2	2	17.2	Bereitstellung von Ressourcen	0=nein	1=ja		
resmat2	2	17.2.1	Material	0=nein	1=ja		
rescap2	2	17.2.2	Kapital	0=nein	1=ja		
resinfr2	2	17.2.3	Infrastruktur	0=nein	1=ja		
resemp2	2	17.2.4	Personal	0=nein	1=ja		
resoth2	2	17.2.5	Sonstiges	0=nein	1=ja		
resotsp2	2	17.2.5.1	Bitte spezifizieren Sie:	Offen			
sup2	2	17.3	Unterstützung	0=nein	1=ja		
supem2	2	17.3.1	Emotionale Unterstützung	0=nein	1=ja		
supins2	2	17.3.2	Institutionelle Unterstützung	0=nein	1=ja		
oth2	2	17.4	Sonstiges	0=nein	1=ja		
othspe2	2	17.4.1	Bitte spezifizieren Sie:	Offen			
frequ2	2	18	Wie intensiv war ihr Kontakt?	1=einmalig/ sporadisch (jährlich)	2=gelegentlich (monatlich)	3=häufig (wöchentlich)	4=ständig (fast täglich)
part3	3	21.2	Gründungspartner	0=nein	1=ja		
sexa3	3	25	Geschlecht	1=männlich	2=weiblich		

Variable	Phase	Frage Nr.	Frage	Antwortmöglichkeiten	
fam3	3	26.1	Familie	0=nein	1=ja
friend3	3	26.2	Freund	0=nein	1=ja
acad3	3	26.3	Akademia	0=nein	1=ja
disca3	3	26.3.1	Gleiche Disziplin?	0=nein	1=ja
discoth3	3	26.3.1.1	Andere Disziplin:	Offen	
inst3	3	26.3.2	Eigenes Institut?	0=nein	1=ja
ind3	3	26.4	Industrie	0=nein	1=ja
funds3	3	26.5	Fördergeber	0=nein	1=ja
others3	3	26.6	Sonstiges	0=nein	1=ja
othspec3	3	26.6.1	Sonstiges spezifiziere:	Offen	
info3	3	27.1	Informationsaustausch/ Diskussion	0=nein	1=ja
infolaw3	3	27.1.1	Rechtliche Informationen	0=nein	1=ja
infoeco3	3	27.1.2	Wirtschaftliches Know-how	0=nein	1=ja
infocus3	3	27.1.3	Informationen zu / Vermittlung von potentiellen Kunden	0=nein	1=ja
infoemp3	3	27.1.4	Informationen zu / Vermittlung von Personal	0=nein	1=ja
infooth3	3	27.1.5	Sonstiges	0=nein	1=ja
infotsp3	3	27.1.5.1	Bitte spezifizieren Sie:	Offen	
res3	3	27.2	Bereitstellung von Ressourcen	0=nein	1=ja
resmat3	3	27.2.1	Material	0=nein	1=ja
rescap3	3	27.2.2	Kapital	0=nein	1=ja
resinfr3	3	27.2.3	Infrastruktur	0=nein	1=ja
resemp3	3	27.2.4	Personal	0=nein	1=ja
resoth3	3	27.2.5	Sonstiges	0=nein	1=ja
resotsp3	3	27.2.5.1	Bitte spezifizieren Sie:	Offen	
sup3	3	27.3	Unterstützung	0=nein	1=ja
supem3	3	27.3.1	Emotionale Unterstützung	0=nein	1=ja
supins3	3	27.3.2	Institutionelle Unterstützung	0=nein	1=ja
oth3	3	27.4	Sonstiges	0=nein	1=ja
othspe3	3	27.4.1	Bitte spezifizieren Sie:	Offen	

Variable	Phase	Frage Nr.	Frage	Antwortmöglichkeiten			
				1=einmalig/ sporadisch (jährlich)	2=gelegentlich (monatlich)	3=häufig (wöchentlich)	4=ständig (fast täglich)
frequ3	3	28	Wie intensiv war Ihr Kontakt?				

8.2.3 Alter-Alter

Variable	Phase	Frage Nr.	Frage	Antwortmöglichkeiten	
EGOID			ID Ego		
ALTERID			ID Alter		
mainalt1	1	3	Hauptanstoß zur Chancenerkennung	0=nein	1=ja
sexa1	1	4	Geschlecht	1=männlich	2=weiblich
fam1	1	5.1	Familie	0=nein	1=ja
friend1	1	5.2	Freund	0=nein	1=ja
acad1	1	5.3	Akademia	0=nein	1=ja
disc1	1	5.3.1	Gleiche Disziplin?	offen	
discoth1	1	5.3.1.1	Andere Disziplin:	offen	
inst1	1	5.3.2	Eigenes Institut?	0=nein	1=ja
ind1	1	5.4	Industrie	0=nein	1=ja
funds1	1	5.5	Fördergeber	0=nein	1=ja
others1	1	5.6	Sonstiges	0=nein	1=ja
othspec1	1	5.6.1	Sonstiges spezifiziere:	Offen	
info1	1	6.1	Informationsaustausch / Diskussion	0=nein	1=ja
infolaw1	1	6.1.1	Rechtliche Informationen	0=nein	1=ja
infoeco1	1	6.1.2	Wirtschaftliches Know-how	0=nein	1=ja
infocus1	1	6.1.3	Informationen zu / Vermittlung von potentiellen Kunden	0=nein	1=ja
infoemp1	1	6.1.4	Informationen zu / Vermittlung von Personal	0=nein	1=ja
infooth1	1	6.1.5	Sonstiges	0=nein	1=ja
infotsp1	1	6.1.5.1	Bitte spezifizieren Sie:	Offen	
res1	1	6.2	Bereitstellung von Ressourcen	0=nein	1=ja
resmat1	1	6.2.1	Material	0=nein	1=ja
rescap1	1	6.2.2	Kapital	0=nein	1=ja
resinfr1	1	6.2.3	Infrastruktur	0=nein	1=ja
resemp1	1	6.2.4	Personal	0=nein	1=ja
resoth1	1	6.2.5	Sonstiges	0=nein	1=ja
resotsp1	1	6.2.5.1	Bitte spezifizieren Sie:	Offen	
sup1	1	6.3	Unterstützung	0=nein	1=ja
supem1	1	6.3.1	Emotionale Unterstützung	0=nein	1=ja

				0=nein	1=ja		
supins1	1	6.3.2	Institutionelle Unterstützung	0=nein	1=ja		
oth1	1	6.4	Sonstiges	0=nein	1=ja		
othspe1	1	6.4.1	Bitte spezifizieren Sie:	Offen			
frequ1	1	7	Wie intensiv war ihr Kontakt?	1=einmalig / sporadisch (jährlich)	2=gelegentlich (monatlich)	3=häufig (wöchentli ch)	4=ständig (fast täglich)
sexa2	2	15	Geschlecht	1=männlich	2=weiblich		
fam2	2	16.1	Familie	0=nein	1=ja		
friend2	2	16.2	Freund	0=nein	1=ja		
acad2	2	16.3	Akademia	0=nein	1=ja		
disc2	2	16.3.1	Gleiche Disziplin?	0=nein	1=ja		
discoth2	2	16.3.1.1	Andere Disziplin:	Offen			
inst2	2	16.3.2	Eigenes Institut?	0=nein	1=ja		
ind2	2	16.4	Industrie	0=nein	1=ja		
funds2	2	16.5	Fördergeber	0=nein	1=ja		
others2	2	16.6	Sonstiges	0=nein	1=ja		
othspec2	2	16.6.1	Sonstiges spezifiziere:	Offen			
info2	2	17.1	Informationsaustausch / Diskussion	0=nein	1=ja		
infolaw2	2	17.1.1	Rechtliche Informationen	0=nein	1=ja		
infoeco2	2	17.1.2	Wirtschaftliches Know-how	0=nein	1=ja		
infocus2	2	17.1.3	Informationen zu / Vermittlung von potentiellen Kunden	0=nein	1=ja		
infoemp2	2	17.1.4	Informationen zu / Vermittlung von Personal	0=nein	1=ja		
infooth2	2	17.1.5	Sonstiges	0=nein	1=ja		
infotsp2	2	17.1.5.1	Bitte spezifizieren Sie:	Offen			
res2	2	17.2	Bereitstellung von Ressourcen	0=nein	1=ja		
resmat2	2	17.2.1	Material	0=nein	1=ja		
rescap2	2	17.2.2	Kapital	0=nein	1=ja		
resinfr2	2	17.2.3	Infrastruktur	0=nein	1=ja		
resemp2	2	17.2.4	Personal	0=nein	1=ja		
resoth2	2	17.2.5	Sonstiges	0=nein	1=ja		
resotsp2	2	17.2.5.1	Bitte spezifizieren Sie:	0=nein	1=ja		
sup2	2	17.3	Unterstützung	0=nein	1=ja		
supem2	2	17.3.1	Emotionale Unterstützung	0=nein	1=ja		

supins2	2	17.3.2	Institutionelle Unterstützung	0=nein	1=ja	
oth2	2	17.4	Sonstiges	0=nein	1=ja	
othspe2	2	17.4.1	Bitte spezifizieren Sie:	Offen		
frequ2	2	18	Wie intensiv war ihr Kontakt?	1=einmalig / sporadisch (jährlich)	2=gelegentlich (monatlich)	3=häufig (wöchentlich)	4=ständig (fast täglich)
part3	3	21.2	Gründungspartner	0=nein	1=ja	
sexa3	3	25	Geschlecht	1=männlich	2=weiblich	
fam3	3	26.1	Familie	0=nein	1=ja	
friend3	3	26.2	Freund	0=nein	1=ja	
acad3	3	26.3	Akademia	0=nein	1=ja	
disca3	3	26.3.1	Gleiche Disziplin?	0=nein	1=ja	
discoth3	3	26.3.1.1	Andere Disziplin:	Offen		
inst3	3	26.3.2	Eigenes Institut?	0=nein	1=ja	
ind3	3	26.4	Industrie	0=nein	1=ja	
funds3	3	26.5	Fördergeber	0=nein	1=ja	
others3	3	26.6	Sonstiges	0=nein	1=ja	
othspec3	3	26.6.1	Sonstiges spezifiziere:	Offen		
info3	3	27.1	Informationsaustausch / Diskussion	0=nein	1=ja	
infolaw3	3	27.1.1	Rechtliche Informationen	0=nein	1=ja	
infoeco3	3	27.1.2	Wirtschaftliches Know-how	0=nein	1=ja	
			Informationen zu / Vermittlung von potentiellen			
infocus3	3	27.1.3	Kunden	0=nein	1=ja	
infoemp3	3	27.1.4	Informationen zu / Vermittlung von Personal	0=nein	1=ja	
infooth3	3	27.1.5	Sonstiges	0=nein	1=ja	
infotsp3	3	27.1.5.1	Bitte spezifizieren Sie:	Offen		
res3	3	27.2	Bereitstellung von Ressourcen	0=nein	1=ja	
resmat3	3	27.2.1	Material	0=nein	1=ja	
rescap3	3	27.2.2	Kapital	0=nein	1=ja	
resinfr3	3	27.2.3	Infrastruktur	0=nein	1=ja	
resemp3	3	27.2.4	Personal	0=nein	1=ja	
resoth3	3	27.2.5	Sonstiges	0=nein	1=ja	
resotsp3	3	27.2.5.1	Bitte spezifizieren Sie:	Offen		
sup3	3	27.3	Unterstützung	0=nein	1=ja	

				0=nein	1=ja		
supem3	3	27.3.1	Emotionale Unterstützung	0=nein	1=ja		
supins3	3	27.3.2	Institutionelle Unterstützung	0=nein	1=ja		
oth3	3	27.4	Sonstiges	0=nein	1=ja		
othspe3	3	27.4.1	Bitte spezifizieren Sie:	Offen			
frequ3	3	28	Wie intensiv war ihr Kontakt?	1=einmalig / sporadisch (jährlich)	2=gelegentlich (monatlich)	3=häufig (wöchentlich)	4=ständig (fast täglich)

297

8.3 Questionnaire

Online Fragebogen zum Sozialen Kapital universitärer Unternehmensgründer in Österreich

Sehr geehrte(r) Unternehmensgründer(in),

der folgende Fragebogen wurde im Rahmen meiner Dissertation an der Wirtschaftsuniversität Wien, Institut für Marketing-Management, zum Thema Akademische Spin-off Gründungen in Österreich entwickelt.

Zielsetzung meiner Arbeit ist es, die Einflussfaktoren einer erfolgreichen Ausgründung aus dem universitären Umfeld mit speziellem Fokus auf die Kontakte und das personelle Umfeld (soziales Netzwerk) des Unternehmensgründers zu beleuchten.

Eine grundlegende Annahme der Arbeit ist, dass sich der Ausgründungsprozess idealtypisch in drei Phasen gliedern lässt, wie in nachfolgender Graphik dargestellt:

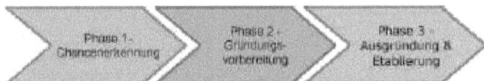

Phase 1: Chancenerkennung
In dieser Phase sind Sie als Unternehmensgründer(in) das erste Mal auf die Idee gestoßen, die dann die Grundlage für die spätere Geschäftstätigkeit dargestellt hat. Sie haben das Für und Wider dieser Idee abgewogen und sich eventuell mit anderen Personen darüber ausgetauscht sowie zusätzliche Informationen unterschiedlichster Natur (technisch, wirtschaftlich, rechtlich, etc.) eingeholt.

Phase 2: Gründungsvorbereitung
Der Entschluss zu gründen ist gefallen. In Phase 2 setzen Sie die notwendigen Schritte zur Vorbereitung der späteren Unternehmensgründung. Das beinhaltet unter anderem die Entwicklung des Geschäftskonzepts sowie die Suche nach Kapital, Personal und / oder Infrastruktur. Notwendige Ressourcen müssen beschafft werden. Damit verbunden ist der Aufbau eines Kontaktnetzes zu zukünftigen Marktpartnern wie Kapitalgebern, Lieferanten und Kunden.

Phase 3: Gründung des Spin-off und Etablierung am Markt
Diese Phase beginnt üblicherweise mit der formalen Gründung. Die konkrete Verwirklichung der Unternehmensidee erfolgt. Sie präsentieren das neu gegründete Unternehmen als neuen Marktteilnehmer. Dazu müssen Sie Bekanntheitsgrad und Renommee aufbauen, was kommunikationspolitische Maßnahmen und die Akquisition von Referenzkunden impliziert.

In diesem Zusammenhang, möchte ich Sie, als Gründer(in) eines Universitäts-Spin-offs, bitten, mir bei der Realisierung meiner Arbeit zu helfen und nachfolgenden Fragebogen auszufüllen. Die Befragung ist entlang der einzelnen Phasen des Gründungsprozesses gegliedert. Die Fragen wiederholen sich zum Teil, was daran liegt, dass in Abhängigkeit von der jeweiligen Phase unterschiedlichen Antworten möglich sind.

Bitte versuchen Sie, sich an den durch die jeweilige Phase repräsentierten Zeitraum zu erinnern und in die damalige Zeit hineinzuversetzen, und die Fragen im Hinblick auf die damalige (Kontakt-)Situation zu beantworten. So kann es z.B. sein, dass eine Person, mit der Sie in der Phase der Ideenfindung nur losen Kontakt hatten, später zu einem engen Freund wurde, so dass Sie für Phase 1 und Phase 2 unterschiedliche Angaben zur selben Person machen.

Die Beantwortung des Fragebogens wird in etwa 20 Minuten dauern. Selbstverständlich ist die Befragung anonym und die Daten werden absolut vertraulich behandelt.

Figure 8-1: Start Page of Online Questionnaire

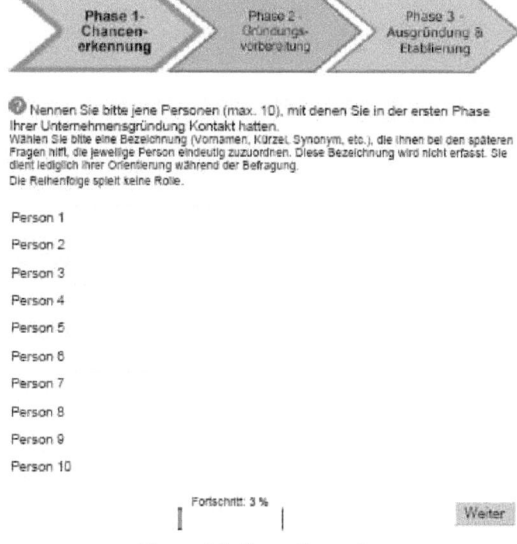

Figure 8-2: Name Generator

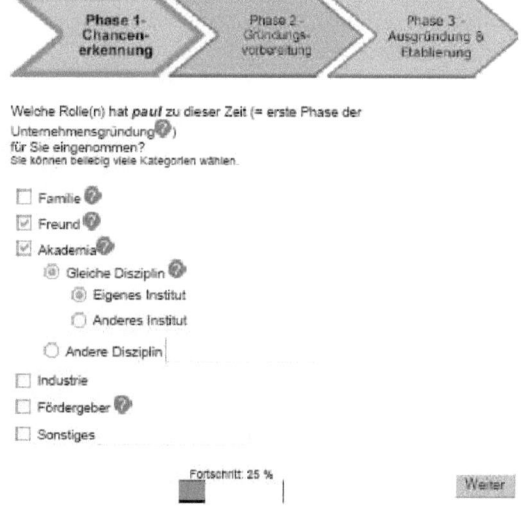

Figure 8-3: Question on Relational Types

Figure 8-4: Question on the Content of Relations

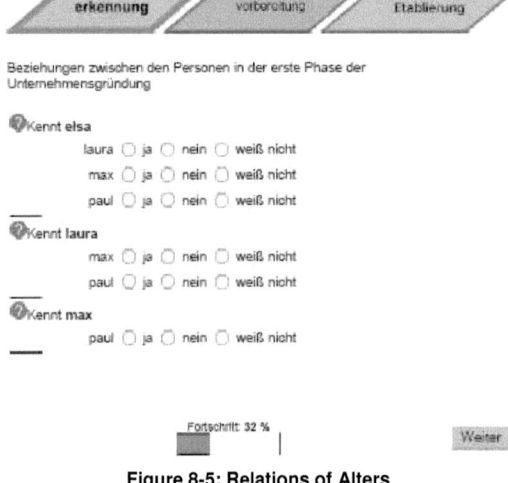

Figure 8-5: Relations of Alters

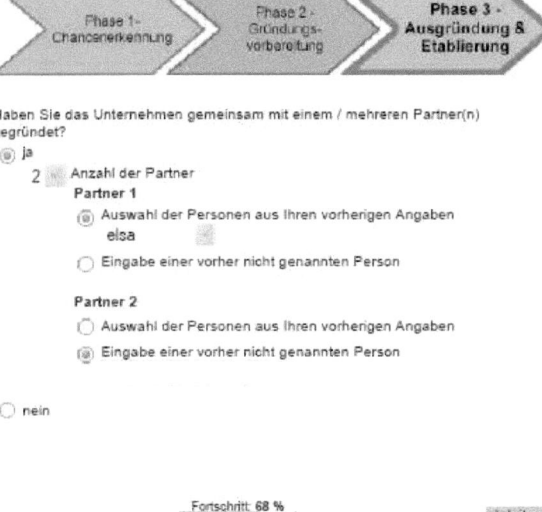

Figure 8-6: Questions on Co-founders

VDM Verlagsservicegesellschaft mbH

Die VDM Verlagsservicegesellschaft sucht für wissenschaftliche Verlage abgeschlossene und herausragende

Dissertationen, Habilitationen, Diplomarbeiten, Master Theses, Magisterarbeiten usw.

für die kostenlose Publikation als Fachbuch.

Sie verfügen über eine Arbeit, die hohen inhaltlichen und formalen Ansprüchen genügt, und haben Interesse an einer honorarvergüteten Publikation?

Dann senden Sie bitte erste Informationen über sich und Ihre Arbeit per Email an *info@vdm-vsg.de*.

Sie erhalten kurzfristig unser Feedback!

VDM Verlagsservicegesellschaft mbH
Dudweiler Landstr. 99   Telefon +49 681 3720 174
D - 66123 Saarbrücken   Fax   +49 681 3720 1749
www.vdm-vsg.de

Die VDM Verlagsservicegesellschaft mbH vertritt

Printed by Books on Demand GmbH, Norderstedt / Germany